"This book is a fresh, timely and thought-provoking analysis of this challenging dimension, in particular the policy and institutional innovations required. It is a very welcome and valuable resource."

Ambassador David Donoghue, *Permanent Representative of Ireland to the UN 2013-2017 and co-facilitator of the UN negotiations on the 2030 Agenda for Sustainable Development*

"The book provides an excellent analysis of governance mechanisms and policy processes that effectively promote sustainable development. This is extremely important for advancing action on the SDGs, despite the Covid-19 pandemic and increasing pressures of climate change."

Prof. Dr. Imme Scholz, *Co-Chair of the UN Independent Group of Scientists for the Global Sustainable Development Report 2023 and Co-Chair of the German Sustainability Council (RNE)*

"This book helps to deepen our understanding of the governance and policy innovations that are needed to support policy coherence across sectors and actors at different levels to deliver on Agenda 2030."

Alina Rocha Menocal, *Principal Research Fellow at the Overseas Development Institute (ODI) and Director of the global Thinking and Working Community of Practice (TWP CoP)*

GOVERNING THE INTERLINKAGES BETWEEN THE SDGS

Governing the Interlinkages between the SDGs: Approaches, Opportunities and Challenges identifies the institutional processes, governance mechanisms and policy mixes that are conducive to devising strategies of integrated Sustainable Development Goal (SDG) implementation.

The book edited by Anita Breuer, Daniele Malerba, Srinivasa Srigiri, and Pooja Balasubramanian examines the dedicated policies targeting the SDGs, as well as political and institutional drivers of synergies and trade-offs between the SDGs in selected key areas – both cross-nationally and in specific country contexts. Their analysis moves beyond the focus on links between SDG indicators and targets. Instead, the book takes advantage of recent evidence from the initial implementation phase of the SDGs and each chapter explores the question of which political-institutional prerequisites, governance mechanisms, and policy instruments are suited to accelerate the implementation of the SDGs. The findings presented are intended to both inform high-level policy debates and to provide orientation for practitioners working on development cooperation.

This volume will be of great interest to practitioners and policy makers in the field of sustainable development, as well as academics in the fields of sustainability research, political science, and economics.

Anita Breuer is a senior researcher at the German Institute of Development and Sustainability (IDOS), leading the inter-disciplinary research project "Implementing the 2030 Agenda: Integrating Growth, Environment, Equality and Governance". Her primary research interests involve sustainability governance, democracy promotion, accountability, and participatory politics, as well as the impact of digitalization on political behavior and collective action.

Daniele Malerba is a senior researcher at IDOS in the inter-disciplinary research project focused on the implementation of the 2030 Agenda. His research focuses on just transitions, socio-ecological transitions, and how to address the distributional implications of climate policies with the use of social protection.

Srinivasa Srigiri is a senior researcher within the Research Programme on Environmental Governance at IDOS. His primary research interests involve governance of natural resources, sustainability transformations in rural and agricultural sectors, strategies for climate adaptation at sub-national and local levels, and the water-energy-food nexus.

Pooja Balasubramanian is a researcher at IDOS and part of an inter-disciplinary research project on the implementation of the 2030 Agenda. She focuses on the interlinkages between poverty, inequality and growth.

GOVERNING THE INTERLINKAGES BETWEEN THE SDGS

Approaches, Opportunities and Challenges

Edited by Anita Breuer, Daniele Malerba, Srinivasa Srigiri and Pooja Balasubramanian

Routledge
Taylor & Francis Group
LONDON AND NEW YORK

First published 2023
by Routledge
4 Park Square, Milton Park, Abingdon, Oxon OX14 4RN

and by Routledge
605 Third Avenue, New York, NY 10158

Routledge is an imprint of the Taylor & Francis Group, an informa business

British Library Cataloguing-in-Publication Data
A catalogue record for this book is available from the British Library

Library of Congress Cataloging-in-Publication Data
A catalog record has been requested for this book

ISBN: 978-1-032-18469-2 (hbk)
ISBN: 978-1-032-18465-4 (pbk)
ISBN: 978-1-003-25468-3 (ebk)

DOI: 10.4324/9781003254683

Typeset in Bembo
by Taylor & Francis Books

CONTENTS

ILLUSTRATIONS

Figures

Tables

ACKNOWLEDGMENTS

This book is a truly collaborative work. We are grateful to all the contributors for their dedication, professionalism, and expertise. Our thanks also go to the Routledge team for their confidence in our project and for their help and support throughout the editing process, in particular to our editorial assistant, Jyotsna Gurung. We would also like to thank the anonymous reviewers of the original book proposal; we are most appreciative of their insightful comments. Colleagues from the administrative and service departments at the German Institute of Development and Sustainability, Bonn, have made every effort to enable us to work with and around constraints imposed by the COVID-19 pandemic. Our project coordinator Dr. Sonja Packschies deserves special mention in this regard. Christopher Ihinegbu provided attentive assistance in reference management and in preparing the individual chapters for publication. This project has been partly funded by the German Federal Ministry for Economic Cooperation and Development/ Bundesministerium für wirtschaftliche Zusammenarbeit und Entwicklung.

CONTRIBUTORS

Cameron Allen is a Post-Doctoral Fellow at the University of New South Wales and Senior Advisor to the UN Sustainable Development Solutions Network. He has almost years of professional experience with research institutions, non-governmental organizations (NGO) and the Australian Government. Cameron supported the negotiation process of the Sustainable Development Goals (SDG) and worked as a UN advisor on green economy and environmental management.

Pooja Balasubramanian is a researcher at the German Institute of Development and Sustainability (IDOS), part of an inter-disciplinary research project on the implementation of the 2030 Agenda. She focuses on the interlinkages between poverty, inequality, and growth. In addition, her research focusses on the gendered impacts of structural change, economic growth, and indebtedness.

Mairon G. Bastos Lima is a Research Fellow at the Stockholm Environment Institute (SEI), assessing leverage points and institutional innovations for sustainability transitions in land use, particularly around tropical ecosystems. Mairon is an Earth System Governance fellow and has collaborated with the UN Development Programme (UNDP) and UN Research Institute for Social Development.

Therese Bennich is a Research Fellow at the Stockholm Environment Institute. Her research focuses on integrated assessments, policy analysis, and decision support in relation to sustainability initiatives and frameworks such as the 2030 Agenda and the Paris Agreement. She is also the Project Manager of Mistra Geopolitics, a research programme that examines the dynamics of geopolitics, human security, and environmental change.

Anita Breuer is a senior researcher at IDOS, leading the inter-disciplinary research project "Implementing the 2030 Agenda: Integrating Growth, Environment,

Equality and Governance". Her primary research interests involve sustainability governance, democracy promotion, accountability, and participatory politics, as well as the impact of digitalization on political behavior and collective action.

Henrik Carlsen has been a Senior Research Fellow at Stockholm Environment Institute since 2014. His main research focus is on applying methodologies and approaches, mainly from systems theory on problems in sustainability science. Currently he acts as Deputy Programme Director for Mistra Geopolitics, a research programme that examines the dynamics of geopolitics, human security, and environmental change.

Ines Dombrowsky heads the Research Programme 'Environmental Governance and Transformation to Sustainability' at IDOS. Her research, grounded in institutional economics and political sciences, focuses on coordination and cooperation in environmental governance across levels, sectors, actor types, and scales. A particular thematic focus lies on water and the water-energy-food-climate nexus.

Johannes Emmerling is a Scientist at the European Institute on Economics and the Environment and co-leads the Low Carbon Pathways unit. He is co-leading the development of the integrated assessment model WITCH. His main areas of research include climate change and energy economics, risk and uncertainty, welfare economics, and development. He is also an Associate Editor of the *Public Finance Review*.

Arvinn Gadgil is the director of UNDP's Oslo Governance Centre. He has been engaged in governance and development policy for almost 20 years. Arvinn's positions prior to joining UNDP include, among others, Senior Director of the Norwegian Refugee Council and Deputy Minister and Junior Minister for international development, climate, and environment in Norway.

Sarah Khan joined the University of Göttingen as a postdoctoral research fellow under Stephan Klasen in November 2015. Her research focuses primarily on gender issues in developing countries, and she has been involved in various projects related to female empowerment in developing countries, including two meta-analysis analyzing the impact of female empowerment interventions.

Julia Kercher is an independent consultant with a focus on SDGs, human rights, and governance. For more than 15 years Julia has worked for different UN agencies and international NGOs and was closely involved in the drafting and negotiation process of the SDGs. She holds law degrees and an MSc in 'Development Practice'.

Julia Leininger has been head of the department "Transformation of political (dis-) order" at IDOS since 2014. She has also been leading the inter-disciplinary research projects "Social cohesion in Africa". Her primary research interests involve democracy promotion, social cohesion, sustainability governance, future scenarios, and religion in democratization.

Daniele Malerba is a senior researcher at IDOS in the inter-disciplinary research project focused on the implementation of 2030 Agenda. His research focuses on just transitions, socio-ecological transitions, and how to address the distributional implications of climate policies with the use of social protection.

Louis Meuleman has some 40 years of public administration experience at subnational, national, European, and UN level. He is a researcher, consultant, and trainer at Public Strategy for Sustainable Development (PS4SD) (Brussels). Among other appointments, he is a member of the UN Committee of Experts on Public Administration and visiting professor public governance at Leuven University (Belgium).

Yannick Oswald recently completed his PhD at the University of Leeds, in the School of Earth and Environment. His research focuses on people's expenditure patterns across products to identify consumption patterns that are unsustainable and to suggest how to satisfy people's needs with different consumption patterns and more sustainable provisioning systems.

Waltina Scheumann is an associated researcher with IDOS in Bonn and the Humboldt University Berlin. Her research interests are cooperation on trans-boundary water bodies, governance and management of irrigation and drainage systems, implementation of international social and environmental standards of hydropower projects, and analysis of cross-sectoral coordination of water, agriculture, energy and environment.

Karen Siegel is head of the research group "Transformation and Sustainability Governance in South American Bioeconomies" at the University of Münster. Her research interests include the politics of sustainable development and sustainability transitions with a focus on inclusiveness, natural resource governance in the global political economy, and regionalism in South America.

Srinivasa Srigiri is a senior researcher within the Research Programme on Environmental Governance at IDOS. His primary research interests involve governance of natural resources, sustainability transformations in rural and agricultural sectors, strategies for climate adaptation at sub-national and local levels, and the water-energy-food nexus.

Nina Weitz is a Research Fellow at the Governance and Institutions Unit at SEI. Nina leads and executes projects focused on how policy and institutions can respond more effectively and in sustainable and just ways to the increasingly interlinked challenges that societies face. Since 2018 she has managed SEI's portfolio on 2030 Agenda implementation.

1

GOVERNING THE INTERLINKAGES BETWEEN THE SDGS

Approaches, Opportunities and Challenges

Anita Breuer, Daniele Malerba, Srinivasa Srigiri and Pooja Balasubramanian

Introduction

In 2015, the UN member states adopted the 2030 Agenda for Sustainable Development, which sets out a 15-year plan to achieve 17 Sustainable Development Goals (SDGs) with 169 sub-targets by 2030. Essentially, the SDGs and their targets constitute a universal call to end poverty, protect the planet and improve the lives and livelihoods of everyone, everywhere.[1]

The 2030 Agenda reflects a new understanding of global development problems that differs from preceding global development frameworks such as the World Bank's and International Monetary Fund's Poverty Reduction Strategy Papers and the UN Millennium Development Goals (MDG), in several ways. A first defining feature of the 2030 Agenda is its aspiration to universality. The MDGs were mainly conceived as an agenda focused on achieving a set of basic minimum living standards in lower income countries. By contrast, the 2030 Agenda is universal in scope and commits high-income and low-income nations alike to contribute to the efforts to achieve global sustainability. Another distinctive characteristic is the 2030 Agenda's strong emphasis on inclusiveness. This becomes manifest in the overarching principle to "Leave No One Behind" (LNOB) and in the pledge to "reach the furthest behind first", which are both enshrined in the Agenda's preamble (§4). A novelty also consists in the strong emphasis on multi-level governance. References to the importance of implementing the SDGs at all levels – from global over national to sub-national level – can be found throughout the text, for example in SDG 16.7, which strives to "ensure responsive, inclusive, participatory and representative decision-making at all levels". The adoption of SDG 16 on "Peace, Justice and Strong Institutions" is one of the most substantial differences between the SDGs and the MDGs. Importantly, issues of good governance, peace and human rights are not only presented as goals in and of themselves but also as enablers for the achievement of all other goals. Finally, and most

DOI: 10.4324/9781003254683-1

importantly for this book, the 2030 Agenda represents a paradigm shift from previous development approaches in its recognition of the indivisibility of the social, economic, and ecological dimensions of sustainable development.

As a result, the 17 SDGs and their 169 targets constitute a network of development objectives that are fundamentally interdependent. Evidently, the practical implementation of such an ambitious development vision poses new challenges to political institutions and processes. Achieving the SDGs simultaneously, will require an integrated[2] implementation of the 2030 Agenda and increased policy coherence. Such policy coherence across different policy sectors, government levels, and societal actors can only be achieved through the dismantling of traditional silos. To this end, the adoption of innovative governance approaches that facilitate the harnessing of synergies and mitigation of trade-offs between and within the SDGs will be necessary (Biermann et al., 2014; Breuer, Leininger, and Tosun, 2019; International Science Council, 2017; Organisation for Economic Co-operation and Development, 2019; Tosun and Leininger, 2017).

The COVID-19 pandemic that erupted throughout the world early in 2020 once more cast a spotlight on the systemic risks that arise from the close interconnection between socio-economic and human development on one side and the natural environment and resources on the other (Mechler, Stevance, and Deubelli, 2020; European Commission, 2021; Mechler, 2020). At the same time, the global crisis sparked by the pandemic opened a window of opportunity in the sense of a strengthened consensus on the imperative of integrated sustainability governance to minimize systemic risks in the future. Amid the ongoing pandemic, the global community renewed its recognition of the need for structural reform and innovative governance approaches with bilateral and multi-lateral donors stressing the need to align strategies for post-pandemic recovery and future resilience building with the principles of the 2030 Agenda (Bundesministerium für Umwelt, Naturschutz und nukleare Sicherheit and Bundesministerium für wirtschaftliche Zusammenarbeit und Entwicklung, 2020; Inter-American Development Bank, 2020; ODI, 2020; Organisation for Economic Co-operation and Development, 2020; UN Development Programme, 2020; World Bank, 2020).

In response to the above challenges, in recent years a growing body of literature has proposed a variety of methods and tools to identify and assess interlinkages between the SDGs, both globally and in individual country contexts. Critical reviews of this body of literature are provided by Miola, Borchardt, and Buscaglia (2019); Breuer, Janetschek, and Malerba (2019); Pham-Truffert, Metz, Fischer, Rueff, and Messerli (2020); and the contribution by Weitz and Carlsen in Part I of this book.

Despite this growing literature, as of yet, there is little academic and political agreement about how SDG interlinkages, once identified, can best be tackled in practice, both in terms of policy mixes and governance structures. Six years into the adoption of the 2030 Agenda, a mixed picture also emerges regarding the degree to which signatory states have established the 2030 Agenda as a reference framework for their national development planning and cross-sectoral and cross-level policy coordination. In high- and lower income countries alike, integrated implementation of the SDGs continues to pose enormous challenges for political, societal, and business actors. Important knowledge gaps persist regarding how to best support policy integration across all levels and sectors of government in varying country contexts. Clearly, thus, there is a need for science to undertake research that supports the development of evidence-based strategies to support implementation of the SDGs (Allen, Metternicht, and Wiedmann, 2021). Such research must be cognizant of the fact that the SDGs cannot be implemented in isolation from existing political social, and economic realities, and that consequently there can be no universal recipe for their implementation. Instead, successful localization of the 2030 Agenda will only be possible if mechanisms and policies to govern its implementation are carefully tailored to country and local contexts.

To this end, it needs to be established which governance mechanisms and policy instruments are suitable for effectively addressing interactions between the SDGs in a given national or sub-national context. This also implies, that the political-institutional factors that potentially affect SDG implementation, such as political regime type or state capacity, must be analysed and taken into consideration, as must the vested interests and power constellations that might hamper integrated implementation.

Furthermore, sustainability research should increasingly strive to generate insights on how the principle of LNOB can be realized in different contexts. To provide one example, achieving environmental goals while simultaneously eradicating poverty is a challenging policy objective; designing policy packages to realize environmental objectives that are socially acceptable to a large section of the population and protect the needs of vulnerable groups requires political negotiation processes. The nature and outcome of these processes will necessarily differ between high-income countries (with mostly urban, highly educated and skilled but aging populations) and low-income countries (with predominantly rural populations, lower levels of educational attainment and job skills, high levels of self-employment in the informal sector, and large youth populations).

Against this background, the present book pursues the following key questions:

1. What do we know about the most important interlinkages between the SDGs?
2. What governance mechanisms and policy processes are needed to address power and capacity asymmetries between stakeholders form different sectors and levels in order to promote an inclusive, coherent and participatory governance of SDG interlinkages?
3. Which policy mixes to increase policy coherence in the implementation of different interlinked SDGs have been found to be effective, socially just and acceptable and leaving no one behind?
4. What political-institutional preconditions are conducive to the establishment of effective governance mechanisms to manage SDG interactions?

In addressing the above questions, particular attention will be paid to three governance dimensions of SDG implementation, which serve as research heuristics underlying contributions to this book:

i the role of political-institutional preconditions (e.g. regime type, state capacity, and the quality of governance);
ii governance mechanisms as formats of political decision-making (e.g. inter-ministerial coordination mechanisms, commissions, consultation forums and platforms, mechanisms for the management of natural resources, etc.);
iii policy instruments and policy mixes to steer the implementation of politically set goals such as the combination of environmental measures with social policies; or the need to accompany economic growth with redistribution.

In the following sub-sections, we elaborate on these three governance dimensions and their discussion in social science literature in more detail.

Political-institutional Preconditions: Democracy, Good Governance and State Capacity

With regards to political-institutional preconditions, it is assumed that factors such as regime type, the quality of governance and state capacity influence governance for SDG implementation.

Regarding *regime type*, several studies point to the positive effects of democratic institutions on economic and social development (Das and Das, 2018; Fossati, 2016; Glassa and Newiga, 2019; Halperin, Siegle, and Weinstein, 2010; Nwobashi and Itumo, 2017). Furthermore, empirical evidence suggests that civil liberties improve income and equality (Li, Squire, and Zou, 1998), and countries with higher degrees of civil liberties and political freedoms enjoy better environmental quality (Barrett and Graddy, 2000; Dasgupta and De Cianac, 2018). Yet, these claims are not uncontested. It is frequently argued that input legitimation via free and fair election creates incentives for democratic governments to favour the preferences of today's (voting) citizens over those of future generations, thus creating a trade-off between democracy and sustainable development (Landman and Lauth, 2019; Wurster, 2013). Another frequently made argument is that the decentralization of power that is inherent to democracies leads to slow and potentially inefficient decision-making processes (e.g., Stasavage, 2020). At the same time, various autocratic regimes in South and East Asia, where strong forms of developmental state capacity are exercised towards meeting transformative sustainable development goals, have (at least area-specific) successes to show for. This begs the question whether the "Churchill hypothesis" of democracies' relative advantage over autocracies can be affirmed in terms of their sustainability performance (Wurster, 2013).

The concept of "democracy" is often conflated with the concept of "good governance" although the two concepts, in fact, rely on very different criteria. One of the most widely recognized definitions of democracy proposed by Dahl (1989) relies entirely on procedural indicators of electoral democracy whereby universal suffrage, elections that reliably reflect voter preferences, and a fair choice between alternatives constitute the basis on which political power is exercised. Whether the way in which power is exercised can be described as "good governance" is not part of this definition. The popular definition of governance by Fukuyama (2013), in turn, refers to governance as "*the government's ability to make and enforce rules, and to deliver services*" irrespective of whether the government in question is democratically elected or not (ibid., p. 3). However, while governance clearly embraces government institutions, it also subsumes informal, non-governmental institutions operating within the public realm (Bøås, 1998). According to Fukuyama (2013), governance refers to the ability to make progress towards objectives and ambitions that derive from the dynamic interaction and power relations among actors of the state and society (2013).

Debates about good governance have been decisively shaped by intergovernmental organizations in the context of development cooperation. Essentially, actions to foster good governance focus on mitigating two undesirable traits: the unrepresentative character of governments and the inefficiency of non-market systems (Albritton and Bureekul, 2009). As Bøås (1998, p. 119) states: "*the World Bank operationalized 'bad governance' as personalisation of power, lack of human rights, endemic corruption and un-elected and unaccountable governments. And so, 'good governance' must be the natural opposite*". Indeed, most definitions of good governance today include representative and participatory decision-making, accountability, and transparency as its key characteristics, which have also been enshrined in SDG 16 on Peace and Strong Institutions (SDG 16.6 and 16.7). Good governance has become an important component of the international agenda, and is often mentioned as political and economic conditionality for the provision of development assistance or investments from international lending agencies. Although the debate about the precise components of good governance continues, the concept clearly goes beyond competitive multiparty elections and the existence of an independent judiciary and parliament, which have been emphasized as primary symbols of Western-style democracy (Weiss, 2000). Consequently, democracy alone is not a sufficient cause of good governance and elements of good governance can also be present in autocratic regimes.

Concerns about *state capacity* and the related effectiveness of states in achieving the goals of economic and social development have been present in social science literature since the 1980s (Cireno Fernandes, de Moura Palotti, and Christo Fernandes, 2017). Early research on the subject focused on gaining a better knowledge of the processes that lead to the establishment of national states, as well as aspects of state capacity building (e.g., Mann, 1998). Over the course of the 1990s, the focus of analyses turned to state capacity as a prerequisite for implementing plans, programs, and public policies, as well as the provision of goods and services (de Ávila Gomide, de Sá e Silva, and Pires, 2014; Grindle, 1996). By the 2010s, state capability had become regarded as a necessary component of effective governance (Cingolani, 2013; Jreisat, 2012; Savoia and Sen, 2014) and the protagonism of the state came to be perceived as a crucial factor for the success of economic development (Cireno Fernandes et al., 2017). Several studies put forward arguments in favour of the so-called 'developmental state', which plays an active, interventionist, and leadership role in business sectors, in contrast to a minimalist view of the state's function. The phenomenon of the developmental state, which is characterized by the prominent role of bureaucratic elites with technical skills who are recruited through meritocratic systems and occupy key positions to carry out high-level economic development strategies, is particularly evident in the so called "Asian-Tiger" economies (Cireno Fernandes et al., 2017; Leftwich, 1995). However, the focus of literature has since shifted to a new model of state action that emerged towards the end of the twentieth century in reaction to globalization, and related increased competition, the emergence of new governance standards, and pressures for democratization and transparency. While social science research continues to identify a strong role for the state, it recognizes that this role is performed through flexible structures that mobilize non-state actors in innovation networks and connect local and international spaces (O'Riain, 2004).

The above development in literature reflects the growing understanding that policymaking to achieve sustainable development is a multidimensional, cross-cutting endeavour that needs to resolve conflicts between ecological, social and economic policy goals as productively as possible. In practical implementation, this implies a highly complex task, the fulfilment of which requires the application of different types of instruments that are adapted to the respective context, and therefore a high capacity of the state. One political science assumption in sustainability research is that, depending on regime type and state capacity, different expectations must be formulated for different types of sustainability instruments. Some studies appear to suggest, that democratic states with a high state capacity apply a more balanced mix of instruments than autocratic states with a low state capacity, in which fewer planning and participation-oriented sustainability instruments are used (Chimhowu, Hulme, and Munro, 2019; Swanson, Pintér, Bregha, Volkery, and Jacob, 2004). However, this view is not uncontested (e.g., Pickering, Bäckstrand, and Schlosberg, 2020).

Governance Mechanisms

The term *governance mechanisms* has different meanings in the literature. In this book, it refers to different formats and processes of political decision-making that complement or supplement traditional representative and bureaucratic institutions by opening public action to the participation of different stakeholder groups and civil society organizations (Bandeira and Ferraro, 2017). In that sense, governance mechanisms are an important part of the interaction in which the public sector engages with "*the environment around it and the society it is dealing with to produce results*" (Fukuyama, 2013, p. 355). These interactions result in different outcomes such as, for instance, the quality of and access to public services like health and education, or the sustainable or unsustainable use of a natural resource.

Over the last few decades, there has been a growing demand for broadening participation in public governance. The arguments behind this demand are varied and perspectives differ between academic disciplines. Democratic theorists stress a lack of input legitimacy, by pointing to the democratic deficits that traditional hierarchical institutions are suffering from (e.g., Goodhart et al., 2012). Scholars of public administration, in turn, focus on throughput and output problems by emphasizing that growing policy complexity and the ensuing government overload necessitate more participatory and interactive governance mechanisms (Torfing, Peters, Pierre, and Sørensen, 2012). Finally, among economists, market failure is seen as the main explanation behind "*a growing fascination with governance mechanisms*" (Jessop, 1998). The above problems resonate with and are reflected in the 2030 Agenda's demands for inclusive, participatory and representative decision-making at all levels (SDG 16.7), multi-stakeholder partnerships (SDG 17.16 and 17.17), as well as in the Agenda's emphasis of the integrated and indivisible nature of the SDGs and the related need for enhanced policy coordination and coherence for sustainable development (SDGs 17.4 and 17.14).

The governance mechanisms adopted to work towards these aspirations can take different forms such as government appointed commissions to prepare policy proposals (Strøm, Müller, and Bergman, 2003)[3]; issue-specific multi-stakeholder forums (e.g., Londres, Larson, and Sarmiento Barletti, 2021; Parker, Coleman, Manyindo, Mukuru, and Schultz, 2019); public-private partnerships (e.g., Beisheim and Campe, 2012; Prats, 2021; Xiong, Chen, Wang, and Zhu, 2018); or inter-ministerial coordination committees with non-state actor involvement (Bouckaert, Peters, and Verhoest, 2010). A good example for the latter are the designated SDG bodies set up in many countries to coordinate and oversee the implementation of the 2030 Agenda across government ministries, departments and organizations (see Breuer, Leininger, and Malerba in this book). For another example, in the water sphere river basin organizations can principally provide platforms for coordination among state and non-state actors using the resource (Dombrowsky and Hensengerth, 2018; Huitema and Meijerink, 2017; Pahl-Wostl, Dombrowsky, and Naho, 2021, and Srigiri and Scheumann in this book).

Which kind of governance mechanism is adopted will not only depend on the policy problem at hand but also on the forms of actor coordination that exist in a society and that are usually referred to as governance modes (e.g., Héritier, 2002; Pahl-Wostl, 2015) or governance styles (Meuleman, 2018). Following Ostrom, Tiebout, and Warren (1961), the political economy school of thought distinguishes between three ideal-typical governance modes: markets, hierarchies and networks that produce different patterns of interaction between state and non-state actors. While hierarchical governance tends to produce coercion as a pattern of interaction, competition is more likely to be found under market-oriented governance, and cooperation will more likely occur under network-based governance (e.g., Ostrom et al., 1961; Stephan, Marshall, and McGinnis, 2019). Against the backdrop of the problems of input legitimacy mentioned above, it could be argued that societies, dominated by a hierarchical governance mode would generally benefit from more participatory and interactive governance mechanisms. By contrast, in societies where networked governance is already the dominant mode the adoption of further participatory mechanisms might make decision-making unnecessarily slow and ineffective. Furthermore, which governance modes (or combinations thereof) are dominant in a given society will, among other factors, again depend on the political-institutional preconditions in place. The evolution of governance mechanisms thus depends on the broader social, political, and institutional context in a society. In how far (and which) different contexts provide an enabling environment for the emergence of effective governance mechanisms for achieving coherent sustainability policies is a matter of inquiry that will be addressed in this book.

Policy Coherence and Policy Mixes

As the SDGs comprise many different goals and targets, which are interlinked, it is obvious that policies implemented to address one specific goal might affect a number of other goals and targets. This challenge and on how to avoid contradiction (trade-offs) and foster mutual support (synergies) between interdependent policy objectives has received increased attention in current research in political science and environmental economics (Cejudo and Michel, 2017; Lenschow, Bocquillon, and Carafa, 2018; Nilsson et al., 2012; Srigiri and Dombrowsky, 2022).

The literature often uses the terms *coordination, policy coherence*, and *policy integration* interchangeably. Policy coherence and policy integration are often seen as loosely equivalent terms and understood as types of coordination that seek to achieve compatibility among the objectives of different policy areas (Cejudo and Michel, 2017). In this book, however, we adopt a conceptualization that defines all three terms as processes rather than outcomes (e.g., Cejudo and Michel, 2017; Tosun and Lang, 2017). *Coordination* describes the process of sharing knowledge and information between different decision-making centers (e.g. of government agencies or programs). *Policy coherence* builds on coordination, as it describes the process of designing policies, which are complementary in their objective and can potentially "*reinforce each other so that they solve, together, a greater and more complex problem*" (Cejudo and Michel, 2017, p. 756). Thus, policy coherence is a process that generally occurs during the initial design phase and describes the setting up of policies (Cejudo and Michel, 2017). Finally, *policy integration* refers to the entire process of making strategic and administrative decisions aimed at solving a complex problem. Solving this complex problem is an objective that encompasses – but exceeds – the objectives of individual decision-making centers. Instead, under policy integration, at every moment of the policy process decision-making centers make decisions based on a new collective logic that aims at addressing the complex problem at hand (ibid. p. 755). Obviously then, addressing such complex problems cannot be achieved through the simultaneous implementation of individual policy instruments but instead requires the combination of different instruments in so-called *policy mixes* (del Rio and Howlett, 2013; Kirschke and Kosow, 2021; Kosow, Weimer-Jehle, León, Minn, and Zahumensky, 2020).

Against this background, over the past decade, the term 'policy-mix' has been increasingly taken up by both policy makers and scholars in the field of environmental economics (e.g., Lehmann, 2012), innovation studies (e.g., Nauwelaers et al., 2009), and policy analysis (e.g., Flanagan, Uyarra, and Laranja, 2011; Howlett and Rayner, 2007). Normative expectations about policy mixes include demands for them to be 'appropriate', 'effective', 'well-functioning' or 'balanced', the achievement of which is generally perceived as a challenge of 'coherence' and 'coordination' (e.g. Guy et al., 2009; CREST Policy Mix Expert Group, 2008; Rosenow et al., 2016; Guy, Boekholt, Cunningham, and Rammer, 2009; Rosenow, Fawcett, Eyre, and Oikonomou, 2016). The underlying premise of these perceptions appears to be that policymakers fail to use the full array of instruments that are theoretically available to them, which, in turn, is a negative thing. More recently, a set of more comprehensive studies has broadened the focus to also include the policy processes that lead to the adoption of policy mixes (Tosun and Koch, 2021). In this book, we draw on the concept of 'policy mixes' put forward by Rogge and Reichhardt (2016), who define the policy mix as a combination of the three building blocks: (i) elements, (ii) processes, and (iii) characteristics, with contributions to this books particularly focussing on the elements and characteristics of policy mixes.

Elements of policy mixes include a *policy strategy* to achieve objectives that are underpinned by long-term targets (such as national development plans, roadmaps, framework conventions and guidelines) and *policy instruments* as tools or techniques adopted by a governing body to achieve these policy objectives (such as regulations and standards, subsidies or market incentives, voluntary agreements, tariffs, and taxes, and charges).

Characteristics of policy mixes include the *consistency of their elements* and the *coherence of processes*. Consistency captures how well the elements of the policy mix are aligned with each other, thereby allowing policy objectives to be met. The degree of consistency could range from the absence of trade-offs to the existence of synergies within and across the elements of the policy mix. Coherence of processes, in turn, refers to synergistic and systematic policy making and implementation that contributes towards the achievement of policy objectives. Such synergistic and systematic policy processes might be achieved through structural and procedural mechanisms, including for example strategic planning, coordinating structures and communication networks. The achievement of coherence of policy processes across different policy sectors and governance levels requires advanced organizational capacities, such as, for example, the ability to assemble related knowledge from diverse sources, to build networks with all relevant actors, or to engage with multiple stakeholders (Rogge and Reichardt, 2016).

This Book's Structure and Content

This book is the result of the combined efforts of an international and interdisciplinary cast of senior researchers, with significant professional experience in policy consultancy. We hope that, beyond their relevance for the academic community, findings presented in this book will inform high-level policy debates about sustainability transformation and provide orientation for practitioners in the field of international development cooperation.

Following this introductory chapter, the book is structured as follows:

Part I sets the scene by presenting an overview of the academic 'state-of-the-art" research around SDG interlinkages and implementation. The two chapters in this

Part offer methodological approaches to navigate the rapidly growing body of literature in this field in a way that allows for the extraction of actionable policy implications.

Part II comprises two chapters, which focus on the institutions and governance structures that are needed to achieve coherent implementation of the SDGs across different policy areas and levels of government. They further investigate the degree to which existing institutional set-ups answer to those needs.

Part III adopts a problem-oriented approach. The five chapters comprising this Part delve into the analysis of interdependencies and trade-offs between specific selected goals and targets and critically discuss the capability and limitations of existing governance mechanisms to address the resulting challenges. They do so based on evidence from individual and comparative qualitative case studies from varying social, political, and economic contexts, as well as from cross-national quantitative analyses.

Part IV takes a solution-oriented stance. The three chapters in this Part focus on policy mixes and policy processes that have demonstrated potential in effectively addressing specific SDG interlinkages and contributing to the integrated implementation of the 2030 Agenda.

Finally, **Part V**, presents conclusions based on key findings of the individual chapters and indicates avenues for future research.

Notes

1 The idea for this book emerged from the research project "Implementing the 2030 Agenda: Integrating Growth, Environment, Equality and Governance", which is being conducted since 2017 at IDOS (formerly German Development Institute/Deutsches Institut für Entwicklungspolitik, DIE) with financial support from the German Federal Ministry for Economic Cooperation and Development. The introduction to this book is partly based on the conceptual analytical framework of this project, to the development of which our colleagues Ines Dombrowsky (IDOS) and Julia Leininger (IDOS) made substantial contributions.

2 Literature on policy integration usually refers to "integration" as a dimension on which policies in a specific issue area can be assessed as being more or less coherent. Integration can thus be conceptualized as a continuum that ranges from "least coherent" to "fully coherent" (see, for instance, UN [United Nations], 2018).

3 In 2018, for example, the German Federal Government appointed the 'Coal Commission' (officially Commission on Growth, Structural Change and Employment) to develop a proposal for phasing out the extraction and use of coal. The body was made up of 31 individuals from the spheres of politics, business, science and civil society and presented its final report in 2019.

References

Albritton, R.B. and Bureekul, T. (2009). Are Democracy and "Good Governance" Always Compatible? Competing Values in the Thai Political Arena. Retrieved from: https://kipdf.com/working-paper-series-no-47_5b18104f7f8b9ac0708b45ca.html.

Allen, C., Metternicht, G., and Wiedmann, T. (2021). Priorities for science to support national implementation of the sustainable development goals: A review of progress and gaps. *Sustainable Development*, 1–18. doi:10.1002/sd.2164.

Bandeira, P. and Ferraro, A. (2017). Integrating participatory institutions into the traditional representative and bureaucratic model of public governance. *International Political Science Review*, 38(5), 642–658. doi:10.1177/0192512116641815.

Barrett, S. and Graddy, K. (2000). Freedom, growth, and the environment. *Environment and Development Economics*, 5(4), 433–456. doi:10.1017/S1355770X00000267.

Beisheim, M. and Campe, S. (2012). Transnational Public–Private Partnerships' Performance in Water Governance: Institutional Design Matters. *Environment and Planning C: Government and Policy*, 30(4), 627–642. doi:10.1068/c1194.

Biermann, F. et al. (2014). Integrating Governance into the Sustainable Development Goals. In *POST2015/UNU-IAS Policy Brief #3*. Tokyo: United Nations University Institute for the Advanced Study of Sustainability.

Bundesministerium für Umwelt, Naturschutz und nukleare Sicherheit and Bundesministerium für wirtschaftliche Zusammenarbeit und Entwicklung. (2020). UN-Nachhaltigkeitsforum: Deutschland fordert nachhaltigen Weg aus der Corona-Krise. Ministerkonferenz überprüft Fortschritt bei der Umsetzung der UN-Ziele für nachhaltige Entwicklung im Lichte der COVID-19-Pandemie. *Pressemitteilung Nr. 114/20*. Retrieved from: www.bmu.de/pressemitteilung/un-nachhaltigkeitsforum-deutschland-fordert-nachhaltigen-weg-aus-der-corona-krise.

Bøås, M. (1998). Governance as multilateral development bank policy: The cases of the African development bank and the Asian development bank. *The European Journal of Development Research*, 10(2), 117–134. doi:10.1080/09578819808426720.

Bouckaert, G., Peters, B.G., and Verhoest, K. (2010). *The Coordination of Public Sector Organizations. Shifting Patterns of Public Management*. Palgrave Macmillan.

Breuer, A., Janetschek, H., and Malerba, D. (2019). Translating Sustainable Development Goal (SDG) interdependencies into policy advice. *Sustainability*, 11(7). doi:10.3390/su11072092.

Breuer, A., Leininger, J., and Tosun, J. (2019). Integrated Policymaking. In *Choosing an Institutional Design for Implementing the Sustainable Development Goals (SDGs)*. Bonn: Deutsches Institut für Entwicklungspolitik.

Cejudo, G.M. and Michel, C.L. (2017). Addressing fragmented government action: coordination, coherence, and integration. *Policy Sciences*, 50(4), 745–767. doi:10.1007/s11077-017-9281-5.

Chimhowu, A.O., Hulme, D., and Munro, L.T. (2019). The 'New' national development planning and global development goals: Processes and partnerships. *World Development*, 120, 76–89. doi:10.1016/j.worlddev.2019.03.013.

Cingolani, L. (2013). The State of State Capacity: a review of concepts, evidence and measures. In *UNU-MERIT Working Papers No*. 053. Maastricht: Maastricht University.

Cireno Fernandes, F., de Moura Palotti, P.L., and Christo Fernandes, C.C. (2017). Measuring and comparing state capacities to achieve the Sustainable Development Goals (SDGs) in an emerging economy: challenges and research proposal. In *3rd International Conference on Public Policy (ICPP3)*. Vaulx-en-Velin: International Public Policy Association.

CREST Expert Group. (2008). *Policy Mix Peer Reviews: Country Report AUSTRIA*. Retrieved from: https://ec.europa.eu/invest-in-research/pdf/download_en/pol_mix_at.pdf.

Dahl, R. (1989). *Democracy and its Critics*. London: Yale University Press.

Das, S.C. and Das, G. (2018). Public Resource Allocation through Grassroots Democratic Institutions: Evidence from Assam, India. *International Journal of Public Administration*, 41 (16), 1325–1337. doi:10.1080/01900692.2017.1387918.

Dasgupta, S. and De Cianac, E. (2018). The influence of institutions, governance, and public opinion on the environment: Synthesized findings from applied econometrics studies. *Energy Research & Social Science*, 43, 77–95. doi:10.1016/j.erss.2018.05.023.

de Ávila Gomide, A., de Sá e Silva, F., and Pires, R.R.C. (2014). Capacidades estatais e políticas públicas: passado, presente e futuro da ação governamental para o desenvolvimento. In L. Monteiro Monasterio, M. Côrtes Neri, and S.S. Dillon Soares (Eds.), *Brasil em desenvolvimento 2014: estado, planejamento e políticas públicas* (pp. 231–246). Brasília: Instituto de Pesquisa Econômica Aplicada.

del Rio, P. and Howlett, M. (2013). Beyond the "Tinbergen Rule" In *Policy Design: Matching Tools and Goals in Policy Portfolios* (Vol. 1, pp. 1–16).

Dombrowsky, I. and Hensengerth, O. (2018). Governing the Water-Energy-Food Nexus Related to Hydropower on Shared Rivers—The Role of Regional Organizations. *Frontiers in Environmental Science*, 6(153). doi:10.3389/fenvs.2018.00153.

European Commission. (2021). *Strategic Foresight Report The EU's capacity and freedom to act.* Retrieved from: https://ec.europa.eu/info/sites/default/files/foresight_report_com750_en.pdf.

Flanagan, K., Uyarra, E., and Laranja, M. (2011). Reconceptualising the 'policy mix' for innovation. *Research Policy*, 40(5), 702–713. doi:10.1016/j.respol.2011.02.005.

Fossati, D. (2016). Is Indonesian local government accountable to the poor? Evidence from health policy implementation. *Journal of East Asian Studies*, 16, 307–330. doi:10.1017/jea.2016.17.

Fukuyama, F. (2013). What Is Governance? *Governance: An International Journal of Policy, Administration, and Institutions*, 26(3), 347–368. doi:10.1111/gove.12035.

Glassa, L.-M. and Newiga, J. (2019). Governance for achieving the Sustainable Development Goals: How important are participation, policy coherence, reflexivity, adaptation and democratic institutions? *Earth System Governance*, 2. doi:10.1016/j.esg.2019.100031.

Goodhart, M. et al. (2012). *Democratic Imperatives: Innovations in Rights, Participation, and Economic Citizenship*. Washington, DC: American Political Science Association.

Grindle, M.S. (1996). *Challenging the State: Crisis and Innovation in Latin America and Africa.* Cambridge: Cambridge University Press.

Guy, K., Boekholt, P., Cunningham, P., and Rammer, C. (2009). *Designing policy mixes: enhancing innovation system performance and R&D investment levels*. Retrieved from: www.researchgate.net/publication/270798985.

Halperin, M., Siegle, J., and Weinstein, M. (2010). *The Democracy Advantage. How Democracies Promote Prosperity and Peace.* London: Routledge.

Héritier, A. (2002). New Modes of Governance in Europe: Policy Making without Legislating? Retrieved from: www.ihs.ac.at/publications/pol/pw_81.pdf.

Howlett, M. and Rayner, J. (2007). Design Principles for Policy Mixes: Cohesion and Coherence in 'New Governance Arrangements'. *Policy and Society*, 26(4), 1–18. doi:10.1016/S1449-4035(07)70118-2.

Huitema, D. and Meijerink, S. (2017). The politics of river basin organizations: institutional design choices, coalitions, and consequences. *Ecology and Society*, 22(2). doi:10.5751/ES-09409-220242.

Inter-American Development Bank. (2020). IDB-AFD Strengthen Partnership in Times of COVID-19 Crisis and Beyond. Retrieved from: www.iadb.org/en/news/idb-afd-strengthen-partnership-times-covid-19-crisis-and-beyond.

International Science Council. (2017). *A guide to SDG interactions: From science to implementation.* Retrieved from: https://sdgintegration.undp.org/guide-sdg-interactions-science-implementation?utm_source=EN&utm_medium=GSR&utm_content=US_UNDP_PaidSearch_Brand_English&utm_campaign=CENTRAL&c_src=CENTRAL&c_src2=GSR&gclid=Cj0KCQjwntCVBhDdARIsAMEwACkyEOyWSNwYflT60UxvM0SbIsHFotS3dtLb4ZgR8R4Y4doHlY9Q6CYaApCaEALw_wcB.

Jessop, B. (1998). The rise of governance and the risks of failure: the case of economic development. *International Social Science Journal*, 50, 29–45. Retrieved from: https://onlinelibrary.wiley.com/doi/epdf/10.1111/issj.12186?saml_referrer.

Jreisat, J.E. (2012). Rethinking Administrative Capacity Development: The Arab States. *Public Organization Review*, 12, 139–155. doi:10.1007/s11115-011-0164-5.

Kirschke, S. and Kosow, H. (2021). *Designing policy mixes for emerging wicked problems. The case of pharmaceutical residues in freshwaters*. Paper presented at the The 5th International Conference on Public Policy (ICPP5), Barcelona.

Kosow, H., Weimer-Jehle, W., León, C., Minn, F., and Zahumensky, Y. (2020). *Designing synergetic and sustainable policy mixes – a new methodology*. Paper presented at the Paper presented at the iEMSs 2020. http://trust-grow.de/wp-content/uploads/2020/09/E3_-Kosow-et-al-2020-Designing-synergetic-and-sustainable-policy-mixes-20200914.pdf.

Landman, T. and Lauth, H.-J. (2019). Political Trade-Offs: Democracy and Governance in a Changing World. *Politics and Governance*, 7(4), 237–242. doi:10.17645/pag.v7i4.2642.

Leftwich, A. (1995). Bringing politics back in: towards a model of the developmental state. *Journal of Development Studies*, 31(3), 400–427. doi:10.1080/00220389508422370.

Lehmann, P. (2012). Justifying a Policy Mix for Pollution Control: A Review of Economic Literature. *Journal of Economic Surveys*, 26(1), 71–97. doi:10.1111/j.1467-6419.2010.00628.x.

Lenschow, A., Bocquillon, P., and Carafa, L. (2018). Understanding coherence between policy spheres. *Environmental Policy and Governance*, 28(5), 323–328. doi:10.1002/eet.1818.

Li, H., Squire, L., and Zou, H.-F. (1998). Explaining International and Intertemporal Variations in Income Inequality. Retrieved from: https://EconPapers.repec.org/RePEc:cuf:wpaper:73.

Londres, M., Larson, A.M., and Sarmiento Barletti, J.P. (2021). The costs of elite-oriented multi-stakeholder forums to address deforestation: the case of the Green Municipalities Program in the Brazilian Amazon. *International Forestry Review* 23(S1), 76–89. Retrieved from: www.cifor.org/knowledge/publication/8061.

Mann, M. (1998). *States, War and History*. Oxford: Blackwell Publishers.

Mechler, S.a.D. (2020). Bouncing Forward Sustainably: Pathways to a post-COVID World Governance for Sustainability. Retrieved from: https://stories.council.science/iiasa-isc/.

Meuleman, L. (2018). *Metagovernance for Sustainability: A Framework for Implementing the Sustainable Development Goals*: Routledge.

Miola, A., Borchardt, S., and Buscaglia, D. (2019). Interlinkages and policy coherence for the Sustainable Development Goals implementation: An operational method to identify trade-offs and co-benefits in a systemic way. Retrieved from: https://publications.jrc.ec.europa.eu/repository/bitstream/JRC115163/sdg_interlinkages_jrc115163_final_on_line.pdf.

Nauwelaers, C., Boekholk, P., Mostert, B., Cunningham, P., Guy, K., Hofer, R., and Rammer, C. (2009). Policy Mixes for R&D in Europe. Retrieved from: https://op.europa.eu/en/publication-detail/-/publication/b3a5d015-05aa-45f0-95c0-c03535fca99f/language-en/format-PDF/source-68821682.

Nilsson, M., Zamparutti, T., Petersen, J.E., Nykvist, B., Rudberg, P., and McGuinn, J. (2012). Understanding Policy Coherence: Analytical Framework and Examples of Sector–Environment Policy Interactions in the EU. *Environmental Policy and Governance*, 22(6), 395–423. doi:10.1002/eet.1589.

Nwobashi, H.N. and Itumo, A. (2017). Democracy, Town Unions and Provision of Health Facilities in Rural Communities of Ebonyi State: Evaluation of Some Selected Communities. *The Turkish Online Journal of Design, Art and Communication*, 7, 496–506. doi:10.7456/1070ASE/051.

O'Riain, S. (2004). *The Politics of High Tech Growth. Developmental Network States in the Global Economy*. Cambridge: Cambridge University Press.

ODI. (2020). Four ways to prevent the Sustainable Development Goals becoming a casualty of Covid-19 responses. Retrieved from: https://odi.org/en/insights/four-ways-to-prevent-the-sustainable-development-goals-becoming-a-casualty-of-covid-19-responses.

Organisation for Economic Co-operation and Development. (2019). Policy Coherence for Sustainable Development 2019: Empowering People and Ensuring Inclusiveness and Equality. Retrieved from: www.oecd.org/publications/policy-coherence-for-sustainable-development-2019-a90f851f-en.htm.

Organisation for Economic Co-operation and Development. (2020). *Global Outlook on Financing for Sustainable Development 2021: A New Way to Invest for People and Planet.* Paris: OECD Publishing.

Ostrom, V., Tiebout, C.M., and Warren, R. (1961). The organization of government in metropolitan areas: a theoretical inquiry. *The American political science review*, 55(4), 831–842. doi:10.1017/S0003055400125973.

Pahl-Wostl, C. (2015). Governance Modes. In C. Pahl-Wostl (Ed.), *Water Governance in the Face of Global Change: From Understanding to Transformation* (pp. 85–98). Cham: Springer International Publishing.

Pahl-Wostl, C., Dombrowsky, I., and Naho, M. (2021). Water governance and policies. In J. J. Bogardi et al. (Eds), *Handbook of Water Resources Management: Discourses, Concepts and Examples* (pp. 253–272). Springer.

Parker, R., Coleman, E., Manyindo, J., Mukuru, E., and Schultz, B. (2019). Impacts of community stakeholder engagement interventions in Ugandan oil extractives. Retrieved from: www.3ieimpact.org/evidence-hub/publications/impact-evaluations/impacts-community-stakeholder-engagement-interventions.

Pham-Truffert, M., Metz, F., Fischer, M., Rueff, H., and Messerli, P. (2020). Interactions among Sustainable Development Goals: Knowledge for identifying multipliers and virtuous cycles. *Sustainable Development*, 28, 1236–1250. doi:10.1002/sd.2073.

Pickering, J., Bäckstrand, K., and Schlosberg, D. (2020). Between environmental and ecological democracy: theory and practice at the democracy-environment nexus. *Journal of Environmental Policy and Planning*, 22(1), 1–15. doi:10.1080/1523908X.2020.1703276.

Prats, J. (2021). *The Governance of Public –Private Partnerships. A Comparative Analysis.* Washington, DC: Inter-American Development Bank.

Rogge, K.S. and Reichardt, K. (2016). Policy mixes for sustainability transitions: An extended concept and framework for analysis. *Research Policy*, 45(8), 1620–1635. doi:10.1016/j.respol.2016.04.004.

Rosenow, J., Fawcett, T., Eyre, N., and Oikonomou, V. (2016). Energy efficiency and the policy mix. *Building Research & Information*, 44(5–6), 562–574. doi:10.1080/09613218.2016.1138803.

Savoia, A. and Sen, K. (2014). Measurement, evolution, determinants, and consequences of state capacity: a review of recent research. *Journal of Economic Surveys*, 29(3), 441–458. doi:10.1111/joes.12065.

Srigiri, S.R. and Dombrowsky, I. (2022). Analysing the Water-Energy-Food Nexus from a polycentric governance perspective: Conceptual and methodological framework. *Frontiers in Environmental Science. Environmental Economics and Management*, February 2022. Retrieved from: www.frontiersin.org/articles/10.3389/fenvs.2022.725116/abstract.

Stasavage, D. (2020). Democracy, Autocracy, and Emergency Threats: Lessons for COVID-19 From the Last Thousand Years. *International organization*, 74(S1), E1–E17. doi:10.1017/S0020818320000338.

Stephan, M., Marshall, G.R., and McGinnis, M.D. (2019). An Introduction to Polycentricity and Governance. In A. Thiel, W.A. Blomquist, and D.E. Garrick (Eds),

Governing Complexity: Analysing and Applying Polycentricity (pp. 21–44). Cambridge: Cambridge University Press.

Strøm, K., Müller, W.C., and Bergman, T. (2003). *Delegation and Accountability in Parliamentary Democracies*. Oxford: Oxford University Press.

Swanson, D., Pintér, L., Bregha, F., Volkery, A., and Jacob, K. (2004). *National strategies for sustainable development. Challenges, approaches and innovations in strategic and co-ordinated action*. Winnipeg, Bonn: International Institute for Sustainable Development, Deutsche Gesellschaft für Technische Zusammenarbeit.

Torfing, J., Peters, G., Pierre, J., and Sørensen, E. (2012). *Interactive Governance: Advancing the Paradigm*. Oxford: Oxford University Press.

Tosun, J. and Koch, M. A. (2021). Policy mixes for biodiversity: a diffusion analysis of state-level citizens' initiatives in Germany. *Journal of Environmental Policy & Planning*, 1–13. doi:10.1080/1523908X.2021.1992265.

Tosun, J. and Lang, A. (2017). Policy integration: mapping the different concepts. *Policy Studies*, 38(6), 553–570. doi:10.1080/01442872.2017.1339239.

Tosun, J. and Leininger, J. (2017). Governing the Interlinkages between the Sustainable Development Goals: Approaches to Attain Policy Integration. *Global Challenges*, 1(9), 1700036. doi:10.1002/gch2.201700036.

United Nations. (2018). *World Public Sector Report 2018*. Retrieved from: www.un.org/development/desa/publications/world-public-sector-report-2018.html.

United Nations Development Programme. (2020). *Beyond Recovery: Towards 2030*. New York: United Nations Development Programme.

Weiss, T.G. (2000). Governance, good governance and global governance: conceptual and actual challenges. *Third World Quarterly*, 21(5), 795–814. Retrieved from: https://library.fes.de/libalt/journals/swetsfulltext/11375717.pdf.

World Bank. (2020). *World Bank Group Partnership Fund for the Sustainable Development Goals. Annual Report 2020*. Washington, DC: World Bank.

Wurster, S. (2013). Comparing ecological sustainability in autocracies and democracies. *Contemporary Politics*, 19(1), 76–93. doi:10.1080/13569775.2013.773204.

Xiong, W., Chen, B., Wang, H., and Zhu, D. (2018). Governing public–private partnerships: A systematic review of case study literature. *Australian Journal of Public Administration*, 78, 95–112. doi:10.1111/1467-8500.12343.

2

SCIENTIFIC APPROACHES TO SDG INTERACTIONS ANALYSES

The State of Play

Therese Bennich, Nina Weitz, and Henrik Carlsen

Introduction

The 2030 Agenda should be treated as an indivisible whole, meaning that the implementation process must consider interactions between its goals, targets, and indicators. Failing to account for these interactions in implementing strategies could hamper the attainment of the 2030 Agenda and associated Sustainable Development Goals (SDG) in several ways. First, important trade-offs might be overlooked, giving rise to unintended consequences of actions aimed to promote the SDGs. Second, opportunities to leverage synergies between goals that are mutually supporting might be lost. Third, failing to recognize the interactions might lead to policy resistance, i.e., situations where actions to implement the SDGs trigger indirect effects that cause policies to fail.

As highlighted in the introduction to this volume, the practical implementation of the ambitious and integrated vision of the 2030 Agenda brings new challenges to policymakers. There is no political consensus on how to best identify and govern interactions between the SDGs, and the formulation of the 2030 Agenda itself offers little guidance. The present volume on the governance of the SDGs focuses on several critical questions; asking what we know about the most important interactions between the SDGs, what governance mechanisms are needed to manage them, and how to strengthen the capacities of different stakeholders to take an integrated approach to SDG implementation. This chapter focuses specifically on how the scientific literature addresses these questions to support integrated governance of the SDGs.

The number of publications on what we refer to as "SDG interactions" is growing rapidly, in what seems to be an exponential fashion (Figure 2.1). Using the search string ("sustainable development goals" OR SDGs) AND (interlinkages OR interlinked OR interactions OR interconnected OR interconnections OR integrated) in the Scopus database generates 2,641 hits. Using the same search

DOI: 10.4324/9781003254683-2

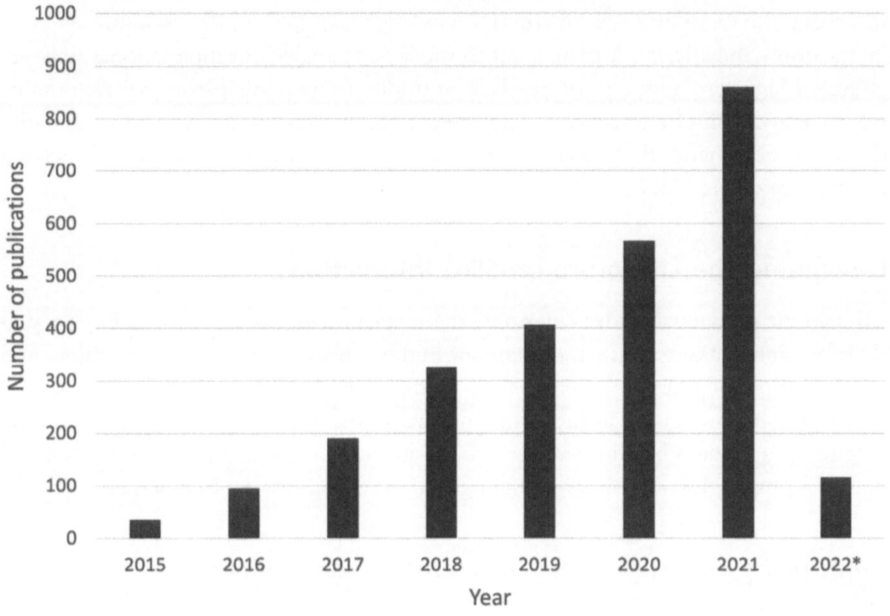

FIGURE 2.1 The number of publications on SDG interactions in the Scopus database
*Cut-off date: February 9, 2022.
Source: Author's own elaboration

string in Google Scholar generates close to 25,000 hits. The present chapter provides a starting point for navigating this vast body of literature. It draws on previous research conducted by the authors of this chapter in combination with a review of recently published literature on SDG interactions. More specifically, this chapter:

1. Introduces a reading guide to the literature on SDG interactions, based on a review and mapping of studies published between 2015 and 2019, originally presented in Bennich, Weitz, and Carlsen (2020).
2. Revisits the literature on SDG interactions, reviewing work published between 2019 and 2022. The aim is to analyze if some of the knowledge gaps identified in the early literature have received attention in recent work.

The present chapter is structured as follows. First, it presents the reading guide to the literature on SDG interactions. The reading guide, originally presented in Bennich et al. (2020), was developed based on a literature review of 70 scientific articles published between 2015 and early 2019. The analysis focused specifically on identifying what policy challenges the studies responded to, how SDG interactions were conceptualized (what is connected to what and how?), and what data sources and methods were used. The analysis did also include a mapping of what approaches can be used to address different policy and governance challenges in SDG implementation. Second, this chapter presents a review of critical knowledge gaps in the early literature on SDG interactions. Specifically, it focuses on (i) the lack of analysis of international spillover effects of SDG progress; and (ii) the lack of studies using participatory research methods to analyze SDG interactions. The present chapter concludes with recommendations for improving the ability of the scientific literature to support integrated governance of the SDGs.

Navigating the Literature on SDG Interactions

To help the individual reader determine the scope and usefulness of a specific study to their decision-making context, consistency in the conceptualization of interactions and categorization of policy challenges and methods are helpful. For this purpose, we present the reading guide in the following five sub-sections to navigate the literature. The reading guide is based on a set of questions. For a more detailed presentation and illustrative examples of studies responding to these questions, see Bennich et al. (2020).

What policy challenges does the literature respond to?

A defining feature of the literature on SDG interactions is that it aims to be relevant to policymaking. The first question to ask when approaching the literature on SDG interactions is: **To what policy challenge(s) does this study respond?** In the early stages of SDG implementation, studies on SDG interactions typically focused on i) policy integration and coherence; ii) policy innovation; iii) contextualizing SDG interactions; iv) policy prioritization; v) realizing new stakeholder perspectives; and vi) monitoring and evaluation.

Policy integration and coherence: coherent policies are needed to generate efficient and sustainable outcomes. Studies addressing this challenge focus on how more integrated decision-making can be realized, for example by informing new ways of collaborating across government levels and societal sectors. Another way the scientific literature supports policy integration and coherence is by providing evidence on how the SDGs (all or a sub-set) interact in terms of synergies and trade-offs (see e.g., Scharlemann et al., 2020 and Warchold, Pradhan, and Kropp, 2021).

Policy innovation: Researchers have been questioning the outputs of traditional policymaking, stating that the 2030 Agenda requires new uses of existing policy instruments or that completely new approaches are developed. Studies contributing to policy innovation often focus on structural change in specific sectors, e.g., the business sector (Dahlmann, Stubbs, Griggs, and Morrell, 2019). However, there are also studies offering more general analyses of how to turn trade-offs into synergies (Kroll, Warchold, and Pradhan, 2019).

Contextualizing SDG interactions: While the 2030 Agenda is globally focused, its implementation requires that goals and their interactions are understood in context (Nilsson et al., 2018; Nilsson, Griggs, and Visbeck, 2016). Studies aiming to contextualize SDG interactions span governance levels (including analysis at the local (Bennich et al., 2021), national (Weitz, Carlsen, Nilsson, and Skånberg,

2018), and regional levels (Allen et al., 2017)) and societal sectors (e.g., the forest sector; see Matsumoto, Hasegawa, Morita, and Fujimori, 2019).

Policy prioritization: Some policy options might be prerequisites for other policies to succeed. Other policies might deliver similar outcomes but might be more or less feasible to pursue due to contextual political, financial, or technological factors. Therefore, making priorities constitutes a key challenge in the implementation of the 2030 Agenda. Studies address this challenge in various ways, for example by providing knowledge about systemic multiplier effects (Pham-Truffert, Metz, Fischer, Rueff, and Messerli, 2020) and by offering frameworks and tools to guide priority setting (Weitz et al., 2018).

Integrated perspective: Stakeholders from a range of sectors need to take an active part in the implementation of the 2030 Agenda. Hence, the ability of these stakeholders to take an integrated perspective is key. Some studies aim to support this by, for example, sharing practices for teaching sustainability and systems thinking in the context of SDGs (Weber, Lindenmeyer, Liò, and Lapkin, 2021), and by reflecting on practical experiences of taking an integrated approach to SDG implementation (Krantz and Gustafsson, 2021).

Monitoring and evaluation: Studies address this challenge in multiple ways. For example, they explore accountability regimes (Karlsson-Vinkhuyzen, Dahl, and Persson, 2018), review indicators (Giles-Corti, Lowe, and Arundel, 2020), and identify challenges and ways forward when it comes to measuring progress on integrated goals (Biggeri, Clark, Ferrannini, and Mauro, 2019).

What is an SDG interaction?

Scientific and policy debates often stress that the 2030 Agenda is integrated and indivisible. But what does this mean? Given the lack of conceptual clarity, the second question to ask when reading a study on SDG interactions is: **How are the interactions conceptualized?** To break this question down further, it might be useful to ask what "interaction entities" a given study focuses on, e.g., the goals, targets, or indicators of the 2030 Agenda. It might also be useful to ask what information is provided about the relationship between the interaction entities, here referred to as the "interaction qualifiers."

Interaction entities

Some studies focus solely on interactions within the 2030 Agenda itself. The interaction entities in such studies are commonly the goals, targets, or indicators of the 2030 Agenda. The analysis might focus on goal-goal interactions, target-target interactions, indicator-indicator interactions, or interactions across all these levels. There are also studies examining interactions across the policies used to support SDG implementation.

In addition to analyses of interactions internal to the 2030 Agenda, numerous studies explore interactions that are part of the broader policy landscape and context, here referred to as "external entities." External entities could be goals belonging to other global sustainability agendas, such as the Paris Agreement. They could also be policies and practices belonging to a wide range of areas, such as land management, business, and technology development.

Interaction qualifiers

The interaction qualifiers provide information about the relationships between two interaction entities. Many studies are limited to stating that a relationship exists, e.g., that SDG 1 is connected to SDG 4. Such studies do not provide information about the nature of that connection. Other studies guide the reader further by offering more detail. The information in such studies can be descriptive, for example labeling an interaction as neutral, a trade-off, or co-beneficial. Other authors make explicit if they identify correlations or claimed causalities. Finally, the interaction qualifiers can be semi-quantitative or fully quantified. The use of a numerical scale to assess the nature of pairwise interactions is an example of the former, while studies using integrated assessment models to explore interactions between SDG indicators provide examples of the latter.

What data sources underpin SDG interactions?

The third question to ask when approaching the literature on SDG interactions is: **What data sources are used to underpin the existence of SDG interactions?** Given the broad range of scientific disciplines engaging with the topic of the 2030 Agenda, the data come from diverse sources. In the early stages of SDG implementation, many studies performed literature reviews to identify SDG interactions. Hence, the most common data source was the scientific literature. The second most common data source was official databases such as the World Bank compilation of cross-country development data (World Bank, 2021). The third most used data source was expert and stakeholder knowledge. For example, experts and stakeholders have often been consulted in studies using numerical scales or labels to describe the relationships between selected SDGs. Other data sources used include models and spatial maps.

What methods and tools can be used to identify and analyze SDG interactions?

The fourth question to ask when approaching the literature on SDG interactions is: **What methods or tools are used to identify and analyze SDG interactions?** There are several methods to choose from, each with its strengths and weaknesses in terms of its ability to carry out integrated analyses of the 2030 Agenda. In the early stages of implementation, the most common method was document analysis. This was followed by different forms of qualitative and quantitative systems modeling methods. Other methods employed include network analysis, cross-impact analysis, participatory methods, statistical analysis, qualitative scenario analysis, and multi-criteria analysis.

Connecting scientific approaches and policy challenges

We now turn to the question of whether certain methods and tools are better equipped to address specific policy challenges than others. As a policymaker or practitioner, what tools are best suited to your specific needs and context? One way to answer this question is to map the scientific approaches commonly used to address certain challenges, for example using network analysis methods and tools. The network analysis presented in Bennich et al. (2020) combined a modularity-based technique and mapping for visualization (Newman, 2006; Newman and Girvan, 2004; Waltman, van Eck, and Noyons, 2010). This to identify how policy challenges, conceptualizations, data sources, and methods appear together in the early literature on SDG interactions. Three distinct clusters of co-occurrences were identified (Figure 2.2).

The yellow cluster addresses the challenge of policy prioritization. The methods and tools used belong to the quantitative modeling family, including integrated

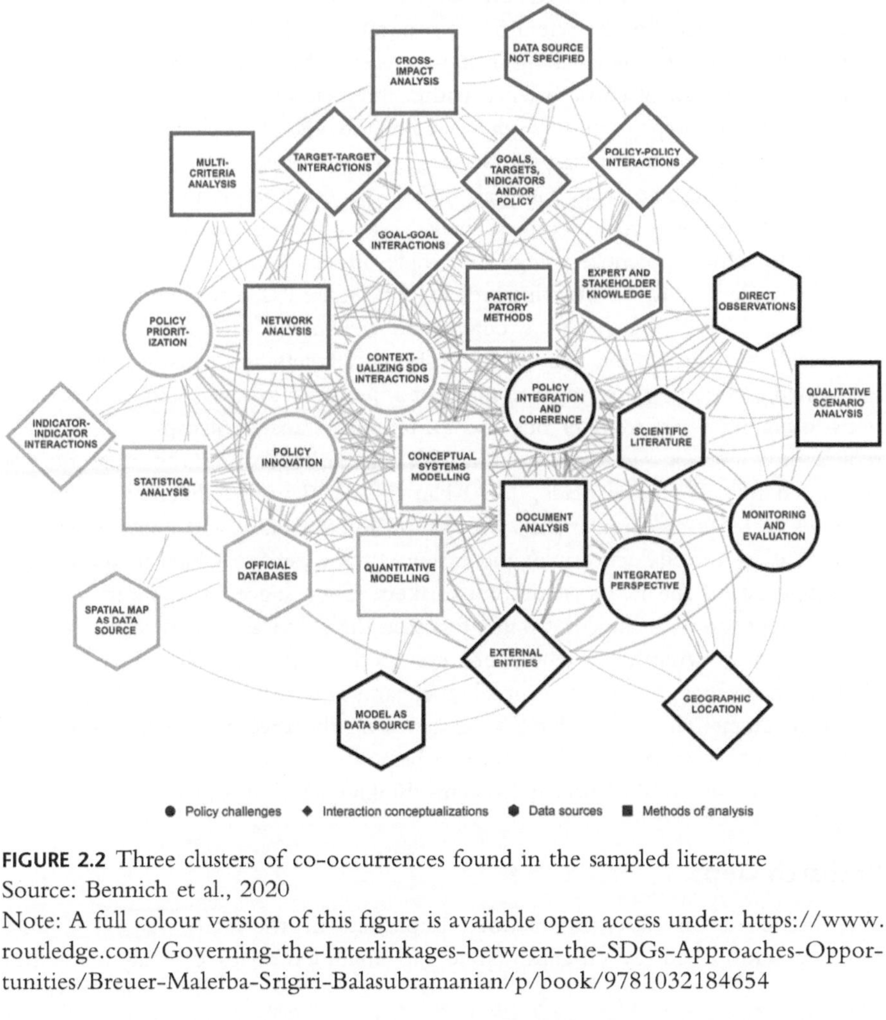

FIGURE 2.2 Three clusters of co-occurrences found in the sampled literature
Source: Bennich et al., 2020
Note: A full colour version of this figure is available open access under: https://www.routledge.com/Governing-the-Interlinkages-between-the-SDGs-Approaches-Opportunities/Breuer-Malerba-Srigiri-Balasubramanian/p/book/9781032184654

assessment models and system dynamics models. Another method used to respond to the challenge of making policy priorities include statistical analysis. Commonly used data sources in this cluster include official databases. The analysis is often detailed and place-based, hence also responding to the challenge of translating the global 2030 Agenda to specific regional, national, and local decision-making contexts.

The red cluster addresses challenges of policy integration and coherence. It also aims to help stakeholders take an integrated perspective. Methods and tools commonly used are qualitative, including literature reviews and scenario-building exercises. The studies belonging to this cluster often rely on the scientific literature and direct observations for data collection.

The green cluster encompasses methods and tools such as network analysis, cross-impact analysis, and participatory methods. Data is collected from expert and stakeholder consultations and engagement. However, it is not evident from the literature what policy challenges this cluster addresses. Hence, the green cluster highlights the need for the scientific community to clearly articulate if and how their studies support policymaking in the context of SDG implementation.

The identification of co-occurrence within three clusters, as shown in Figure 2.2, might inform the design of future studies based on the type of policy challenge one seeks to address. For example, if facing the challenge to make policy priorities, the co-occurrence network suggests that statistical analysis or quantitative modeling could be employed. If the policy challenge is to achieve policy coherence, reviewing the scientific literature to identify potential synergies and trade-offs could instead prove more useful.

As a complement to the mapping presented in Figure 2.2, recent studies provide an in-depth analysis of the specific strengths and weaknesses of different methods for assessing SDG interactions. For example, Breuer, Janetschek, and Malerba (2019) performed a review of a selected number of approaches for analyzing SDG interactions, outlining their strengths but also a number of analytical gaps (e.g., issues of replicability, context-sensitivity, and a lack of clarity around the time horizon accounted for). Di Lucia, Slade, and Khan (2021) identified different qualities of methods valued by decision-makers (e.g., simplicity, flexibility, and ability to generate results that are directly actionable) and explored the extent to which method developers understand the needs of decision-makers. They suggest that future methods development should include a stronger focus on the practical value methods bring to decision-makers. Another methods-oriented study is provided by Horvath et al. (2022). The authors provide a comprehensive and systematic review of the literature on SDG interactions, identifying and categorizing the methods used, and assessing their suitability to analyze SDG interactions based on a number of criteria (e.g., if the method promotes collaboration and systems thinking and if it is practical to use).

Research Gaps

The literature on SDG interactions published in the early stages of SDG implementation had several gaps. These gaps included a lack of analysis focusing on policy innovation, monitoring and evaluation, interactions between SDG indicators, interactions between actors, coherence across global agendas (e.g., analyzing interactions across the 2030 Agenda and the Paris Agreement), international spillovers, interactions between the full set of SDGs, and research using participatory methods (Bennich et al., 2020). We now revisit some of these gaps to explore to what extent the recent literature has made progress in terms of addressing them.

International spillovers

National and local SDG implementation can create international spillovers, i.e., impacts that could support or hamper SDG progress elsewhere. For example, because of climate change, harvest failures are expected to become increasingly common in the decades to come. Harvest failures create global disruptions through food shortages and spikes in food prices, affecting both export- and import-dependent countries. Responding to such disruptions is necessary to realize the SDGs on the national level (e.g., SDG 2 on zero hunger). However, some responses, such as stockpiling commodities or limiting exports, could worsen the situation for many countries, specifically low-income countries dependent on imports. In this way, local climate change impacts create responses that result in disturbances and damage far away from the original source (Benzie and Harris, 2020).

Recognizing international spillovers is imperative to the design of efficient implementation strategies, as it makes it possible to reinforce positive spillovers while ensuring that one country does not make progress on the SDGs in a way that is harmful to other countries. Thus, accounting for international spillovers speaks to both the universal and transformative principles of the 2030 Agenda. The importance of identifying, measuring, and ultimately managing international spillovers has been recognized in SDG reporting (Sachs, Schmidt-Traub, Kroll, Durand-Delacre, and Teksoz, 2017). Dedicated indicators have been proposed to track countries' international spillovers in multiple areas, including (i) the environmental and social spillovers embodied in trade; (ii) direct cross-border flows, (iii) international economic and financial flows; and (iv) peacekeeping and security spillovers (Sachs, Traub-Schmid, Kroll, Lafortune, and Fuller, 2021). Nevertheless, the analysis of such spillovers is not straightforward. For example, the available data is scarce and incomplete, there is a lack of frameworks to guide data collection, and tracing international spillovers might be politically sensitive (Sachs et al., 2017).

Given these challenges, the scientific community could play an important role in SDG implementation by expanding the knowledge base on international spillovers. Yet, this topic was poorly covered by the scientific literature on SDG interactions in the early stages of SDG implementation. Our recent review of literature on SDG interactions indicates that the topic has gained some attraction. We conducted a Scopus search for papers containing combinations of the keywords "sustainable development goals" or SDGs and international/cross-border/transboundary impacts/effects/spillovers in the title, keywords, or abstracts. The search included all work published between 2019 and 2022 and generated 99 hits. Only papers that i) explicitly address the SDGs (and not merely mention them or use them as a research framing); and b) analyze the SDGs in an integrated manner were included in the final sample. An initial screening and subsequent read of selected publications resulted in a final sample of five papers.

Some of the papers in the final sample look specifically at the international spillovers related to trade. Amos and Lydgate (2020) focus on SDG 12 on responsible consumption and production. They highlight the failure of high-consumption countries to account for the transboundary impacts of their resource demands. However, they also emphasize that accounting for transboundary impacts might offer opportunities for more efficient cooperation and moving beyond narrow problem definitions. Further, Malik et al. (2021) offer an analysis of the negative spillovers created by textile supply chains. Using multi-regional input-output analysis, the authors attribute fatal and non-fatal accidents in the global textile industry to the consumption of textile products within the EU. They find that some 80% of the accidents are associated with demands for textiles originating from EU countries, calling for both SDG policies and EU strategies that reduce such negative spillovers.

Zeng, Runting, Watson, and Carrasco (2022) explore human-nature interactions across distant systems, referring to such interactions as telecouplings (based on the metacoupling framework, see e.g., Liu, 2018). They stress that the effects of telecoupling will be persistent in SDG implementation, yet we know very little about these effects beyond a few selected environmental impacts. Using network analysis, the authors find that most SDG indicators are telecoupled with indicators of environmental impacts. They also find that accounting for such telecouplings might reduce trade-offs between environmental and development targets in low-income countries. Also, Zhao et al. (2021) base their work on the metacoupling framework to study the spatial interactions of the SDGs. They develop a new theoretical framework to guide the identification of SDG interactions within and across boundaries. They then demonstrate how it works by applying it to the case of tourism and nature conservation.

Finally, Engström et al. (2021) take a slightly different approach, looking at SDG implementation at the city level and its potential international spillovers. The authors emphasize that this scale of analysis is useful for several reasons. First, actions on the city level often differ from national-scale decision-making. Second, decisions made on the city level have global impacts. Third, city strategies for SDG implementation are only beginning to emerge. The authors propose a research agenda to support SDG governance at the city level, stressing the importance of adapting approaches to local realities, increasing access and comparability of data and tools, and reconciling existing knowledge to guide priority setting.

The scientific literature aside, valuable contributions to the knowledge about international spillovers can be found in the gray literature. An in-depth review of the gray literature is beyond the scope of the present chapter. However, examples of recent contributions include analysis proposing methods and tools to measure transboundary impacts (Organisation for Economic Co-operation and Development and Joint Research Centre, European Commission, 2021), research suggesting how to overcome data gaps and limitations (Schmidt-Traub, Hoff, and Bernlöhr, 2019), global analysis of the impacts of international spillovers on national SDG attainment (Li, Wiedmann, Fang, and Hadjikakou, 2021), and analysis of specific supply-chains (Hoff et al., 2019).

Participatory approaches

Participatory research refers to approaches that engage stakeholders in the research process. In the context of sustainability science, such efforts are generally promoted as means to gain new knowledge about both problems and solutions to sustainability issues. Further, they are seen as a way to increase collaboration between stakeholders and to create ownership and uptake of research results (Blackstock, Kelly, and Horsey, 2007; Lang et al., 2012; Sedlacko, Martinuzzi, Røpke, Videira, and Antunes, 2014). Thus, participatory approaches for analyzing SDG interactions could provide insight into the complex links between goals, targets, indicators, and policies. However, such approaches could also ensure science-based decision-making (Buyana et al., 2020) and that stakeholders from all parts of society take ownership of the 2030 Agenda, as called for by its formulation (United Nations, 2015).

Participatory approaches were scarce in the early literature on SDG interactions. There were studies engaging with experts and stakeholders, but primarily with the aim of collecting data, not involving stakeholders in the research process to a greater extent (Bennich et al., 2020). Revisiting the literature, with a specific focus on work published between 2019 and 2022, indicates that this gap remains. A Scopus-based search, using the search string ("sustainable development goals" OR SDGs) AND (interlinkages OR interlinked OR interactions OR interconnected OR interconnections OR integrated) AND (participatory OR transdisciplinary) generated 119 document results. Screening for papers explicitly using participatory approaches to analyze SDG interactions resulted in a final sample of three papers.

Schwindenhammer and Gonglach (2021) focus on emerging technologies in the agricultural sector. The authors engage with key stakeholders to explore how technology innovation might support goal attainment while at the same time also creating new governance challenges. Aguiar et al. (2020) present an approach to co-designing target-seeking scenarios with stakeholders. The approach aims to capture different perspectives on pathways for sustainability, analyzing how these perspectives converge or diverge across regional and global scales. Finally, Thiam et al. (2021) reflect on transdisciplinary research projects carried out at the city level in five different African countries. The paper focuses specifically on the interactions between the SDGs assessed in the five projects, but also on how the projects have contributed to SDG progress.

Conclusion

The successful implementation of the SDGs hinges on better understanding SDG interactions and dedicated strategies to govern these interactions in practice, or at least make policy and decision-making processes more responsive to them. The early stages of SDG implementation have seen exponential growth in the scientific literature on SDG interactions. Our review of this literature finds that it in many ways respond to key governance issues, such as how to achieve policy coherence, how to build the capacity of stakeholder to take an integrated perspective, and how to break down institutional and scientific "silos" to ensure integrated implementation of the SDGs. In addition, the knowledge gaps identified in the early stages of SDG implementation are now beginning to be addressed. For example, our knowledge about the international spillovers of SDG progress is expanding. Addressing such gaps is essential, not only to adhere to the "indivisible" principle of the 2030 Agenda, but also to realize its universal and transformative aspirations.

As a final remark, given the rapid growth in the literature, it is worth reiterating that not only new knowledge is needed to achieve integrated governance of the SDGs. In addition, the knowledge uptake among those responsible for implementation must be ensured. It should be easy to access, understand, and compare the existing knowledge on SDG interactions. Based on our review of the literature, participatory research is still relatively scarce in the study of SDG interactions. This is unfortunate, as engaging key stakeholders in the research process could be one way to ensure that the research informs governance in practice. Another promising suggestion, as previously emphasized by several authors in the field, is to create a data repository synthesizing knowledge on SDG interactions (Allen, Metternicht, and Wiedmann, 2021; Messerli et al., 2019; Nilsson et al., 2018; Pham-Truffert et al., 2020). To our knowledge, a comprehensive and continuously updated repository does not yet exist. The reading guide presented in this chapter could be one way to structure the literature in the development of such a repository.

References

Aguiar, A. P. D. et al. (2020). Co-designing global target-seeking scenarios: A cross-scale participatory process for capturing multiple perspectives on pathways to sustainability. *Global Environmental Change*, 65, 102198. doi:10.1016/j.gloenvcha.2020.102198.

Allen, C., Metternicht, G., and Wiedmann, T. (2021). Priorities for science to support national implementation of the sustainable development goals: A review of progress and gaps. *Sustainable Development*, 29(4), 635–652. doi:10.1002/sd.2164.

Allen, C., Nejdawi, R., El-Baba, J., Hamati, K., Metternicht, G., and Wiedmann, T. (2017). Indicator-based assessments of progress towards the sustainable development goals (SDGs): a case study from the Arab region. *Sustainability Science*, 12(6), 975–989. doi:10.1007/s11625-017-0437-1.

Amos, R. and Lydgate, E. (2020). Trade, transboundary impacts and the implementation of SDG 12. *Sustainability Science*, 15(6), 1699–1710. doi:10.1007/s11625-019-00713-9.

Bennich, T., Belyazid, S., Stjernquist, I., Diemer, A., Seifollahi-Aghmiuni, S., and Kalantari, Z. (2021). The bio-based economy, 2030 Agenda, and strong sustainability – A regional-scale assessment of sustainability goal interactions. *Journal of Cleaner Production*, 283, 125174. doi:10.1016/j.jclepro.2020.125174.

Bennich, T., Weitz, N., and Carlsen, H. (2020). Deciphering the scientific literature on SDG interactions: A review and reading guide. *Science of the Total Environment*, 728, 138405. doi:10.1016/j.scitotenv.2020.138405.

Benzie, M. and Harris, K. (2020). *Transboundary climate risk and adaptation. Science for Adaptation Policy Brief* 2. The World Adaptation Science Programme Secretariat, UNEP, Nairobi.

Biggeri, M., Clark, D.A., Ferrannini, A., and Mauro, V. (2019). Tracking the SDGs in an 'integrated' manner: A proposal for a new index to capture synergies and trade-offs between and within goals. *World Development*, 122, 628–647. doi:10.1016/j.worlddev.2019.05.022.

Blackstock, K.L., Kelly, G.J., and Horsey, B.L. (2007). Developing and applying a framework to evaluate participatory research for sustainability. *Ecological Economics*, 60(4), 726–742. doi:10.1016/j.ecolecon.2006.05.014.

Breuer, A., Janetschek, H., and Malerba, D. (2019). Translating Sustainable Development Goal (SDG) interdependencies into policy advice. *Sustainability*, 11(7). doi:10.3390/su11072092.

Buyana, K., et al. (2020). *Advancing the 2030 Agenda in African cities through knowledge co-production.* Retrieved from: https://council.science/wp-content/uploads/2020/04/LIRA-2030-report-2020-04-29.pdf.

Dahlmann, F., Stubbs, W., Griggs, D., and Morrell, K. (2019). Corporate actors, the UN Sustainable Development Goals and Earth System Governance: A research agenda. *The Anthropocene Review*, 6(1–2),167–176. doi:10.1177/2053019619848217.

Di Lucia, L., Slade, R., and Khan, J. (2021). Decision-making fitness of methods to understand Sustainable Development Goal interactions. *Nature Sustainability*, 5, 131–138. doi:10.1038/s41893-021-00819-y.

Engström, R.E. et al. (2021). Succeeding at home and abroad: accounting for the international spillovers of cities' SDG actions. *npj Urban Sustainability*, 1(1), 18. doi:10.1038/s42949-020-00002-w.

Giles-Corti, B., Lowe, M., and Arundel, J. (2020). Achieving the SDGs: Evaluating indicators to be used to benchmark and monitor progress towards creating healthy and sustainable cities. *Health Policy*, 124(6), 581–590. doi:10.1016/j.healthpol.2019.03.001.

Hoff, H. et al. (2019). *International Spillovers in SDG Implementation: The Case of Soy from Argentina.* Retrieved from www.sei.org/publications/spillovers-sdg-implementation-soy-argentina.

Horvath, S.-M. et al. (2022). Handling a complex agenda: A review and assessment of methods to analyse SDG entity interactions. *Environmental Science & Policy*, 131, 160–176. doi:10.1016/j.envsci.2022.01.021.

Karlsson-Vinkhuyzen, S., Dahl, A.L., and Persson, Å. (2018). The emerging accountability regimes for the Sustainable Development Goals and policy integration: Friend or foe? *Environment and Planning C: Politics and Space*, 36(8), 1371–1390. doi:10.1177/2399654418779995.

Krantz, V. and Gustafsson, S. (2021). Localizing the sustainable development goals through an integrated approach in municipalities: early experiences from a Swedish forerunner. *Journal of Environmental Planning and Management*, 64(14), 2641–2660. doi:10.1080/09640568.2021.1877642.

Kroll, C., Warchold, A., and Pradhan, P. (2019). Sustainable Development Goals (SDGs): Are we successful in turning trade-offs into synergies? *Palgrave Communications*, 5(1), 140. doi:10.1057/s41599-019-0335-5.

Lang, D. J. et al. (2012). Transdisciplinary research in sustainability science: practice, principles, and challenges. *Sustainability Science*, 7(1), 25–43. doi:10.1007/s11625-011-0149-x.

Li, M., Wiedmann, T., Fang, K., and Hadjikakou, M. (2021). The role of planetary boundaries in assessing absolute environmental sustainability across scales. *Environ Int*, 152, 106475. doi:10.1016/j.envint.2021.106475.

Liu, J. (2018). An Integrated Framework for Achieving Sustainable Development Goals Around the World. *Ecology, Economy and Society–the INSEE Journal*, 1(2). doi:10.37773/ees.v1i2.32.

Malik, A., Lafortune, G., Carter, S., Li, M., Lenzen, M., and Kroll, C. (2021). International spillover effects in the EU's textile supply chains: A global SDG assessment. *Journal of Environmental Management*, 295, 113037. doi:https://doi.org/10.1016/j.jenvman.2021.113037.

Matsumoto, K. i., Hasegawa, T., Morita, K., and Fujimori, S. (2019). Synergy potential between climate change mitigation and forest conservation policies in the Indonesian forest sector: implications for achieving multiple sustainable development objectives. *Sustainability Science*, 14(6), 1657–1672. doi:10.1007/s11625-018-0650-6.

Messerli, P. et al. 2019). Expansion of sustainability science needed for the SDGs. *Nature Sustainability*, 2(10), 892–894. doi:10.1038/s41893-019-0394-z.

Newman, M.E.J. (2006). Modularity and community structure in networks. *Proceedings of the National Academy of Sciences*, 103(23), 8577–8582. doi:10.1073/pnas.0601602103.

Newman, M.E.J. and Girvan, M. (2004). Finding and evaluating community structure in networks. *Physical Review E*, 69(2), 026113. doi:10.1103/PhysRevE.69.026113.

Nilsson, M. et al. (2018). Mapping interactions between the sustainable development goals: lessons learned and ways forward. *Sustainability Science*, 13(6), 1489–1503. doi:10.1007/s11625-018-0604-z.

Nilsson, M., Griggs, D., and Visbeck, M. (2016). Policy: Map the interactions between Sustainable Development Goals. *Nature*, 534, 320–322. doi:10.1038/534320a.

Organisation for Economic Co-operation and Development and Joint Research Centre, European Commission. (2021). *Understanding the Spillovers and Transboundary Impacts of Public Policies*.

Pham-Truffert, M., Metz, F., Fischer, M., Rueff, H., and Messerli, P. (2020). Interactions among Sustainable Development Goals: Knowledge for identifying multipliers and virtuous cycles. *Sustainable Development*, 28, 1236–1250. doi:10.1002/sd.2073.

Sachs, J., Schmidt-Traub, G., Kroll, C., Durand-Delacre, D., and Teksoz, K. (2017). *SDG Index and Dashboards Report 2017*. Retrieved from

Sachs, J., Traub-Schmid, G., Kroll, C., Lafortune, G., and Fuller, G. (2021). The Sustainable Development Goals and Covid-19 Includes the SDG Index and Dashboards. Retrieved from: www.cambridge.org/de/academic/subjects/economics/economic-development-and-growth/sustainable-development-report-2020-sustainable-development-goals-and-covid-19-includes-sdg-index-and-dashboards?format=PB.

Scharlemann, J.P.W. et al. (2020). Towards understanding interactions between Sustainable Development Goals: the role of environment–human linkages. *Sustainability Science*, 15(6), 1573–1584. doi:10.1007/s11625-020-00799-6.

Schmidt-Traub, G., Hoff, H., and Bernlöhr, M. (2019). *International spillovers and the Sustainable Development Goals (SDGs) Measuring how a country's progress towards the SDGs is affected by actions in other countries*. Retrieved from

Schwindenhammer, S. and Gonglach, D. (2021). SDG Implementation through Technology? Governing Food-Water-Technology Nexus Challenges in Urban Agriculture. *Politics and Governance*, 9(1), 176–186. doi: 10.17645/pag.v9i1.3590.

Sedlacko, M., Martinuzzi, A., Røpke, I., Videira, N., and Antunes, P. (2014). Participatory systems mapping for sustainable consumption: Discussion of a method promoting systemic insights. *Ecological Economics*, 106, 33–43. doi:10.1016/j.ecolecon.2014.07.002.

Thiam, S., Aziz, F., Kushitor, S.B., Amaka-Otchere, A.B.K., Onyima, B.N., and Odume, O. N. (2021). Analyzing the contributions of transdisciplinary research to the global sustainability agenda in African cities. *Sustainability Science*, 16(6), 1923–1944. doi:10.1007/s11625-021-01042-6.

United Nations. (2015). Transforming our world: the 2030 Agenda for Sustainable Development. Retrieved from: www.un.org/ga/search/view_doc.asp?symbol=A/RES/70/1&Lang=E.

Waltman, L., van Eck, N.J., and Noyons, E.C.M. (2010). A unified approach to mapping and clustering of bibliometric networks. *Journal of Informetrics*, 4(4), 629–635. doi:10.1016/j.joi.2010.07.002.

Warchold, A., Pradhan, P., and Kropp, J.P. (2021). Variations in sustainable development goal interactions: Population, regional, and income disaggregation. *Sustainable Development*, 29(2), 285–299. doi:10.1002/sd.2145.

Weber, J.M., Lindenmeyer, C.P., Liò, P., and Lapkin, A.A. (2021). Teaching sustainability as complex systems approach: a sustainable development goals workshop. *International Journal of Sustainability in Higher Education*, 22(8), 25–41. doi:10.1108/IJSHE-06-2020-0209.

Weitz, N., Carlsen, H., Nilsson, M., and Skånberg, K. (2018). Towards systemic and contextual priority setting for implementing the 2030 Agenda. *Sustainability Science*, 13(2), 531–548. doi:10.1007/s11625-017-0470-0.

World Bank. (2021). World Development Indicators. Retrieved from https://datatopics.worldbank.org/world-development-indicators.

Zeng, Y., Runting, R.K., Watson, J.E.M., and Carrasco, L.R. (2022). Telecoupled environmental impacts are an obstacle to meeting the sustainable development goals. *Sustainable Development*, 30(1), 76–82. doi:10.1002/sd.2229.

Zhao, Z. et al. (2021). Synergies and tradeoffs among Sustainable Development Goals across boundaries in a metacoupled world. *Science of the Total Environment*, 751, 141749. doi:10.1016/j.scitotenv.2020.141749.

3

THE ROLE OF GOOD GOVERNANCE IN REDUCING POVERTY AND INEQUALITY[1]

Evidence from a scoping review of interlinkages between SDGs 16, 10, and 1

Cameron Allen, Anita Breuer, Julia Kercher, Pooja Balasubramanian, Julia Leininger and Arvinn Gadgil

Introduction

Recognizing that governance is a critical means to achieve sustainable development, the 2030 Agenda includes a standalone SDG 16 on peace, justice and inclusive institutions. Targets under SDG 16 highlight several concepts that are viewed as key institutional principles of good governance, such as transparency, accountability, and inclusiveness of decision-making, and are seen as key enablers for all other SDGs (UN Department of Economic and Social Affairs, 2019). Despite their systemic importance, recent global studies on SDG interlinkages have either excluded or provided limited coverage of SDG 16 targets in their analyses (IGS, 2019; International Council for Science, 2017; Pham-Truffert, Metz, Fischer, Rueff, and Messerli, 2020).

To fill this gap, this chapter presents the results and key findings from a scoping review of literature dealing with interlinkages between institutional aspects that are targets under SDG 16 and the achievement of poverty reduction (SDG 1) and reduced inequalities (SDG 10). This chapter responds to key research questions for this volume relating to knowledge on interlinkages between the SDGs and the political-institutional preconditions to achieve sustainable development. The chapter briefly outlines the methods and research design used for the scoping review, then presents and discusses key findings, and finally offers some concluding remarks and insights for future research.

Methods and Research Design

The study adopted a scoping literature review approach to synthesize evidence on whether and how progress on institutional aspects of SDG 16 impact upon the achievement of poverty reduction (SDG 1) and reduced inequalities (SDG 10). The study involved several steps. First, a conceptual framework for evaluating

DOI: 10.4324/9781003254683-3

interlinkages was devised, followed by the development of the query protocol for literature retrieval and screening criteria, and ultimately a review and synthesis of relevant literature. These steps are summarized below.

Conceptual framework and approach for evaluating interlinkages

Many previous assessments of interlinkages between the SDGs rely on an evaluation of target-to-target interactions (Allen, Metternicht, and Wiedmann, 2019; Nilsson, Griggs, and Visbeck, 2016; Weitz, Carlsen, Nilsson, and Skånberg, 2017). Important initial steps include the identification of SDG targets that are of interest for the analysis, the directionality of impacts being evaluated, and a conceptual framework and methodology for the analysis.

Targets under SDG 16 are a blend of objectives relating to peace, justice and strong institutions. Similarly, SDG 1 and SDG 10 incorporate a range of targets covering aspects relating to poverty, social protection systems, access to basic services and social and economic inclusion. Of primary interest for this review were SDG 16 targets relating to key institutional principles of good governance: participation and inclusion, accountability, and transparency. These principles are reflected in several targets of SDG 16, which were selected as the entry points for this analysis (namely 16.5, 16.6, 16.7, 16.9 and 16.10). Key concepts of primary interest with regard to SDG 1 and SDG 10 ('impact goals') corresponded to nine human development-related targets across both SDGs (1.1, 1.2, 1.3, 1.4, 1.5, 10.1, 10.2, 10.3, 10.4).

In many instances, the SDG targets of interest either combined multiple principles or included overlapping concepts across multiple targets, which created impediments to adopting a target-to-target approach for the analysis. To provide conceptual clarity for the evaluation of interlinkages, the study therefore adopted a clustering approach, which separated distinct concepts and grouped closely related concepts. The logic for the clustering for SDG 16 (three entry clusters) and SDGs 1 and 10 (three impact clusters) is presented in Figure 3.1. The guiding question for the evaluation of interlinkages was:

> "Based on the evidence, does an increase/improvement in entry cluster X have an enabling/constraining/neutral/inconclusive impact on impact cluster Y?"

The primary scale of interest of our review related to public decision-making at national and sub-national levels. During the review, additional information on the specific key terms or sub-categories for each entry and exit cluster was also collected (e.g. specifically relating to 'poverty' or 'income inequality' within the impact cluster 'A. Reduce Poverty'), along with explanations provided in the respective article regarding the potential causal factors and pathways for impacts. A number of reviewed articles also analysed aspects that related to SDG 16 more generically (e.g. 'good governance') or that were beyond primary focus of the analysis (e.g. 'government effectiveness', 'political stability'). These interlinkages were also captured during the review process by allocating them to an additional entry cluster '4. Good Governance'. Additional information was also collected on the study methods and the type of evidence (e.g. quantitative, qualitative, etc.)

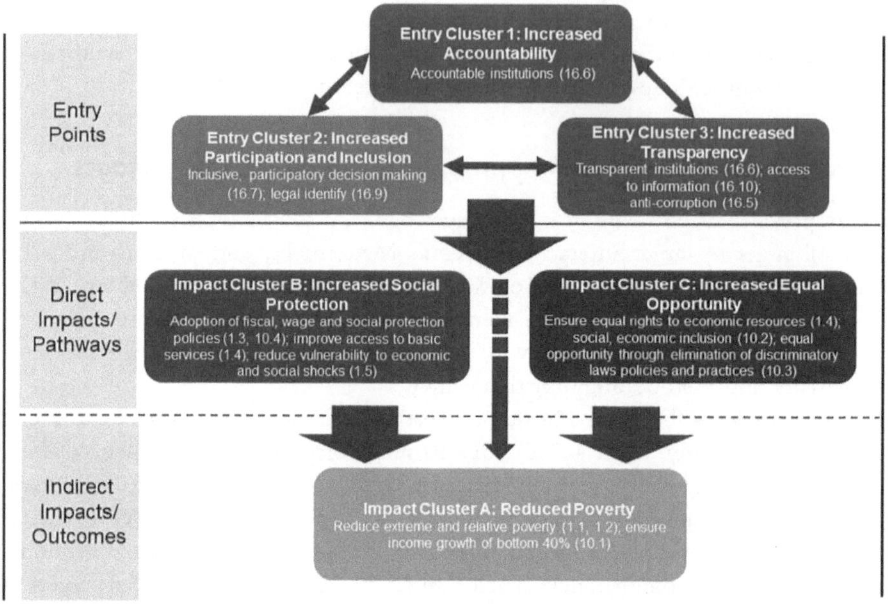

FIGURE 3.1 Framework for clustering key concepts from SDG targets into three main entry clusters and impact clusters

Source: Author's own elaboration

Note: A full colour version of this figure is available open access under: https://www.routledge.com/Governing-the-Interlinkages-between-the-SDGs-Approaches-Opportunities/Breuer-Malerba-Srigiri-Balasubramanian/p/book/9781032184654

as well as the geographic scale or scope of the analysis (multi-country, national, subnational) and the number of countries included in the sample for each study. This information was used to indicate the type and coverage of evidence supporting the interlinkages identified in the studies.

Query protocol, criteria and literature retrieval

To identify relevant literature, we developed a review protocol based on a standardized set of inclusion/exclusion criteria including query terms and conditions. Key terms used in the query protocol are provided in Table 3.1 corresponding to priority concepts from SDGs 16, 1, and 10 as well as other important terms that helped to refine the scope (e.g. public administration, institutions, government, etc.). The primary method of resource retrieval was based on a Web of Science (WoS) search using a query string conducted in March 2021. To ensure that the obtained number of publications was manageable and scientifically robust we included only peer-reviewed journal articles published since 2015 (i.e. since the adoption of the 2030 Agenda). Articles had to include at least one key word corresponding to the SDG 16 entry clusters, plus at least one keyword corresponding to the SDG 1 and SDG 10 impact clusters, plus at least one relating to the additional scoping terms.

The WoS query delivered 426 articles, which we then screened for relevance based on their title, keywords and abstract. Based on the screening exercise, we retained a total of 58 articles for our detailed review of interlinkages.

Summary Results for Interlinkages between SDGs 16, 1 and 10

Close to half of the articles reviewed focused at the national or subnational scale while the remainder were multi-country studies ranging from three to 176 countries. In terms of the geographic scope[2], the majority of the studies focused on lower income countries in Sub-Saharan Africa (SSA, 25%) and Central and Southern Asia (17%), as well as countries in Europe and North America (15%), Latin America and the Caribbean (15%), and East and South-East Asia (10%). About 13% of studies were considered global in scope. Of the studies reviewed, 42 (72%) were considered quantitative analyses (mostly relying on panel data), while 16 (28%) were considered qualitative (mostly drawing on comparative case studies).

We identified a total of 83 interlinkages between the three primary entry clusters (1. Increased Accountability; 2. Increased Participation and Inclusion; 3. Increased Transparency) and three impact clusters (A. Reduced Poverty; B. Increased Social Protection; C. Increased Equal Opportunity). In addition, we

TABLE 3.1 Query terms used in the protocol

Key concepts	*Query terms*
1. Accountability	"accountable institutions"; "accountability", "accountable governance"
2. Participation and inclusion	"inclusive decision-making"; "participatory decision-making"; "inclusive institutions"; "participatory institutions"; "political inclusion"; "public participation"; "public consultation"; "public engagement"; "legal identity"; "civil registration"; "participatory governance"; "inclusive governance"; "civic engagement"; "democratic governance"
3. Transparency	"transparent institutions"; "transparency"; "access to information"; "freedom of information"; "right to information"; "open government data"; "transparent governance"; "anti-corruption"
	AND
A. Poverty	"poverty"; "income equality"; "income inequality";
B. Social Protection	"social protection"; "access to services"; "access to basic services"; "economic vulnerability"; "social vulnerability"; "vulnerability to shocks"
C. Equal Opportunity	"social inclusion"; "economic inclusion"; "equal opportunity"; "discriminatory laws"; "discriminatory policies"; "social inequality"; "political inequality"
	AND
Additional key terms	"institutions"; "public sector"; "government"; "public administration"; "governance"

Source: Author's own elaboration

identified a further 39 interlinkages corresponding to additional attributes of good governance (e.g. 'government effectiveness', 'political stability', 'good quality governance', etc.) and the three impact clusters. Of these, 88% were interlinkages from the entry to the impact cluster while the remainder were reverse interlinkages from impact to entry clusters. The subsequent analysis focuses on interlinkages from entry to impact clusters.

Overall, evidence in the literature testified to many more enabling effects (50 enabling interlinkages) than constraining effects (four constraining interlinkages), while 15 interlinkages were identified as neutral (little or no impact). The most common enabling effects identified were from '3. Increased Transparency' to 'A. Reduced Poverty' (13 enabling interlinkages) followed by '2. Increased Participation and Inclusion' to 'A. Reduced Poverty' (ten enabling interlinkages) (Figure 3.2). Enabling effects were also higher for '1. Increased Accountability' on both 'B. Increased Social Protection' (six enabling interlinkages) and '2. Increased Participation and Inclusion' on 'B. Increased Social Protection'. Mixed results (both enabling and constraining) were found for the effects of increased accountability and improved transparency on equal opportunity (1 to C, 3 to C). Overall, only four studies identified constraining effects. Note that reverse interlinkages are excluded from these numbers.

Interlinkages were also evaluated at the 'sub-cluster' level, corresponding to specific concepts within each of the entry and impact clusters. Figure 3.3 provides a

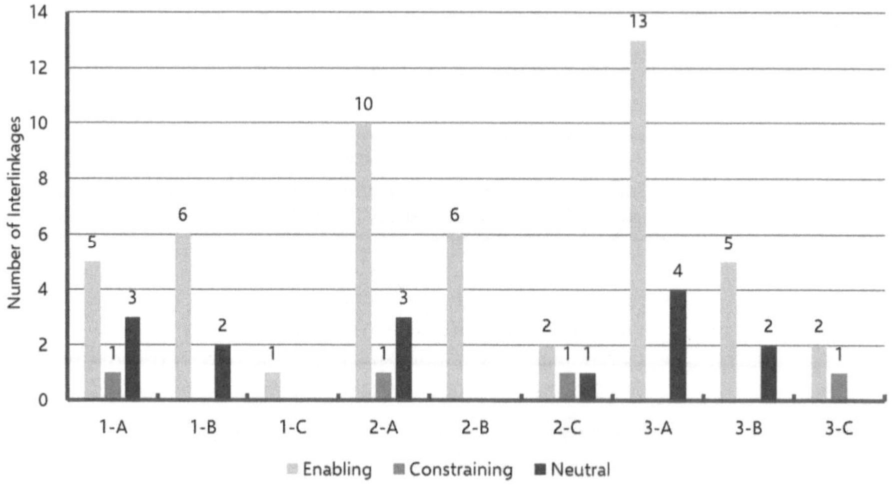

FIGURE 3.2 Number of enabling, constraining and neutral interlinkages identified between the three primary entry clusters and three impact clusters

1. Increased Accountability; 2. Greater Participation and Inclusion; 3. Increased Transparency and A. Reduced Poverty; B. Greater Social Protection; C. Enhanced Equal Opportunity

Source: Author's own elaboration

Note: A full colour version of this figure is available open access under: https://www.routledge.com/Governing-the-Interlinkages-between-the-SDGs-Approaches-Opportunities/Breuer-Malerba-Srigiri-Balasubramanian/p/book/9781032184654

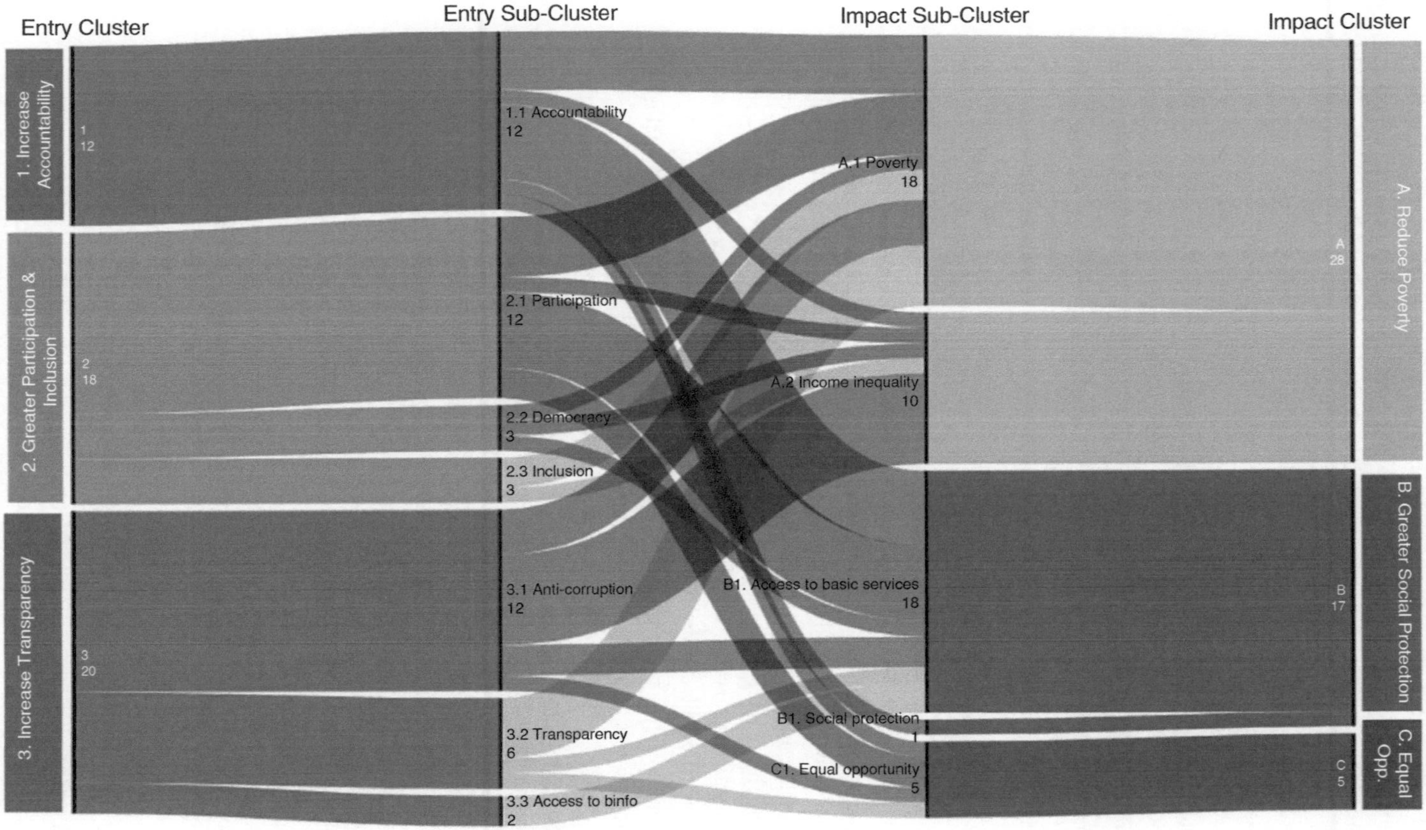

FIGURE 3.3 Flow diagram of enabling interlinkages between the three entry clusters and entry subcategories (left) and the impact subcategories and impact clusters (right). Width of the flows (and numbers inserted) represent the number of positive/enabling interlinkages

Source: Author's own elaboration

Note: A full colour version of this figure is available open access under: https://www.routledge.com/Governing-the-Interlinkages-between-the-SDGs-Approaches-Opportunities/Breuer-Malerba-Srigiri-Balasubramanian/p/book/9781032184654

summary of these more detailed sub-cluster interlinkages in the form of a flow diagram, where the width of the 'flows' corresponds to the number of enabling interlinkages identified. In most cases, each interlinkage corresponds to an individual article, so the values or flows can also be roughly interpreted as the number of articles supporting a particular enabling interlinkage.

For the first entry cluster on increased accountability, no sub-clusters were identified as this cluster incorporates a single SDG target (16.6). Here, out of 12 enabling interlinkages the majority improved access to basic services (five enabling interlinkages) and reduced poverty (four enabling interlinkages) (Figure 3.3). Single enabling interlinkages were also identified for increasing social protection and equal opportunity and reducing income inequality. Overall, positive enabling effects were identified between increasing accountability and all three impact clusters, but primarily 'A. Reduced Poverty' and 'B. Increased Social Protection'. Accountability was also the only entry cluster with an identified enabling effect specifically on the 'social protection' sub-cluster (one enabling interlinkage).

For entry cluster '2. Increased Participation and Inclusion', a total of 18 enabling interlinkages were identified across three sub-clusters, as depicted in Figure 3.3. For the 'participation' sub-cluster, 12 interlinkages were identified, which enabled access to basic services (five), poverty reduction (four), equal opportunity (two), and reduced income inequality (one). For the sub-cluster 'democracy', three interlinkages were identified, which enabled the reduction of income inequality and poverty, as well as greater access to basic services. Finally, for the 'inclusion' sub-cluster, the three interlinkages identified enabled poverty reduction and reduced income inequality. Again, positive enabling effects were identified between greater participation and inclusion and all impact clusters, but in particular impact clusters A and B.

For entry cluster '3. Increased Transparency', 20 enabling interlinkages were identified across two sub-clusters. Most of these related to anti-corruption measures (12 interlinkages), which enabled all the impact clusters to varying degrees. In particular, enabling effects were evident between corruption control and reducing income inequality (six) and poverty (three), as well as improving access to basic services (two) and equal opportunity (one). For the 'transparency' sub-cluster, enabling effects corresponded to reducing poverty (four), as well as improving access to basic services (one) and equal opportunity (one). Finally, for the 'access to information' sub-cluster, enabling effects were evident for access to basic services (2). While most of the interlinkages we found were enabling, four studies showed constraining effects. Their findings imply that judicial accountability increases income inequality, transparency increases inequality in opportunities, democracy promotes income inequality, and participation constrains political equality. The identified interlinkages are described in more detail in the next section.

Important Findings on Interlinkages and Potential Causal Pathways

These results show that there is clear evidence of enabling effects of SDG 16 on SDGs 1 and 10. Yet, it is also important to understand the key mechanisms and pathways underlying these enabling (or constraining) effects. Many of the studies reviewed attempted to explain and interpret their findings, often drawing upon the broader literature to provide insights into potential causal relationships. The review revealed that a combination of both quantitative and qualitative analyses can provide complementary insights into both statistical correlations as well as causal explanations of key pathways to impact.

This section briefly discusses the literature and key findings from individual studies in more detail and summarizes a complex array of causal relationships and dynamics between the selected institutional aspects of SDG 16 and the selected aspects of SDGs 1 and 10. These relationships are often indirect or mediated via a range of other enablers and drivers. In some cases, the effects appear to be conditional upon progress in other areas (e.g. gender equality, GDP growth). Figure 3.4 attempts to capture the relationships identified from the literature in a systems map or 'causal diagram'. To interpret the diagram, note that all blue arrows (+) represent a positive relationship, which can be read as "Increasing/improving variable x is associated with an increase and/or improvement in variable y". In contrast, red arrows (-) represent a negative relationship and should be read as "Increasing/improving variable x is associated with a decrease/decline in variable y". All linkages identified are backed by the literature reviewed in this section. While the resulting diagram is complex, it is still unlikely to capture all the dynamics at play. Nevertheless, it points to some key dynamics and pathways supported by the literature which can assist in developing an overall theory of change.

The colours and shading of variables in Figure 3.4 reflect the three entry clusters used in the conceptual framework for the analysis, however there was some overlap between the relationships and pathways identified in the literature for each cluster. We developed the diagram in three stages, starting with the literature relating to '1. Increased Accountability' (blue variables in Figure 3.4), then adding elements from the literature on '2. Increased Participation and Inclusion' (orange variables), and finishing with elements from the literature relating to '3. Increased Transparency' (green variables).

Accountability effects on impact clusters and review of causal pathways

Five articles found an enabling effect of increased accountability on the reduction of poverty and income inequality (Akobeng, 2016; Chan, 2018; Hill, Byrne, and de Vasconcellos Pegas, 2016; Ramanujam, Caivano, and Agnello, 2019; Workneh, 2020), while one found constraining effects (Berggren and Bjørnskov, 2020). A further seven articles found enabling effects of increased accountability on access to basic services (Andersson and Palacio Chaverra, 2017; Guimarães, Malheiros, and Marques, 2016; Khan, Faguet, and Ambel, 2017; Sukati, Moodley, and Mashige, 2018), social protection (Fossati, 2016), and equal opportunity (Hill et al., 2016; Jones, Abu-Hamad, Pereznieto, and Sylvester, 2016).

With regard to poverty and inequality, the study by Akobeng (2016) finds that accountable (and transparent) institutions increase the poverty-reduction effects of GDP growth in 41 countries in SSA, by enabling economic, property and other rights including respect for contracts which, in turn, provides security and predictability in

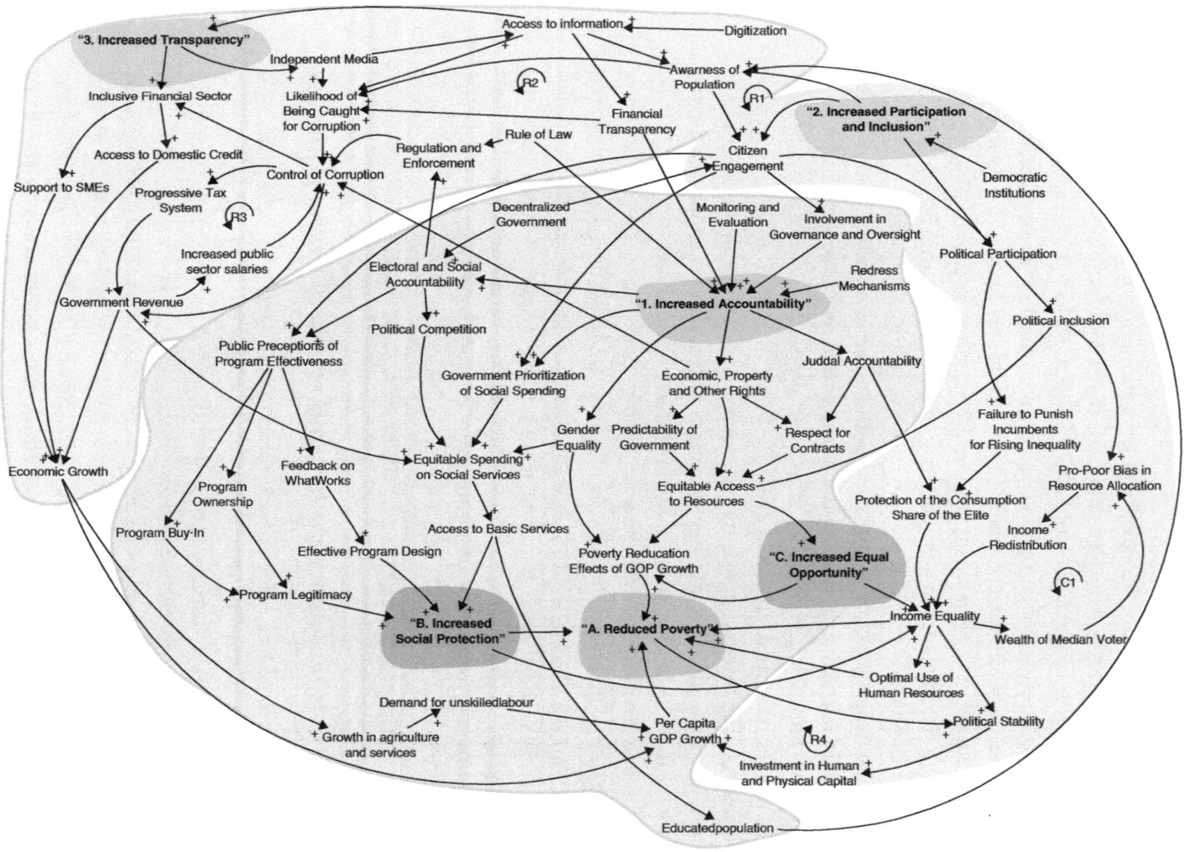

FIGURE 3.4 Capture: Causal Diagram: Causal pathways from SDG 16 Entry Clusters to SDG 1 and 10
Impact Clusters Blue= Increased Accountability; Orange = Increased Participation and Inclusion; Green = Increased Transparency

Source: Author's own elaboration

Note: A full colour version of this figure is available open access under: https://www.routledge.com/Governing-the-Interlinkages-betweenthe-SDGs-Approaches-Opportunities/Breuer-Malerba-Srigiri-Balasubramanian/p/book/9781032184654 government decision-making. In another study on SSA, Workneh (2020) argues that both government accountability and effectiveness play a vital role in improving government service provision and reducing poverty. Other studies suggest that institutions aimed at providing distributive justice will enhance accountability and reduce poverty and inequality (Ramanujam et al., 2019) and that a lack of accountability in resource use governance leads to the persistence of social inequality (Hill et al., 2016). Contrary to these findings on the positive effects of accountability, the global study by Berggren and Bjørnskov (2020) finds that as judicial accountability increases, so does income and consumption inequality.

The study by Sukati et al. (2018) on Swaziland concludes that increasing accountability is expected to improve health system performance and enable increased access to healthcare (specifically eye care). Andersson and Palacio Chaverra (2017) suggest that increasing accountability improves countries' ability to prioritize spending to uphold the social contract with its people and will ultimately increase social spending and access to basic services. Similarly, Guimarães et al. (2016) argue that social participation and accountability will facilitate durable universal access to water and sanitation services for vulnerable populations (slum dwellers) in Brazil. The study by Khan et al. (2017) on Ethiopia suggests that accountability (both upwards to donors and downwards to citizens) improves the equity outcomes associated with grant programs. Fossati (2016) concludes that accountability ensuing from electoral democracy at the local level in Indonesia plays a crucial role for the effective implementation of national social protection programs.

Overall, a total of 12 articles identified enabling effects of '1. Increased Accountability' on SDGs 1 and 10. While the relationships or pathways for these effects were not always explored, some of the key pathways and dynamics are summarized in the causal diagram in Figure 3.4 (highlighted in blue shading). In brief, the articles reviewed highlighted that increasing accountability is associated with an increase in the 'poverty reduction effects' of per capita GDP growth. This was reportedly due to increased security of economic and property rights, respect for contracts, predictability of government decision-making, as well as improvements to gender equality and equitable access to resources. However, the direct mechanisms for these impacts were not always established. Increased social accountability was reported to have an enabling effect on social protection, through improved public perceptions of program effectiveness, legitimacy and buy-in. Increased electoral accountability (through local elections in the context of decentralization) resulted in increased citizen engagement and political competition, which in turn resulted in the prioritization and better targeting of social expenditure and increased access to basic services. A range of measures were identified that increased accountability, including financial transparency, redress mechanisms, citizen engagement, and involvement in governance and oversight. However, increasing judicial accountability was also seen to protect the elite and entrench inequalities in income and consumption.

Increased participation and inclusion effects on impact clusters and review of causal pathways

A selection of primarily qualitative studies provided evidence of enabling interlinkages between greater participation and inclusion and reducing poverty and income inequality (Akobeng, 2016; Andersson and Palacio Chaverra, 2017; Anyanwu, Erhijakpor, and Obi, 2016; Fan, Li, Tao, and Yang, 2020; Hill et al., 2016; Jianu, Dobre, Bodislav, Radulescu, and Burlacu, 2020; Nieto-Aleman, Garcia-Alvarez-Coque, Roig-Tierno, and Mas-Verdú, 2019). A further eight studies identify enabling interlinkages between increased participation and inclusion with access to basic services and equal opportunity. This includes one global study (Elgar, Stefaniak, and Wohl, 2020), a multi-country study (Wickremasinghe et al., 2018) and national and sub-national studies (Das and Das, 2018; Guimarães et al., 2016; Lindström, 2020; Nwobashi and Itumo, 2017; Ye and Yang, 2020).

Anyanwu et al. (2016) find that increasing democracy levels correspond to reduction of income inequality in West African countries. They also find that increasing democracy improves the marginal effect of GDP per capita growth on reducing poverty, a finding also supported by Akobeng (2016) for countries in SSA. Studying countries in the EU, Jianu et al. (2020) find that inequality is less persistent in countries with inclusive institutions, and that social policy instruments and social spending are more effective in countries with inclusive institutions. The global study by Andersson and Palacio Chaverra (2017) also identifies enabling interlinkages between increasing inclusion and reducing poverty. Contrary to other studies, Hicks, Jacobs, and Matthews (2016) suggest that democratic voting increases income inequality in advanced democracies as voters fail to punish rising inequality (up to a point).

Several studies highlight the enabling effects of participation and inclusion on access to services. Wickremasinghe et al. (2018) find that civil society engagement and participation improve access to health care in Ethiopia, Nigeria, and India. Nwobashi and Itumo (2017) find that democratic governance has enhanced the provision of health facilities and promoted programs in rural communities in Nigeria. Guimarães et al. (2016) argue that social participation facilitates durable universal access to water and sanitation services in Brazil. The global study by Elgar et al. (2020) finds that social capital derived from civic engagement is associated with lower levels of mortality from COVID-19. Using local level data from India, the study by Das and Das (2018) concludes that grass-roots democracy, decentralization and participation amplifies the voice of citizens in policy making and enables the efficient distribution of benefits of public welfare schemes and programs.

Overall, a total of 18 studies identified enabling interlinkages between '2. Increased Participation and Inclusion' and the impact clusters from SDGs 1 and 10. Building on the previous dynamics, this incorporates additional linkages and causal pathways (highlighted in orange in Figure 3.4). In particular, the literature highlighted that democratization results in greater political participation and inclusion which in turn reduces income inequality through a pro-poor bias in resource allocation and greater income redistribution. Greater income equality results in greater political stability and investment in human and physical capital and more optimal use of human resources. Increased participation implies greater citizen engagement and more inclusive institutions, which in turn deliver more effective social spending and policy instruments. Democracy and political inclusion are also found to increase the 'poverty reduction effects' of GDP growth. An independent media and education support greater access to information and awareness, which leads to higher political participation. This amplifies the voice of citizens in policy making and enables efficient and equitable distribution of benefits. A constraining effect is also identified, whereby voters in advanced democracies fail to punish incumbents for rising income inequality.

Increased transparency effects on impact clusters and review of causal pathways

The largest share of studies in this cluster (13) testify to the enabling interlinkage between increased transparency and reducing poverty and income inequality. These were predominantly quantitative studies covering a comparatively large sample of countries. This included global studies (Berggren and Bjørnskov, 2020; Chan, 2018), regional studies on SSA (Adams and Klobodu, 2016; Adeleye, Osabuohien, and Bowale, 2017; Kunawotor, Bokpin, and Barnor, 2020), Asia (Warf, 2019) and Latin America (Warf and Stewart, 2016), as well as national and subnational studies on Mexico (Hill et al., 2016), India (Daoud, 2015), Colombia (Bustos and Estupiñán, 2019; Nieto-Aleman et al., 2019), Egypt (Bremer, 2018), and Nigeria (Suleiman and Aminul Karim, 2015). A further seven studies identified enabling interlinkages between increased transparency and access to basic services (Bhat, Holtz, and Avila, 2018; Pinzón-Flórez, Chapman, Cubillos, and Reveiz, 2016; Sukati et al., 2018; Warf and Stewart, 2016; Wickremasinghe et al., 2018) and equal opportunity (Hill et al., 2016; Ye and Yang, 2020).

A key focus of these studies was on the relationship between corruption and income inequality, in particular in SSA. Kunawotor et al. (2020) find the control of corruption and the rule of law to be statistically significant in reducing income inequality in 40 African countries. Adams and Klobodu (2016) find that the control of corruption and transparency in governance are crucial for financial development to reduce income inequality in 21 countries in SSA. The study by Adeleye et al. (2017) in 42 countries in SSA also finds that if corruption is controlled while domestic credit and finance increase, then income inequality will decrease. Studies by Chan (2018) and Warf (2019) suggest that increasing levels of corruption are associated with increasing levels of income inequality.

A range of national and sub-national studies explore the impacts of corruption on poverty. Hill et al. (2016) suggest that ineffective and corrupt governance leads to increased persistence of poverty in their case study analysis in Mexico. Daoud (2015) finds that higher corruption in Indian states results in a higher prevalence of absolute child poverty. Suleiman and Aminul Karim (2015) find that corruption in

Nigeria reduces government revenue and subsequent expenditure on social security which increases poverty and unemployment. Bustos and Estupiñán (2019) analyse data on a multidimensional poverty index and an index of government transparency developed for the study in 23 cities in Colombia over two years and find that increasing government transparency results in poverty reduction. Nieto-Aleman et al. (2019) also find that institutional transparency (through corruption monitoring) and personal safety have been decisive factors for reducing poverty in some regional clusters in Colombia.

Regarding access to basic services, the focus again was primarily on health services. For example, Wickremasinghe et al. (2018) find that increasing access to information helps to scale up health interventions in Ethiopia, while Sukati et al. (2018) find that increasing the control of corruption increases access to healthcare services. The study by Warf and Stewart (2016) in Latin America and the Caribbean suggests that reducing corruption is positively correlated with improved access to education and literacy rates. Ye and Yang (2020) suggest that increasing transparency supports equal opportunity in China, and Hill et al. (2016) find that corrupt governance leads to social inequality in Mexico. Bhat et al. (2018) find that increasing access to information increases access to public health services. Contrary to these findings, Masiero and Maiorano (2018) suggest that the uptake of e-governance to increase accountability in anti-poverty programs in India has reinforced existing power structures that result in unequal access to opportunities.

Finally, for '3. Increased Transparency', evidence on enabling interlinkages with SDGs 1 and 10 were identified across 20 different studies. A key focus of the studies related to the relationship between corruption and inequality and poverty. A range of causal pathways are explored in the literature and are incorporated into the diagram (green variables in Figure 3.4). Corruption creates a biased tax system and supports tax evasion, which reduces revenue and undermines the capacity of governments to fairly redistribute wealth and for spending on social services to reduce poverty. Conversely, inequality motivates corrupt behaviour, and the rich are more able to pay bribes. Enhancing contract enforcement and economic and property rights can help to control corruption and ensure fairer distribution. Increased transparency in governance and the control of corruption are crucial for inclusive financial development, which in turn reduces income inequality. If corruption is controlled while domestic credit and finance increase, then income inequality will decrease. Corruption is less likely to occur when the likelihood of being caught and punished is relatively high, which largely depends on financial transparency, oversight, regulation and enforcement, and access to information. Factors such as education and awareness, an independent media, and higher salaries also inhibit corruption. Controlling corruption results in greater access to health and education services. Increasing access to information also increases awareness of target populations and improves social protection programs.

Unpacking key feedbacks and causal dynamics between SDGs 16, 1 and 10

Figure 3.4 also identifies a selection of feedback loops within the causal diagram. Such feedback loops are a well-known characteristic of complex systems, representing important reinforcing or balancing dynamics, which can lead to complex non-linear behaviour over time (Meadows, 2008). They are important for identifying key entry points, interventions and accelerators which can deliver (or undermine) desirable outcomes. However, as the causal diagram is based only on the relationships identified in the literature reviewed for this study and given the focus on only a selection of goals, it is unlikely to include all relevant dynamics. Nevertheless, it sheds light on some important feedbacks associated with the entry and impact clusters addressed in this study. These are labelled in Figure 3.4 and the following figures as reinforcing loops (R1, R2, R3 and R4), which have positive polarities overall, and counter-balancing loop (C1) which has an overall negative polarity. Again, the colours in the following figures reflect the three main entry clusters used in the study.

First, Figure 3.5 presents a simplified version of Reinforcing Loop 1 (R1 in Figure 3.4). This suggests that increasing awareness of the population leads to increased citizen engagement, which improves government prioritization of social spending

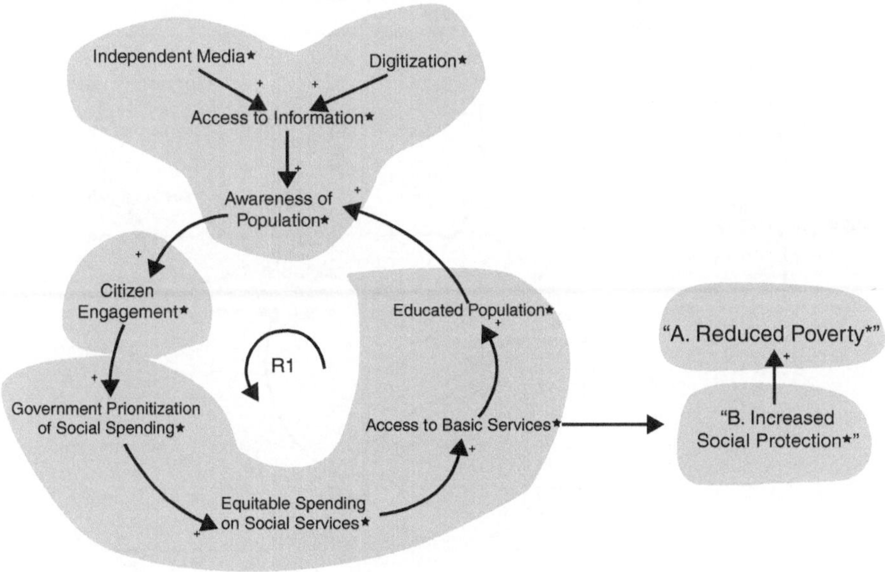

FIGURE 3.5 Reinforcing Feedback Loop R1 "Raising Awareness"
[Blue = Increased Accountability; Orange = Increased Participation and Inclusion; Green = Increased Transparency]

Source: Author's own elaboration

Note: A full colour version of this figure is available open access under: https://www.routledge.com/Governing-the-Interlinkages-between-the-SDGs-Approaches-Opportunities/Breuer-Malerba-Srigiri-Balasubramanian/p/book/9781032184654

and results in more equitable spending on social services. This in turn increases access to basic services including education, which results in a more educated and aware population, continuing the loop around again to citizen engagement. Interventions aimed at the different entry points in this feedback loop would reinforce progress, for example through increased access to information as a result of an independent media and digitization. Increased access to basic services also increases social protection, which reduces poverty.

Second, building on the R1 loop, additional reinforcing loops can be identified from Figure 3.4, which are also associated with raising awareness. Reinforcing feedback loops R2 and R3 highlight additional positive effects of raising awareness on the control of corruption, which further reinforces equitable government spending on social services (Figure 3.6).

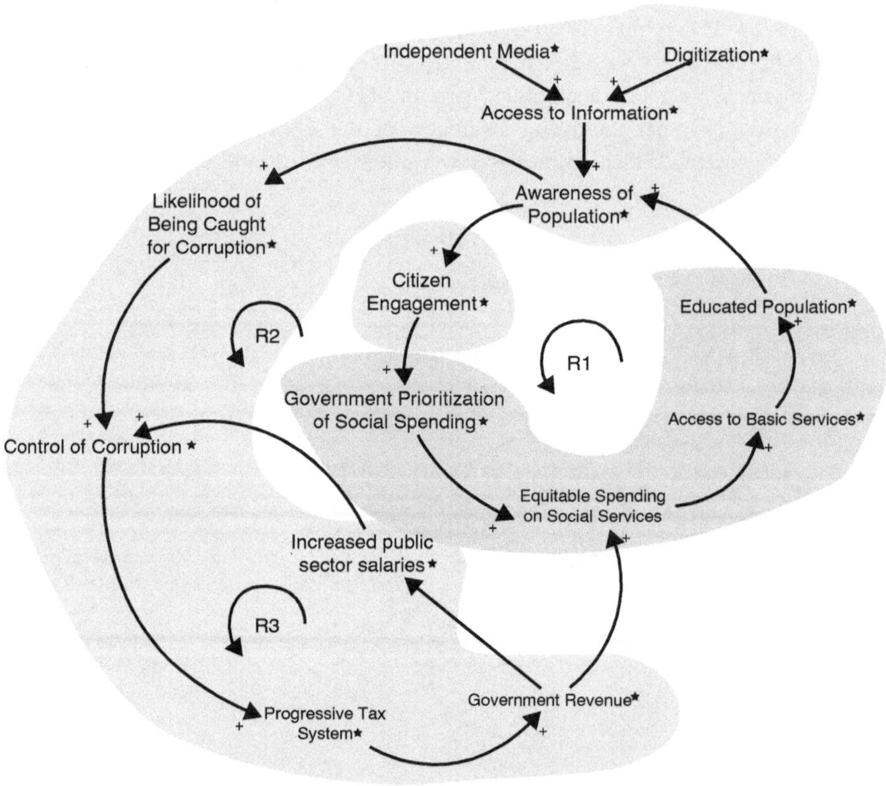

FIGURE 3.6 Reinforcing Feedback Loop R2 and R3: "Control of Corruption" [Blue = Increased Accountability; Orange = Increased Participation and Inclusion; Green = Increased Transparency]

Source: Author's own elaboration

Note: A full colour version of this figure is available open access under: https://www.routledge.com/Governing-the-Interlinkages-between-the-SDGs-Approaches-Opportunities/Breuer-Malerba-Srigiri-Balasubramanian/p/book/9781032184654

Finally, Reinforcing Loop "R4" and Counterbalancing Loop "C1" are summarized in Figure 3.7. The reinforcing loops highlights that per capita GDP growth has an enabling effect on poverty reduction, which increases political stability, increasing investment in human capital and, in turn, increasing per capita GDP growth.

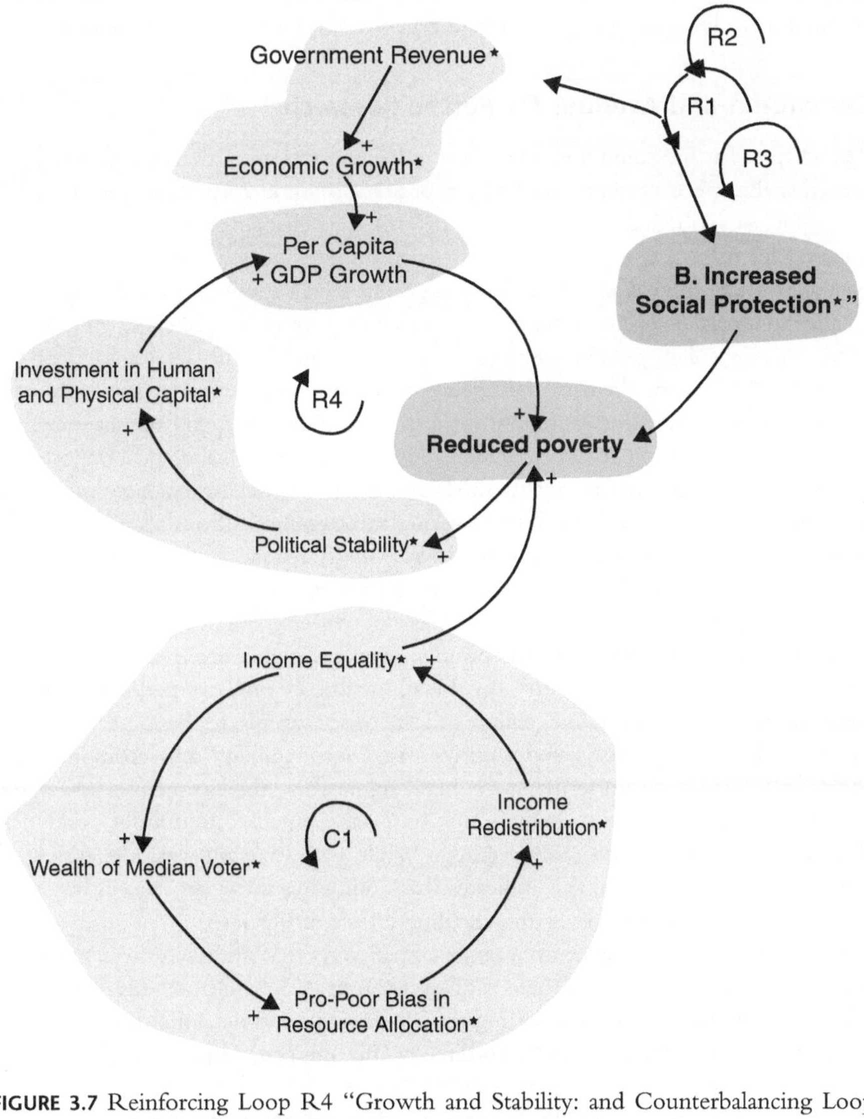

FIGURE 3.7 Reinforcing Loop R4 "Growth and Stability: and Counterbalancing Loop C1 "Median Wealth".
[Blue = Increased Accountability; Orange = Increased Participation and Inclusion; Green = Increased Transparency]
Source: Author's own elaboration
Note: A full colour version of this figure is available open access under: https://www.routledge.com/Governing-the-Interlinkages-between-the-SDGs-Approaches-Opportunities/Breuer-Malerba-Srigiri-Balasubramanian/p/book/9781032184654

However, this feedback loop is offset to some degree by Counterbalancing Loop C1 whereby income redistribution is reduced as increasing income equality raises the wealth of the median voter and reduces the pro-poor bias in resource allocation. This implies that income equality would stabilize over time, also limiting poverty reduction. It would also be possible to link to the other feedback loops (R1, R2 and R3) through variables relating to government revenue and increased social protection.

Conclusion and Avenues for Future Research

This chapter has presented the results from a scoping review of the academic literature related to the effects of institutional aspects of SDG 16 on key aspects of poverty reduction (SDG 1) and reduced inequalities (SDG 10). The chapter responds to the research question for this volume relating to knowledge on interlinkages between the SDGs, as well as the gap in coverage of SDG 16 targets in recent studies on interlinkages. The chapter summarizes recent evidence on the role of governance principles enshrined in SDG 16 as enabling political-institutional preconditions for the achievement of two selected SDGs. At the focus of the review were three institutional principles of good governance, namely participation and inclusion, accountability, and transparency, and their effects on reducing poverty and increasing social protection and equal opportunity.

Overall, the results of our review show empirical evidence from a large number of quantitative and qualitative analyses covering most countries from all regions of the world of the enabling effects of SDG 16 on SDGs 1 and 10 at both national and subnational levels. While we identify enabling interlinkages between all three governance principles and our selected aspects of poverty and inequality reduction, stronger evidence was found for the positive effects of higher levels of "transparency" and "participation and inclusion" on the reduction of poverty, particularly in the form of reduced income inequality and increased access to basic services. The majority of these were associated with increased accountability, anti-corruption, and increased participation. In terms of the SDG framework, these principles correspond to targets 16.5 (reducing corruption), 16.6 (accountable institutions) and 16.7 (inclusive, participatory decision-making). While pathways will vary depending on different country contexts, this indicates that addressing these targets will be critical for countries seeking to harness the enabling effects of SDG 16.

Our review also shed light on a range of pathways that underpin these synergistic impacts. The identification of these pathways constitutes a significant advancement of the theory of change (ToC) regarding the mechanisms and dynamics through which the institutional principles of SDG 16 directly and indirectly impact on the reduction of poverty (SDG 1) and inequalities (SDG 10). While specific pathways will highly depend on the country context and require additional research for any given country, his advancement of the ToC should prove valuable in the readjustment of existing and development of future interventions and programs targeted at the reduction of poverty and inequalities.

The discussion of causal dynamics also pointed to many broader interlinkages with other goals and targets from the SDGs framework, including gender (SDG 5), economic growth (SDG 8), health (SDG 3), and education (SDG 4). Consequently, a further refinement of the theory of change on the effects of SDG 16 on other goals could be achieved by extending the analysis through the inclusion of targets and elements from these goals. Such an extension would enable a more complete picture and allow for the identification of additional important reinforcing and counterbalancing feedback loops. Such feedbacks are a well-known characteristic of complex systems and are important for identifying key leverage points, interventions, and accelerators to achieve desired development outcomes.

Furthermore, although democracy was not at the focus of our review, its findings suggest that it constitutes an enabling political pre-condition for the reduction of poverty and inequalities. For example, evidence was found that electoral democracy contributes to the effectiveness of social protection programs and increased electoral accountability that results from local elections in the context of decentralization resulted in a better targeting of social expenditure and increased access to basic services. While the governance principles of transparency, accountability, and inclusive and participatory decision-making are not exclusive to democratic regimes, they are more frequently found in democratic than in autocratic contexts and are largely considered as key characteristics that distinguish liberal democracies from merely electoral democracies.

In terms of future research directions, priority should therefore be given to broadening the scope of the analysis by including a) democracy as an additional overarching entry cluster and b) additional impact clusters associated with other SDGs.

Finally, while recognizing that exact causal pathways will depend highly on the context and will require additional research for any given country, a number of broad policy implications can be drawn from the review. In particular, there is relatively strong evidence from a wide range of countries that the implementation of measures that increase electoral and social accountability, control corruption, and increase participation and citizen engagement have an enabling effect in reducing inequality and poverty. However, these synergistic effects are generally indirect or manifest in complement to a range of other enablers or drivers. For example, higher levels of democracy and good governance attributes such as accountability and transparency have been shown to increase and sustain the poverty-reduction effects of GDP growth.

Several of the articles reviewed also made specific policy recommendations to improve individual aspects of good governance. Regarding the improvement of accountability, specific measures proposed in the literature included the adoption of redress mechanisms, intensification of monitoring and evaluation, government decentralization, increasing citizen engagement in governance and oversight, legislating economic and property rights, and contract enforcement (particularly where punishment or penalties were adequate). Regarding the improvement of corruption control, proposed measures included the enhancement of contract enforcement, the improvement of transparency through increasing access to information and the creation of an enabling environment for independent media, as well as education and awareness raising.

Notes

1 The original research for this chapter was a collaboration between UN Development Programme (UNDP)'s Oslo Governance Centre, which leads UNDP's policy development and applied research on inclusive governance, and IDOS (previously German Development Institute/Deutsches Institut für Entwicklungspolitik). The contributions of Cameron Allen, Arvinn Gadgil and Julia Kercher were part of their assignments with the Oslo Governance Centre.

2 Regional groups correspond to those used by UN Statistics Division for the Sustainable Development Goals Report: https://unstats.un.org/sdgs/report/2019/regional-groups.

References

Adams, S. and Klobodu, E.K.M. (2016). Financial development, control of corruption and income inequality. *International Review of Applied Economics*, 30(6), 790–808. doi:10.1080/02692171.2016.1208740.

Adeleye, B.N., Osabuohien, E., and Bowale, E.I. (2017). The role of institutions in the finance-inequality nexus in Sub-Saharan Africa. *Journal of Contextual Economics*, 137, 173–192.

Akobeng, E. (2016). Growth and institutions: A potential medicine for the poor in Sub-Saharan Africa. *African Development Review*, 28(1), 1–17. doi:10.1111/1467-8268.12163.

Allen, C., Metternicht, G., and Wiedmann, T. (2019). Prioritising SDG targets: assessing baselines, gaps and interlinkages. *Sustainability Science*, 14(2), 421–438. doi:10.1007/s11625-018-0596-8.

Andersson, M. and Palacio Chaverra, A. (2017). Catch up growth and social capability in developing countries: A conceptual and measurement proposal. *OASIS*, 26. doi:10.18601/16577558.n26.02.

Anyanwu, J.C., Erhijakpor, A.E., and Obi, E. (2016). Empirical analysis of the key drivers of income inequality in West Africa. *African Development Review*, 28(1), 18–38. doi:10.1111/1467-8268.12164.

Berggren, N. and Bjørnskov, C. (2020). Corruption, judicial accountability and inequality: Unfair procedures may benefit the worst-off. *Journal of Economic Behavior & Organization*, 170, 341–354. doi:10.1016/j.jebo.2019.12.010.

Bhat, R., Holtz, J., and Avila, C. (2018). Reaching the missing middle: Ensuring health coverage for India's urban poor. *Health Systems & Reform*, 4(2), 125–135. doi:10.1080/23288604.2018.1445425.

Bremer, J. (2018). Youth unemployment and poverty in Egypt. *Poverty & Public Policy*, 10 (3), 295–316. doi:10.1002/pop4.224.

Bustos, W.O.P. and Estupiñán, J.M.T. (2019). Technical efficiency of local public institutions in Colombia. *Revista de Estudios Regionales*, (115), 15–41.

Chan, S.-G. (2018). The role of country governance on value-added tax and inequality. *E & M: Ekonomie a Management*, 21(4), 79–93. doi:10.15240/tul/001/2018-4-006.

Daoud, A. (2015). Quality of governance, corruption and absolute child poverty in India. *Journal of South Asian Development*, 10(2), 148–167. doi:10.1177/0973174115588844.

Das, S. C. and Das, G. (2018). Public Resource Allocation through Grassroots Democratic Institutions: Evidence from Assam, India. *International Journal of Public Administration*, 41 (16), 1325–1337. doi:10.1080/01900692.2017.1387918.

Elgar, F.J., Stefaniak, A., and Wohl, M.J. (2020). The trouble with trust: Time-series analysis of social capital, income inequality, and COVID-19 deaths in 84 countries. *Social Science & Medicine*, 263, 113365. doi:10.1016/j.socscimed.2020.113365.

Fan, F., Li, M., Tao, R., and Yang, D. (2020). Transfer-based decentralisation, economic growth and spatial inequality: Evidence from China's 2002–2003 tax sharing reform. *Urban Studies*, 57(4), 806–826. doi:10.1177/0042098019856780.

Fossati, D. (2016). Is Indonesian local government accountable to the poor? Evidence from health policy implementation. *Journal of East Asian Studies*, 16(3), 307–330. doi:10.1017/jea.2016.17.

Guimarães, E., Malheiros, T., and Marques, R. (2016). Inclusive governance: New concept of water supply and sanitation services in social vulnerability areas. *Utilities Policy*, 43, 124–129. doi:10.1016/j.jup.2016.06.003.

Hicks, T., Jacobs, A.M., and Matthews, J.S. (2016). Inequality and electoral accountability: Class-biased economic voting in comparative perspective. *The Journal of Politics*, 78(4), 1076–1093. doi:10.1086/686157.

Hill, W., Byrne, J., and de Vasconcellos Pegas, F. (2016). The ecotourism-extraction nexus and its implications for the long-term sustainability of protected areas: what is being sustained and who decides? *Journal of Political Ecology*, 23(1), 308–327. doi:10.2458/v23i1.20219.

IGS. (2019). *Global Sustainable Development Report 2019: The Future is Now–Science for Achieving Sustainable Development*. New York: United Nations.

International Council for Science. (2017). *A Guide to SDG Interactions: From Science to Implementation*. Paris: International Council for Science.

Jianu, I., Dobre, I., Bodislav, D.A., Radulescu, C.V., and Burlacu, S. (2020). The implications of institutional specificities on the income inequalities drivers in European Union. *arXiv preprint arXiv:2007.11436*.

Jones, N., Abu-Hamad, B., Pereznieto, P., and Sylvester, K. (2016). Transforming cash transfers: citizens' perspectives on the politics of programme implementation. *The Journal of Development Studies*, 52(8), 1207–1224. doi:10.1080/00220388.2015.1134772.

Khan, Q., Faguet, J.-P., and Ambel, A. (2017). Blending top-down federalism with bottom-up engagement to reduce inequality in Ethiopia. *World Development*, 96, 326–342. doi:10.1016/j.worlddev.2017.03.017.

Kunawotor, M.E., Bokpin, G.A., and Barnor, C. (2020). Drivers of income inequality in Africa: Does institutional quality matter? *African Development Review*. doi:10.1111/1467-8268.12473.

Lindström, M. (2020). A commentary on "The trouble with trust: Time-series analysis of social capital, income inequality, and COVID-19 deaths in 84 countries". *Social Science & Medicine*, 263, 113386. doi:10.1016/j.socscimed.2020.113386.

Masiero, S. and Maiorano, D. (2018). MGNREGA, power politics, and computerization in Andhra Pradesh. *Forum for Development Studies*, 45(1), 1–24. doi:10.1080/08039410.2017.1345785.

Meadows, D.H. (2008). *Thinking in Systems: A Primer*. White River Junction, VT: Chelsea Green Publishing.

Nieto-Aleman, P.A., Garcia-Alvarez-Coque, J.M., Roig-Tierno, N., and Mas-Verdú, F. (2019). Factors of regional poverty reduction in Colombia: Do institutional conditions matter? *Social Policy & Administration*, 53(7), 1045–1063. doi:10.1111/spol.12474.

Nilsson, M., Griggs, D., and Visbeck, M. (2016). Policy: Map the interactions between Sustainable Development Goals. *Nature*, 534, 320–322. doi:10.1038/534320a.

Nwobashi, H.N. and Itumo, A. (2017). Democracy, Town Unions and Provision of Health Facilities in Rural Communities of Ebonyi State: Evaluation of Some Selected Communities. *The Turkish Online Journal of Design, Art and Communication*, 7, 496–506. doi:10.7456/1070ASE/051.

Pham-Truffert, M., Metz, F., Fischer, M., Rueff, H., and Messerli, P. (2020). Interactions among Sustainable Development Goals: Knowledge for identifying multipliers and virtuous cycles. *Sustainable Development*, 28, 1236–1250. doi:10.1002/sd.2073.

Pinzón-Flórez, C.E., Chapman, E., Cubillos, L., and Reveiz, L. (2016). Prioritization of strategies to approach the judicialization of health in Latin America and the Caribbean. *Revista de Saude Publica*, 50, 56. doi:10.1590/S1518-8787.2016050005728.

Ramanujam, N., Caivano, N., and Agnello, A. (2019). Distributive justice and the sustainable development goals: Delivering agenda 2030 in India. *Law and Development Review*, 12 (2), 495–536. doi:10.1515/ldr-2019-0020.

Sukati, V.N., Moodley, V.R., and Mashige, K.P. (2018). A situational analysis of eye care services in Swaziland. *Journal of Public Health in Africa*, 9(3). doi:10.4081/jphia.2018.892.

Suleiman, M.N. and Aminul Karim, M. (2015). Cycle of bad governance and corruption: The rise of Boko Haram in Nigeria. *Sage Open*, 5(1), 2158244015576053. doi:10.1177/2158244015576053.

UN Department of Economic and Social Affairs. (2019). *Sustainable Development Goal 16: Focus on Public Institutions*. Retrieved from:

Warf, B. (2019). Geographically uneven landscapes of Asian corruption. In *Global Corruption from a Geographic Perspective* (pp. 143–193): Springer.

Warf, B. and Stewart, S. (2016). Latin American corruption in geographic perspective. *Journal of Latin American Geography*, 133–155.

Weitz, N., Carlsen, H., Nilsson, M., and Skånberg, K. (2017). Towards systemic and contextual priority setting for implementing the 2030 Agenda. *Sustainability Science*, 13, 531–548. doi:10.1007/s11625-017-0470-0.

Wickremasinghe, D., Gautham, M., Umar, N., Berhanu, D., Schellenberg, J., and Spicer, N. (2018). "It's About the Idea Hitting the Bull's Eye": How Aid Effectiveness Can Catalyse the Scale-up of Health Innovations. *International Journal of Health Policy and Management*, 7(8), 718.

Workneh, M.A. (2020). Gender Inequality, Governance, and Poverty in Sub-Saharan Africa. *Poverty & Public Policy*, 12(2), 150–174. doi:10.1002/pop4.278.

Ye, L. and Yang, H. (2020). From digital divide to social inclusion: a tale of mobile platform empowerment in rural areas. *Sustainability*, 12(6),2424. doi:10.3390/su12062424.

4

GOVERNANCE MECHANISMS FOR COHERENT AND EFFECTIVE IMPLEMENTATION OF THE 2030 AGENDA

A Cross-national Comparison of Government SDG Bodies

Anita Breuer, Julia Leininger and Daniele Malerba

Policy Integration and Accountable Institutions as Conditions for Coherent and Effective Implementation of the SDGs

As pointed out in the introduction to this volume, the recognition of the interdependent nature of the SDGs is one of the major paradigm shifts that differentiate the 2030 Agenda from previous global development frameworks such as the Millennium Development Goals (MDG). Related to this is the realization that achieving the SDGs simultaneously will require an integrated implementation of the 2030 Agenda and increased policy coherence, which, in turn, will require deep institutional reforms and the adoption of innovative governance approaches (Breuer, Leininger, and Tosun, 2019; International Science Council, 2017; Niestroy, Hege, Dirth, Zondervan, and Derr, 2019; Organisation for Economic Co-operation and Development, 2019; Tosun and Leininger, 2017).

Since the concepts of "policy integration", "policy coherence", and "policy coordination" are ascribed slightly different connotations in the literature (May, Sapotichne, and Workman, 2006), some clarification is due here. Literature on policy integration usually refers to "integration" as a dimension on which policies in a specific issue area can be assessed as being more or less coherent. Integration can thus be conceptualized as a continuum that ranges from "least coherent" to "fully coherent" (United Nations, 2018). In common parlance, coherence implies that various policies go together because they share a set of objectives. As a minimum, policy coherence strives to identify and minimize trade-offs between policies. At a more ambitious level, however, it should also aim to promote synergies and produce mutually reinforcing policies (May et al., 2006). In the context of the 2030 Agenda, enhancing policy coherence implies joint consideration of the economic, social, and environmental dimensions of sustainable development and

DOI: 10.4324/9781003254683-4

related policy sectors on the one hand and of policies at various levels of government on the other hand (UN Environment Programme, 2021). Coordination, in turn, refers to the systematic and regular exchange between actors. In development policy, coordination is a means to divide labour and avoid double work as well as increase policy coherence (Torsvik, 2005). Hence, coordination does not necessarily lead to policy integration and coherence but policy integration and coherence are not possible without coordination (Peters, 2015; Tosun and Lang, 2017).

Another major paradigm shift from the MDGs to the SDGs was the recognition of accountable institutions as a requirement for the effective implementation of the SDGs. In the 2030 Agenda, this is most prominently reflected in SDG 16 that calls for "effective, accountable and inclusive institutions at all levels" .

There is consensus in political science literature that accountable governments are a necessary condition for effective development (e.g., Acemoglu and Robinson, 2012; Fukuyama, 2014; Scharpf, 2003). The term accountability refers to a rule-based system that stimulates or constrains behaviour by holding actors responsible to their actions. Essentially, government accountability requires the presence of three central elements: (i) Information that governments must provide about their activities and performance; (ii) answerability, which implies the capacity to demand explanations from governments and the correspondent duty of governmental actors to justify their conduct; and (iii) sanction, which implies the capacity to either punish governmental misconduct or underperformance, or to reward positive behaviour and good performance (Schedler, 1999; Schmitter, 2004).

However, how to ensure government accountability in the context of the 2030 Agenda remains a subject of debate (Beisheim and Ellersiek, 2017). As is the case with many other multilateral agreements, the SDGs are the product of a complex process of negotiations between UN member states. They are based on the principle of global governance through goal setting (Kanie, Bernstein, Biermann, and Haas, 2017) and there are no formal sanctions for countries who do not comply with the agreement (Chimhowu, Hulme, and Munro, 2019; Young, 2017). Therefore, it is incumbent on signatory states themselves to use existing or establish new governance mechanisms to hold governments accountable for SDG implementation (Breuer and Leininger, 2021).

Since the adoption of the 2030 Agenda, the governance mechanisms adopted for its implementation have been compared and analysed in several international policy reports, which focus on different dimensions of policy integration. However, the evidence base underlying these reports presents certain limitations. While some of them focus on a subset of implementing countries based on their membership in international organizations or unions (e.g., European Commission, 2019; Organisation for Economic Co-operation and Development, 2016), others only compare those countries that presented their VNR in a given year (Canadian Council for International Co-operation, 2018; Gesellschaft für Internationale Zusammenarbeit, 2020; United Nations, 2018).

Yet gaining the most complete possible picture of governance mechanisms for SDG implementation is highly relevant for two reasons. First, it allows us to establish a baseline for future analyses to assess the performance of different institutional models ensuring coherent and effective SDG implementation. Second, the systematic compilation of comparable data on the institutional design of such governance mechanisms from a large sample of countries will also enable the conduction of quantitative analyses that investigate how contextual factors (such as regime type, state capacity, country income group, etc.) shape governments' institutional choices vis-à-vis the attainment of an integrated implementation of SDG policies. Grouping countries according to such socioeconomic and political background characteristics will allow for more efficient and context-sensitive policy advice on SDG implementation (see e.g., Paasi, 2005).

Against this background, and responding to this volume's research question concerning the institutional pre-conditions for achieving integrated SDG implementation, the aim of this chapter is twofold:

Conceptually, it proposes five criteria for assessing the institutional design of national governance mechanisms for SDG implementation regarding their potential to foster coherent and effective implementation of the 2030 Agenda.

Empirically, by applying these criteria to 137 country cases, it paves the way for future empirical analyses, in particular comparative cross-national analysis on institutional factors leveraging coherent and effective sustainability governance.

The remainder of this chapter is structured as follows: In section 2, we introduce five criteria for the assessment of national SDG governance mechanisms. In Section 3, we introduce the data and methods used and discuss the results of the empirical analysis. Section 4 concludes.

Criteria for the Assessment of National Governance Mechanisms to Enhance Coherent and Effective SDG Implementation

Beginning from 2015, countries started to create "designated national SDG bodies" (Transparency, Accountability and Partnership for 2030 Agenda, 2019, p. 15) as coordinating governance mechanisms for the implementation of the SDGs. These bodies come in different forms such as, for example, Inter-Ministerial Committees, National SDG Committees, or High Level Commissions. While some countries established new bodies, others expanded the mandate of existing ones to implement the SDGs. While their names vary, they share similar functions - to coordinate and oversee the implementation of the 2030 Agenda across government ministries, agencies and organizations. For simplicity, we will henceforth refer to these governance mechanisms as "government SDG bodies".

At the time of writing, it is still too early to make empirically substantiated statements about how these institutions are performing in ensuring coherent and effective implementation of the 2030 Agenda and it seems unlikely that a single governance mechanism will prove to be most efficient (Organisation for Economic Co-operation and Development, 2019; Persson, Weitz, and Nilsson, 2016). However, previous research and international policy debates consistently emphasize five factors as crucial for successfully achieving the SDGs:

1. Political leadership by the highest level of government (see Abbott and Bernstein, 2015; United Nations, 2018; Donoghue and Khan, 2019)

2. Horizontal coordination across policy sectors (see UN Development Programme, 2017)
3. Vertical coordination across levels of state and government (see International Science Council, 2017; UN Development Programme, 2017)
4. Horizontal accountability and control between actors of the state (Breuer and Leininger, 2021; Cardinal, Romano, and Sweeney, 2019).
5. Broad societal participation and social accountability (Beisheim and Simon, 2016; Dodds, 2015; Elgin-Cossart and Chandran, 2016; Participate, 2017; Stafford-Smith et al., 2017; Transparency, Accountability and Partnership for 2030 Agenda, 2019; Villalona, 2021)

Building on this literature, we propose five criteria for the assessment of national SDG governance bodies regarding their potential to promote coherent and effective SDG implementation, on which we elaborate below.

Political leadership

The most important place for horizontal policy coordination is assumed to be located at the Centre of Government (CoG), that is, the chief executive and the central agencies serving that executive (Organisation for Economic Co-operation and Development, 2016; Peters, 2015). However, prime ministers and presidents normally do not possess the capacities and time resources to coordinate the abundance of activities under their oversight. Rather than that, they delegate this task to staffs or agencies such as offices of the president or prime minister, or cabinet offices. While structural definitions of CoG only include those institutions that directly support the core executive, functional definitions also include institutions that perform supra-sectoral, whole-of-government functions, such as ministries of planning or finance (Santiso, Lafuente, and Alessandro, 2013). This chapter adopts the latter, functional approach.

Research on the National Councils for Sustainable Development (NCSD), created in the follow-up of the 1992 Rio Earth Summit, found support and leadership by the CoG to be conducive to effective policy coordination. The UN Development Programme (2017), for example, finds that NCSDs that were located within a line ministry often had insufficient political power to effectively coordinate sustainable development matters. By comparison, countries that located their NCSDs under the office of the president or prime minister stated that this high-level position ensured effective coordination. Similarly, Osborn, Cornforth, and Ullah (2014) observed that NCSDs with ministerial members were useful in promoting integrated government approaches to sustainable development when chaired or given strong support by a head of state. Another key role is attributed to ministries of finance and planning. Both the cross-cutting vision and the specific objectives formulated in national sustainable development strategies will likely remain at the margins of government decision-making if they are not adequately considered in budget planning and fiscal priority setting (Cheru, 2006; Swanson, Pintér, Bregha, Volkery, and Jacob, 2004; Whitfield, 2009). On the downside, relying exclusively on the CoG for policy coordination might produce adverse, centralising effects. As Peters (2015) points out: *"The top-down approach common to coordination may often not match the bottom-up reality of the problems"* (ibid, p. 75).

Consequently, to ensure efficient coordination, government SDG bodies should be backed by the highest levels of government while allowing line ministries to contribute both their sectoral expertise and working experience.

Horizontal coordination

Given the interlinked nature of the SDGs, it is often claimed that inter-ministerial, government SDG bodies should be better suited to identify cross-cutting issues and resolve trade-offs between the SDGs than those located in a single line ministry (Organisation for Economic Co-operation and Development, 2016; UN Department of Economic and Social Affairs, 2018; UN Development Programme, 2017).

This claim is supported by research on the 1990s and early 2000s National Sustainable Development Strategies (NSDS). Swanson et al. (2004), for example, observe the development of the NSDS was typically entrusted to environment ministries and consequently often exclusively focused on the environmental dimension while displaying a weaker understanding of the economic and social dimensions. Also, traditionally belonging to the less powerful ministries in many countries, environment ministries often had to court the support of more influential actors which led to a "watering down" of the NSDS (Gjoksi, Sedlacko, and Berger, 2010).

These findings suggest that government SDG bodies will only be able to balance the interdependencies between the three dimensions of the 2030 Agenda if they profit from the expertise and experience of all relevant ministries.

Vertical coordination

Since the majority of SDGs involve both a sub-national and local dimension, there is broad agreement that their coherent implementation will require alignment, integration and coordination between different levels of government (Organisation for Economic Co-operation and Development, 2016; Pisano, Mulholland, and Berger, 2016; UN Department of Economic and Social Affairs, 2018). Local governments play an important role in identifying and formulating policy needs as well as in implementing and delivering services and are therefore essential in linking the global SDGs to the needs of local communities (Ongaro, 2015; Swanson et al., 2004; UN Department of Economic and Social Affairs, 2018).

Notwithstanding, decentralizing the responsibility for SDG implementation to lower levels of government requires the creation of additional structures, legislation, and regulation. Furthermore, the coordination and alignment of SDG action undertaken at different levels of government with diverging policy priorities and implementation capacities might entail the risk of diluting SDG action (UN Department of Economic and Social Affairs, 2018). Concerns about these transaction costs can be reflected in the set-up of government SDG bodies: National governments might be reluctant from granting sub-national governments a formal representation in the body. Alternatively, they may choose to collect the input of sub-national governments by means of technical working groups without formal decision-making power, or they may opt to engage them for a limited period of time or for specific purposes such as consultations to prioritize national SDG targets, awareness raising, or monitoring and evaluation.

To ensure vertical coordination across different levels of government, sub-national governments should be represented in the SDG government body or at least closely collaborate with it.

Horizontal accountability

It has been criticized that despite the adoption of SDG 16 the accountability vision offered by the 2030 Agenda is relatively vague and under ambitious (e.g., Donald, 2016). In fact, the text of the declaration offers no clear guidance to member states on how to set up or adjust their national level accountability frameworks (Breuer and Leininger, 2021).

Given that accountability depicts the relationship between different actors, it is central to ask who is accountable to whom. Strategic actor theory literature has distinguished three main types of accountability (i) vertical accountability, whereby voters hold governments to account through elections (O'Donnell, 2007); (ii) horizontal accountability, exercised by the different branches of power who engage in mutual control and by independent state institutions specialized in oversight like courts, parliaments or audit institutions (ibid.), (iii) social accountability, exercised by civil society organizations (CSO) and independent media (Smulovitz and Peruzzotti, 2000).

Recent research has started to address the related question what processes and formats are suitable to ensure SDG accountability at the national level (e.g., Breuer and Leininger, 2021; Karlsson-Vinkhuyzen, Dahl, and Persson, 2018; Transparency, Accountability and Partnership for 2030 Agenda, 2019). As far as horizontal accountability is concerned, this emerging literature has identified Parliaments and independent oversight agencies, such as Supreme Auditing Institutions (SAI) and National Human Rights Institutions (NHRI) as essential cornerstones for national SDG accountability regimes (INTOSAI Development Initiative, 2019; Inter-Parliamentary Union, 2019; Lafortune and Schmidt-Traub, 2019).

To begin with, Parliaments have several important functions to fulfil in promoting accountability in the SDG process including: law making that is consistent with the 2030 Agenda, overseeing the actions of governments and their agencies in implementing the SDGs, monitoring SDG progress, budget evaluation that ensures an adequate allocation of financial resources to achieve the SDGs, and representing the interests of their constituents in the process of SDG implementation (Chungong, 2018; GLOBE and UN Environment Programme, 2015; Hege and Brimont, 2018; Inter-Parliamentary Union, 2015, 2016).

SAIs, in turn, have the mandate to safeguard rational spending of public money (Nagy, Gál, and Véha, 2012; Organisation for Economic Co-operation and Development, 2016; van Winden, 2017). However, over time, SAIs have evolved from exercising purely fiscal oversight and have increasingly taken a more comprehensive view on the effectiveness and efficiency of policies and programmes (Transparency, Accountability and Partnership for 2030 Agenda, 2019; van Winden, 2017). It has thus been argued that they are also well-equipped to audit the compliance of governments to international commitments such as the 2030 Agenda (e.g., United Nations, 2014; van Leeuwen, 2004).

Finally, NHRIs are state bodies with a legal mandate to hold governments accountable to their international human rights obligations. Their main tasks are to prevent and promote the sanctioning of unlawful actions or omissions of governments and other state agencies and to empower vulnerable people vis-à-vis state (e.g., Moreno and Witmer, 2016). Given this mandate, NHRIs are uniquely positioned to ensure that governments adopt a human-rights based approach of SDG implementation (Danish Institute for Human Rights, 2013).

Considering the above, to ensure horizontal accountability in the SDG implementation process, Parliaments, SAIs, and NHRIs should be represented in or closely collaborate with the government SDG body.

Societal participation and accountability

Since the 1990s, growing consensus emerged that traditional top-down governmental approaches were no longer sufficient in managing sustainable development (Dodds, 2015). As scholars of public administration have pointed out, increasing policy complexity and the resulting government overload, require more participatory and interactive governance mechanisms (Torfing, Peters, Pierre, and Sørensen, 2012).

Today, the call for civil society engagement and social accountability in SDG implementation is practically universal in policy reports dealing with the 2030 Agenda. On the one hand, major societal groups including academia, CSOs and the private sector must be engaged and included in order to establish integrated visions and strategies that support sustainability transformation. On the other hand, social accountability is essential for effective policy implementation. It is typically performed by non-state actors who voice demands regarding the lawfulness of state actions and engage in respective monitoring activities, or perform watchdog functions by exposing and documenting governmental wrongdoing or performance deficits (Peruzzotti and Smulovitz, 2006). In fact, the 2030 Agenda stresses "people" or "citizens", as the main principals in the accountability regime for the SDGs [see 37, paragraphs 47, 73, 79, and 84].

It is crucial to recognize, though, that multi-stakeholder engagement also entails certain risks. Most frequently emphasized in this regard are the potential negative implications of private sector involvement (e.g., Beisheim and Simon, 2016; Pattberg, Biermann, Chan, and Mert, 2012). As ECOSOC (2018, p. 5) puts it: "*Private sector financing of the SDGs has its limits in the profit maximizing rational of private sector activities, as ultimately the SDGs are public goods that cannot become bankable projects.*" Furthermore, and as illustrated by the contribution of Siegel and Bastos Lima in this volume, where partnerships for SDG implementation between governments and powerful private sector actors fail to include civil society representatives and perspectives, pre-existing power asymmetries might be reinforced and important trade-offs between economic and social SDGs remain unaddressed.

The above criteria constitute the basis for our comparative analysis of government SDG bodies, the guiding questions of which are shown in Table 4.1 below.

Clarifications on Data and Methods

The present analysis focuses on the institutional design that governments proposed for their designated SDG bodies. To collect the necessary data, based on the above criteria, we performed a qualitative, directed content analysis (Hsieh & Shannon, 2005) of 137 Voluntary National Reviews (VNR) presented to the UN High-level Political Forum on Sustainable Development between 2016 and 2019.[1] The VNR typically contain a dedicated section on "institutional mechanisms" that provides information on how the country has adapted its institutional framework in light of the 2030 Agenda.

Our descriptive analysis rests on five variables built on the five criteria for the assessment of government SDG bodies presented in section 2.

First, the variable **political leadership** refers to the role of the CoG and ministries in overseeing the implementation of SDGs. In our coding, it takes the values 0 to 2, indicating whether no leadership of the implementation process is assigned (0), leadership is assigned to one or several line ministries (1), or leadership assigned to the CoG or there is a co-leadership by both CoG and a line ministry (2).

Second, the variable **horizontal coordination** refers to the number of line ministries represented in the government SDG body. The variable receives value (0), if no ministry or only one ministry is represented in the body, (1) if at least two ministries are represented, and (2), if three or more ministries are represented. Furthermore, in line with the 2030 Agenda's conceptualization of sustainable global development as integrating the economic, environmental, and social dimensions of development, our analysis

TABLE 4.1 Five criteria for the assessment of national SDG governance mechanism

Criterion	*Guiding question*
Political leadership	Where in the executive is the government SDG body located? Who presides over the body?
Horizontal coordination	How many, and which ministries, are formally represented in the government SDG body?
Vertical coordination	Are sub-national governments formally and permanently represented in government SDG body? Or are they members of permanent working groups that report to the body?
Horizontal accountability	Are Parliaments, SAIs, and NHRIs formally and permanently represented in government SDG body? Or are they members of permanent working groups that report to the body?
Societal participation and accountability	Are non-state stakeholders formally and permanently represented in the national body for SDG implementation? Or are they members of permanent working groups that report to the national body? In addition, which stakeholders are represented?

Source: Author's own elaboration

differentiates between the categories of ministries of the economy, environment, social, and sustainability, as well as foreign ministries and the residual category of other ministries.

Third, the variable **vertical coordination** refers to the relation of subnational entities with the government SDG body. The variable takes value (0) if sub-national governments are neither represented in the SDG government body nor in technical committees or working groups that collaborate with the body, (1) if they are only represented in committees or working groups, and (2), if they are represented in the government SDG body itself.

Fourth, our **variable horizontal accountability** refers to the relation of Parliament and independent oversight agencies with the government SDG body. As with the above variable we distinguish between membership in the government SDG bodies themselves or membership in permanent working groups. The variable receives value (0) if neither Parliament, nor SAIs, nor NHRIs are represented in the body or working groups or committees collaborating with it, value (1), if at least one of these three institutions is represented in a working group or committee and (2), if at least one of the three institutions is represented in the SDG government body.

Fifth, our variable **societal participation and social accountability** refers to the representation of non-state stakeholders in the government SDG body. It takes value (0), if non-state stakeholders are neither represented in the SDG government body nor in technical committees or working groups that collaborate with the body, (1) if they are only represented in committees or working groups, and (2), if they are represented in the government SDG body itself.

Empirical Analysis

Assessment: types of governance mechanisms for SDG implementation

In this section, we use the above criteria to assess national government SDG bodies by using descriptive statistics.

Table 4.2 summarizes the information, by showing how each country scores on each variable using a traffic light colouring scheme. The colour green represents variables coded with 2 as described in the previous section (meaning that the country scores high on the respective criterion); orange represents the value of 1 (corresponding to a medium score); and red represents the value 0 (meaning that the country scores low on this criterion).

As can be seen, only three countries (Mali, Slovakia, and Côte d'Ivoire) score green for all the five assessment criteria. Meanwhile, of the spectrum, six countries score red for all five criteria (Andorra, Latvia, Liechtenstein, Malta, New Zealand, Senegal, and Singapore).

In terms of individual criteria, most countries score high (green) on the criteria *political leadership* and *horizontal coordination* (102 and 115, respectively). By contrast, the majority of countries score low (red) on *vertical coordination* and *horizontal accountability* (115 and 91, respectively). Concerning *societal participation and accountability* the picture is a bit more balanced, with 55 countries scoring red, and 54 scoring green.

TABLE 4.2 Assessment of government SDG bodies according to five criteria
Note: A full colour version of this figure is available open access under: https://www.routledge.com/Governing-the-Interlinkages-betweenthe-SDGs-Approaches-Opportunities/Breuer-Malerba-Srigiri-Balasubramanian/p/book/9781032184654

	Political Leadership	*Horizontal Coordination*	*Vertical Coordination*	*Societal Partici-pation and Accountability*	*Horizontal Accountability*
Afghanistan					
Albania					
Algeria					
Andorra					
Argentina					
Armenia					
Australia					
Azerbaijan					
Bahamas					
Bahrain					
Bangladesh					
Belarus					
Belgium					
Belize					
Benin					
Bhutan					
Bosnia and Herzegovina					
Botswana					
Brazil					
Burkina Faso					
Cabo Verde					
Cambodia					
Cameroon					
Canada					
Central African Republic					
Chad					
Chile					

Colombia					
Congo (Republic of the)					
Costa Rica					
Côte d'Ivoire					
Croatia					
Cyprus					
Czech Republic					
Denmark					
Dominican Republic					
Ecuador					
Egypt					
El Salvador					
Estonia					
Eswatini					
Ethiopia					
Fiji					
Finland					
France					
Georgia					
Germany					
Ghana					
Greece					
Guatemala					
Guinea					
Guyana					
Honduras					
Hungary					
Iceland					
India					
Indonesia					
Iraq					
Ireland					

(*Continued*)

TABLE 4.2 (Cont.)

Israel					
Italy					
Jamaica					
Japan					
Jordan					
Kazakhstan					
Kenya					
Korea, Rep.					
Kuwait					
Laos					
Latvia					
Lebanon					
Lesotho					
Liechtenstein					
Lithuania					
Luxembourg					
Madagascar					
Malaysia					
Maldives					
Mali					
Malta					
Mauritania					
Mauritius					
Mexico					
Monaco					
Mongolia					
Montenegro					
Morocco					
Namibia					
Nepal					
Netherlands					
New Zealand					

Niger					
Nigeria					
Norway					
Pakistan					
Palau					
Panama					
Paraguay					
Peru					
Philippines					
Poland					
Portugal					
Qatar					
Romania					
Rwanda					
Saint Lucia					
Saudi Arabia					
Senegal					
Serbia					
Sierra Leone					
Singapore					
Slovakia					
Slovenia					
South Africa					
Spain					
Sri Lanka					
State of Palestine					
Sudan					
Sweden					
Switzerland					
Tajikistan					
Tanzania					
Thailand					
Timor-Leste					

TABLE 4.2 (Cont.)

Togo					
Tonga					
Tunisia					
Turkey					
Turkmenistan					
UK					
Uganda					
United Arab Emirates					
Uruguay					
Vanuatu					
Venezuela					
Viet Nam					
Zimbabwe					

Source: Author's own elaboration

Note: Low score Medium score High score

From these general findings, two main messages arise. First, judging by the composition of government SDG bodies, the majority of countries displays a high level of government support to the 2030 Agenda by assigning the leadership of the government SDG body to the CoG. Looking at the data in more detail (not presented here for space reasons), 84 out of 137 countries in our sample assigned the leadership to the CoG, 25 to one or multiple line ministries, and 19 assigned it jointly to the CoG and line ministries. Only nine countries fail to clarify the question of political leadership in their VNRs.

Second, the assignment of leadership to the CoG does not necessarily preclude a whole-of-government approach to SDG implementation. In fact, we observe an involvement of line ministries in about 74 per cent of all cases. In most of these cases, it was the ministry of foreign affairs and/or the environment ministry that played a key role. This dominance of the ministries for foreign affairs and the environment is also observed when we inspect the descriptive statistics for the variable political leadership, that is, for those line ministries that are indicated as presiding over the government SDG body.

Going into more detail, it is interesting to investigate which line ministries are most often represented in government SDG bodies. Figure 4.1 provides an overview of how often the individual categories of line ministries are mentioned in VNRs as being members of the national SDG-implementation body, as a percentage share of total VNRs. We can infer from the bar graph that ministries of foreign affairs, of the economy, and of the environment are most frequently mentioned in the VNRs. The assignment of responsibility to ministries of foreign affairs is plausible when bearing in mind that the SDGs are an international policy agenda, which will require a

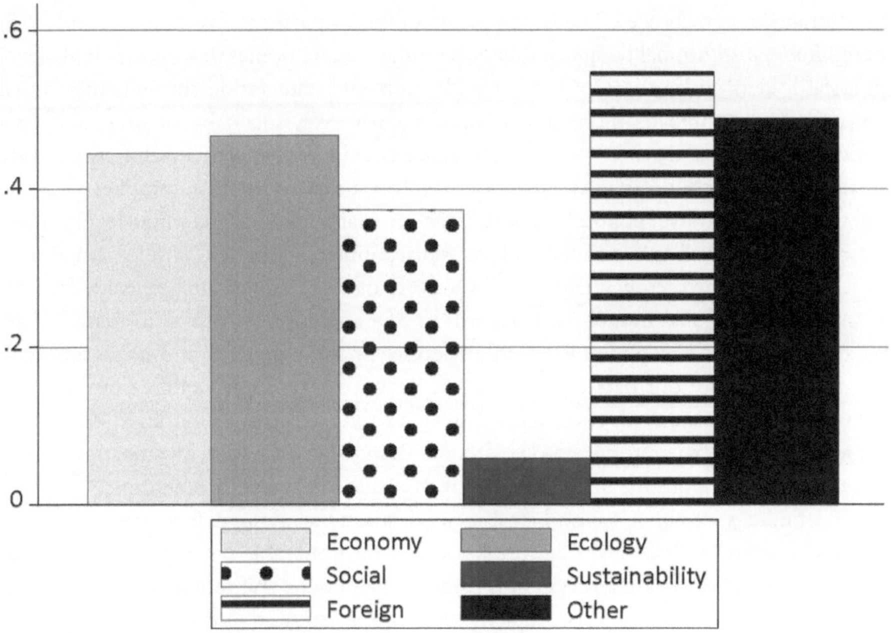

FIGURE 4.1 Share of VNRs mentioning individual categories of line ministries as members of government SDG body

Source: Author's own elaboration

Note: A full colour version of this figure is available open access under: https://www.routledge.com/Governing-the-Interlinkages-between-the-SDGs-Approaches-Opportunities/Breuer-Malerba-Srigiri-Balasubramanian/p/book/9781032184654

considerable degree of coordination among implementing states. However, it raises the question of whether national SDG-implementation bodies are not – to some extent – an effort in showcasing activities related to the 2030 Agenda internationally, rather than an attempt to achieve cross-sectoral policy coordination domestically. Furthermore, the prominent role of environment ministries suggests that the 2030 Agenda is strongly perceived as an environmental agenda, which is interesting considering the wide range of topics covered by the SDGs.

Conclusion

Responding to this volume's research question on adequate institutional pre-conditions for SDG implementation, this exploratory study set out to a) propose criteria for assessing the institutional design of government SDG bodies regarding its potential to promote effective and coherent implementation of the 2030 Agenda; and b) to apply these criteria in a comparative study of existing government SDG bodies.

On the basis of the VNRs submitted to the United Nations between 2016 and 2019 we were able to show that governments across the world have taken different institutional design choices. The largest group of the countries analyzed in this study (15) opted for an institutional design of their government SDG bodies that ensures high-level political leadership by the CoG along with horizontal integration through the formal engagement of three or more line ministries. However, sub-national units and civil society stakeholders are often not formally represented in this approach. Only a relatively small number of countries (six) opted for an institutional design that complies with the call to ensure high-level political support, cross-sectoral horizontal and multi-level vertical coordination, as well as horizontal and societal accountability in SDG implementation.

Evidently, in the present analysis, we were yet not able to discuss the *performance* of the respective institutional models with regards to SDG achievement since in many countries government SDG bodies have only been set up fairly recently and more time will have to pass before our data on institutional design can be meaningfully collated with data on SDG progress. Hence, rather than offering conclusive findings, this study represents a first exploratory venture into the theory-led empirical investigation of the institutional conditions for SDG implementation that are currently emerging.

The findings of this empirical analysis of institutional designs for integrated SDG implementation have various policy implications. One noticeable observation is the strong role of ministries of foreign affairs in government SDG bodies. Although the Agenda 2030 is a global agenda, which is to be implemented in domestic and global arenas, it seems to be primarily perceived as an international agenda by almost half of the governments analysed. Although international action is crucial for addressing global problems formulated in the SDGs, the domestic arena must not be left behind. Environment ministries rank second after foreign affairs in leading government SDG bodies in the countries analysed. The high level of involvement of environment ministries is positive in the sense that it holds potential to increase synergies with global environmental agreements such as the Paris Agreement on Climate Change or the Convention on Biological Diversity. However, there are also downsides to this approach. Environment ministries typically belong to the "weaker" ministries, which have to court the support of stronger ministries. This had led to the watering down of the environmental goals in benefit of economic goals in past sustainability strategies. If cross-sectoral policy coherence is to be improved, government SDG bodies should thus be composed in a way that all three dimensions of the 2030 Agenda (economic, social, and ecological) have ministerial representation.

Another observation is that, so far, both horizontal and social accountability in SDG implementation appear to be under-institutionalized in the majority of countries analysed. Regarding social accountability and participation, our comparative analysis found that while civil society representatives are often integrated in working groups and committees, they are rarely formally represented in government SDG bodies themselves and thus do not have a say in decision-making processes on the national level. However, given the constructive role that non-state actors have been playing in raising awareness for and formulating the SDGs, they should be part of decision-making processes. Furthermore, social accountability has a critical role to play in ensuring effective implementation of the SDGs by monitoring progress, highlighting gaps, preventing governments from backsliding on their commitments and counteracting global tendency of shrinking civil spaces. Yet, regarding horizontal accountability, the findings of our analysis are even more concerning. The number of countries in our sample that report having established robust mechanisms to ensure horizontal government oversight of SDG implementation through parliaments and independent government agencies is negligible. Clearly, serious formal commitment to horizontal accountability in SDG implementation has been a choice of individual governments rather than a standard in national SDG implementation across countries so far. This resonates with earlier criticism concerning the lack of strong accountability in the SDG process at the national level. Addressing this lack will be key to accelerate effective SDG implementation in the future.

Note

1 The coding instructions used for the directed content analysis of the VNR are available from the authors on request. On this occasion, we thank Buğra Ahlatci, Adriana Cassis, Brilliant Dziko, Paula Alejandra González Mateus, Lucas Leopold, Ramona Hägele, Lena Noumi, Julian Rossello, Paul Thalmann, Semyon Pavlenko, and Daniela Zuluaga for their support in coding the data.

References

Abbott, K. W. and Bernstein, S. (2015). The High-Level Political Forum on Sustainable Development: Orchestration by Default and Design. *Global Policy*, 6(3), 222–233.

Acemoglu, D. and Robinson, J.A. (2012). *Why Nations Fail: The Origins of Power, Prosperity, and Poverty*. New York: Random House.

Beisheim, M. and Ellersiek, A. (2017). Partnerships for the 2030 Agenda for Sustainable Development: Transformative, Inclusive and Accountable? Retrieved from: https://sustainabledevelopment.un.org/content/documents/24722017RP14_bsh_elk_01.pdf.

Beisheim, M. and Simon, N. (2016). *Multi-stakeholder partnerships for implementing the 2030 Agenda improving accountability and transparency*. Berlin.

Breuer, A. and Leininger, J. (2021). Horizontal Accountability for SDG Implementation: A Comparative Cross-National Analysis of Emerging National Accountability Regimes. *Sustainability*, 13(13),7002. Retrieved from: www.mdpi.com/2071-1050/13/13/7002.

Breuer, A., Leininger, J., and Tosun, J. (2019). Integrated Policymaking. In *Choosing an Institutional Design for Implementing the Sustainable Development Goals (SDGs)*. Bonn: Deutsches Institut für Entwicklungspolitik.

Cardinal, N., Romano, J., and Sweeney, E. (2019). *SDG Accountability Handbook*. Retrieved from

Canadian Council for International Co-operation. (2018). *Progressing national SDGs implementation: An independent assessment of the voluntary national review reports submitted to the United Nations High-level Political Forum on Sustainable Development*. Retrieved from: https://secureservercdn.net/192.169.222.135/9bz.99d.myftpupload.com/wp-content/uploads/2019/05/SDG-Accountability-Handbook.pdf?time=1653401988.

Cheru, F. (2006). Building and supporting prsps in africa: what has worked well so far? what needs changing? *Third World Quarterly*, 27(2), 355–376. doi:10.1080/01436590500432689.

Chimhowu, A.O., Hulme, D., and Munro, L.T. (2019). The 'New' national development planning and global development goals: Processes and partnerships. *World Development*, 120, 76–89. doi:10.1016/j.worlddev.2019.03.013.

Chungong, M. (2018). *Transforming the SDGs into Everyday Reality: The Role of Parliaments*. Honolulu, HI: East-West Center.

Danish Institute for Human Rights. (2013). *NHRIs' Independence and Accountability*. Retrieved from: www.humanrights.dk/sites/humanrights.dk/files/media/migrated/nhris_independence_and_accountability_publication.docx.pdf.

Dodds, F. (2015). *Multi-stakeholder partnerships: Making them work for the Post-2015 Development Agenda (ECOSOC Policy Brief)*. New York: ECOSOC.

Donald, K. (2016). Promising the World: Accountability and the SDGs. *Health and Human Rights Journal*. Retrieved from: www.hhrjournal.org/2016/01/promising-the-world-accountability-and-the-sdgs.

Donoghue, D. and Khan, A. (2019). Achieving the SDGs and 'leaving no one behind'. Maximising synergies and mitigating trade-offs. ODI Working Paper No. 560. https://cdn.odi.org/media/documents/Achieving_the_SDGs_and_leaving_no_one_behind_maximising_synergies_and_mitigating.pdf.

ECOSOC. (2018). *High level event on financing the SDGs. Breaking the bottlenecks from policy to impact*. Retrieved from: www.un.org/pga/72/es/high-level-event-financing-for-sdgs-breaking-the-bottlenecks-of-investment-from-policy-to-impact-9-2.

Elgin-Cossart, M. and Chandran, R. (2016). *Designing Better Accountability Mechanisms for the 2030 Agenda for Sustainable Development*. Center for American Progress.

European Commission. (2019). The European Green Deal. In COM(2019) 640 Final, 11.12.2019.

Fukuyama, F. (2014). *Political Order and Political Decay: From the Industrial Revolution to the Globalization of Democracy*. New York: Farrar, Straus and Giroux.

Gesellschaft für Internationale Zusammenarbeit. (2020). 2020 Voluntary National Reviews – a snapshot of trends in SDG reporting. Retrieved from: www.partners-for-review.de/wp-content/uploads/2020/12/P4R-2020-VNR-Analysis_FINAL-Nov.-2020.pdf.

Gjoksi, N., Sedlacko, M., and Berger, G. (2010). National sustainable development strategies in Europe: Status quo and recent developments. Retrieved from: www.esdn.eu/fileadmin/ESDN_Reports/2010-September-National_Sustainable_Development_Strategies_in_Europe.pdf.

GLOBE and UN Environment Programme. (2015). Bringing the 2015 Summits Home. In *An Action Agenda for Legislators*. Brussels: UN Environment Programme.

Hege, É. and Brimont, L. (2018). Integrating SDGs into national budgetary processes. *Studies*, 5(18), IDDRI, Paris. Retrieved from: https://www.iddri.org/sites/default/files/PDF/Publications/Catalogue%20Iddri/Etude/201807-ST0518-SDGs-budget-EN_1.pdf.

Hsieh, H.F. and Shannon, S.E. (2005). Three approaches to qualitative content analysis. *Qualitative Health Research*, 15(9), 1277–1288. doi:10.1177/1049732305276687.

INTOSAI Development Initiative. (2019). Are Nations Prepared for Implementation of the 2030 Agenda? Supreme Audit Institutions' Insights and Recommendations.

International Science Council. (2017). *A guide to SDG interactions: From science to implementation*.

Inter-Parliamentary Union. (2015). The Sustainable Development Goals: Turning Words into Action.

Inter-Parliamentary Union. (2016). *Parliaments and the Sustainable Development Goals*.

Inter-Parliamentary Union. (2019). *Institutionalization of the Sustainable Development Goals in the work of parliaments*. Geneva.

Kanie, N., Bernstein, S., Biermann, F., and Haas, P.M. (2017). Introduction: Global Governance through Goal Setting. In N. Kanie and F. Biermann (Eds), *Governing through Goals: Sustainable Development Goals as Governance Innovation* (pp. 1–28). Cambridge, Massachusetts: MIT Press.

Karlsson-Vinkhuyzen, S., Dahl, A.L., and Persson, Å. (2018). The emerging accountability regimes for the Sustainable Development Goals and policy integration: Friend or foe? *Environment and Planning C: Politics and Space*, 36(8), 1371–1390. doi:10.1177/2399654418779995.

Lafortune, G. and Schmidt-Traub, G. (2019). SDG Challenges in G20 Countries. In: *Sustainable Development Goals: Harnessing Business to Achieve the SDGs through Finance, Technology, and Law Reform* (pp. 29–234).

May, P.J., Sapotichne, J., and Workman, S. (2006). Policy Coherence and Policy Domains. *Policy Studies Journal*, 34(3), 381–403. doi:10.1111/j.1541-0072.2006.00178.x.

Moreno, E. and Witmer, R. (2016). The Power of the Pen: Human Rights Ombudsmen and Personal Integrity Violations in Latin America, 1982–2006. *Human Rights Review*, 17(2), 143–164. doi:10.1007/s12142-015-0391-1.

Nagy, S., Gál, J., and Véha, A. (2012). Improving audit functions of supreme audit institutions to promote sustainable development. *Applied Studies in Agribusiness and Commerce*, 6(3–4), 63–69. doi:10.19041/APSTRACT/2012/3-4/9.

Niestroy, I., Hege, E., Dirth, E., Zondervan, R., and Derr, K. (2019). *Europe's approach to implementing the Sustainable Development Goals: Good practices and the way forward*. Brussels: European Union.

O'Donnell, G. (2007). *Dissonances. Democratic Critiques of Democracy*. Notre Dame, IN: University of Notre Dame Press.

Organisation for Economic and Co-operation and Development. (2016). OECD survey on planning and co-ordinating the implementation of the SDGs: First results and key issues. Retrieved from: www.oecd.org/gov/cob-sdg-survey-overview-of-results.pdf.

Organisation for Economic Co-operation and Development. (2019). Policy Coherence for Sustainable Development 2019: Empowering People and Ensuring Inclusiveness and Equality. Retrieved from: www.oecd.org/publications/policy-coherence-for-sustainable-development-2019-a90f851f-en.htm.

Ongaro, E. (2015). Multi-Level Governance: The Missing Linkages. In: *Multi-Level Governance: The Missing Linkages* (Vol. 4, p. i): Emerald Group Publishing.

Osborn, D., Cornforth, J., and Ullah, F. (2014). National Councils for Sustainable Development: Lessons from the past and present. Retrieved from: www.iisd.org/sites/default/files/publications/sdplannet_lessons_from_the_past.pdf.

Paasi, M. (2005). Collective benchmarking of policies: an instrument for policy learning in adaptive research and innovation policy. *Science and Public Policy*, 32(1), 17–27. doi:10.3152/147154305781779669.

Participate. (2017). Participatory Accountability for the SDGs: beyond Social Accountability. Retrieved from www.researchgate.net/publication/344153210_Participatory_Accountability_for_the_SDGs_beyond_Social_Accountability.

Pattberg, P., Biermann, F., Chan, S., and Mert, A. (2012). *Public Private Partnerships for sustainable development. Emergence, influence and legitimacy*. Cheltenham: Edward Elgar.

Persson, Å., Weitz, N., and Nilsson, M. (2016). Follow-up and Review of the Sustainable Development Goals: Alignment vs. Internalization. *Review of European, Comparative & International Environmental Law*, 25(1), 59–68. doi:10.1111/reel.12150.

Peruzzotti, E. and Smulovitz, C. (2006). *Enforcing the rule of law: social accountability in the new Latin American democracies*. Pittsburgh, PA: University of Pittsburgh Press.

Peters, B.G. (2015). *Pursuing horizontal management: The politics of public sector coordination*. Lawrence, KS: University Press of Kansas.

Pisano, U., Mulholland, E., and Berger, G. (2016). *Implementation of the 2030 Agenda for SD and the SDGS in Europe. Stock-taking to share experiences and support peer learning*. Vienna: ESDN.

Santiso, C., Lafuente, M., and Alessandro, M. (2013). *The role of the center of government: A literature review*. Inter-American Development Bank, Technical Note No. IDB-TN-581. Retrieved from: https://publications.iadb.org/publications/english/document/The-Role-of-the-Center-of-Government-A-Literature-Review.pdf

Scharpf, F. (2003). Problem-Solving Effectiveness and Democratic Accountability in the EU. Working Paper 03/1, February 2003.

Schedler, A. (1999). Conceptualizing accountability. In A. Schedler, L. Diamond, and M.F. Plattner (Eds), *The Self-Restraining State. Power and Accountability in New Democracies* (Vol. 13, p. 17). Colorado and London: Lynne Rienner Publishers.

Schmitter, P.C. (2004). The quality of democracy: the ambiguous virtues of accountability. *Journal of Democracy*, 15(4), 47–60.

Smulovitz, C. and Peruzzotti, E. (2000). Societal Accountability in Latin America. *Journal of Democracy*, 11(4), 147–158.

Stafford-Smith, M. et al. (2017). Integration: the key to implementing the Sustainable Development Goals. *Sustainability Science*, 12(6), 911–919. doi:10.1007/s11625-016-0383-3.

Swanson, D., Pintér, L., Bregha, F., Volkery, A., and Jacob, K. (2004). *National strategies for sustainable development. Challenges, approaches and innovations in strategic and co-ordinated action.* Bonn: International Institute for Sustainable Development, Deutsche Gesellschaft für Technische Zusammenarbeit.

Transparency, Accountability and Partnership for 2030 Agenda. (2019). *SDG Accountability Handbook*. Retrieved from: https://secureservercdn.net/166.62.112.219/9bz.99d.myftpupload.com/wp-content/uploads/2019/05/SDG-Accountability-Handbook.pdf?time=1637476956.

Torfing, J., Peters, G., Pierre, J., and Sørensen, E. (2012). *Interactive Governance: Advancing the Paradigm*. Oxford: Oxford University Press.

Torsvik, G. (2005). Foreign economic aid; should donors cooperate? *Journal of Development Economics*, 77(2), 503–515. doi:10.1016/j.jdeveco.2004.05.008.

Tosun, J. and Lang, A. (2017). Policy integration: mapping the different concepts. *Policy Studies*, 38(6), 553–570. doi:10.1080/01442872.2017.1339239.

Tosun, J. and Leininger, J. (2017). Governing the Interlinkages between the Sustainable Development Goals: Approaches to Attain Policy Integration. *Global Challenges*, 1(9), 1700036. doi:10.1002/gch2.201700036.

UN Department of Economic and Social Affairs. (2018).

UN Development Programme. (2017).

UN Environment Programme. (2021). SDG indicator 17.14.1 metadata. Retrieved from https://unstats.un.org/sdgs/metadata/files/Metadata-17-14-01.pdf.

United Nations. (2014). Promoting and fostering the efficiency, accountability, effectiveness and transparency of public administration by strengthening supreme audit institutions. In UN General Assembly A/RES/69/228.

United Nations. (2018). *Working Together: Integration, Institutions and the Sustainable Development Goals, World Public Sector Report 2018*. Retrieved from: https://publicadministration.un.org/publications/content/PDFs/World%20Public%20Sector%20Report2018.pdf.

van Leeuwen, S. (2004). Auditing International Environmental Agreements: The Role of Supreme Audit Institutions. *Environmentalist*, 24(2), 93–99. doi:10.1007/s10669-004-4800-2.

van Winden, E. (2017). *Auditors of Sustainability: Exploring the Role of Supreme Audit Institutions in the Implementation of the Sustainable Development Goals*. The Hague: Utrecht University.

Villalona, C. (2021). Global SDG Accountability Report. Retrieved from: www.partners-for-review.de/wp-content/uploads/2021/06/Global-SDG-Accountability-Report_JUN2021.pdf.

Whitfield, L. (2009). *The Politics of Aid: African Strategies for Dealing with Donors*. Oxford: Oxford University Press.

Young, O.R. (2017) Conceptualization: Goal setting as a strategy for earth system governance. In: Kanie, N. and Biermann, F. (eds.). *Governing Through Goals: Sustainable Development Goals as Governance Innovation. MIT-Press, Cambridge.* (pp. 31–52).

5

A METAGOVERNANCE APPROACH TO MULTILEVEL GOVERNANCE AND VERTICAL COORDINATION FOR THE SDGS

Louis Meuleman

Introduction

The 17 Sustainable Development Goals (SDGs) of the UN 2030 Agenda have been designed as a comprehensive set of goals and targets because the social, economic and environmental systems they aim to transform or preserve, are connected and influencing each other (Kamau, Chasek, and O'Connor, 2018). It is generally understood that these systemic interlinkages should be reflected in how governments at all levels and within a whole-of-government approach relate to each other and collaborate to attain the SDGs. One of the implications is the need to have effective horizontal coordination across policy sectors; another is about effective coordination between the governance frameworks used by public authorities across administrative levels. The latter approach is called multilevel governance (MLG).

This chapter builds on literature and practice examples on MLG from an SDG perspective, and with the conceptual glasses of metagovernance ('governance of governance'). It addresses governance mechanisms between government levels (linked to the second research question of this volume, as regards power asymmetries between stakeholders from different levels), and political-institutional preconditions that are conducive to the establishment of effective governance mechanisms to manage SDG interactions (research question 5), from a MLG perspective.

Conceptual Framework

In order to analyze the different variations of multilevel governance, a conceptual framework is needed based on specific definitions of governance, governance styles, governance frameworks and metagovernance. This framework is presented in Figure 5.1.

DOI: 10.4324/9781003254683-5

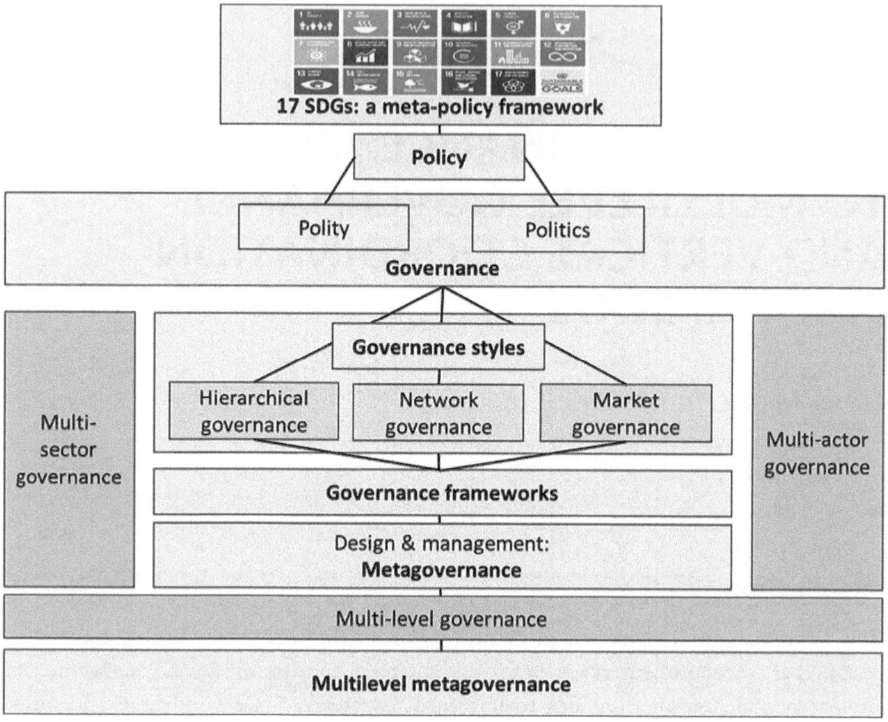

FIGURE 5.1 A conceptual framework: from SDG policies to multilevel metagovernance
Source: Author's own elaboration

In Figure 5.1, the key terms defined above are linked in a conceptual framework. The 17 SDGs cover virtually all areas of public policy, and therefore constitute a policy of policies or 'meta-policy' (Meadowcroft, 2011). In a classical political science tryptic, policy (goals, targets, timelines) is supported by polity (institutions, rules) and politics (processes, actors). Together, polity and politics are covered by the term governance, as will be explained below. Governance for sustainability has a multi- or cross-sector dimension, a multi-actor dimension and a multi-level dimension. Governance frameworks are concrete approaches for specific policies. Design and management of governance frameworks with a sensitivity for the governance environment, including the administrative and societal values, cultures, and traditions in a geographical area, requires a concept 'over and beyond', or 'meta' governance. Multilevel metagovernance is thus about creating actionable mechanisms to foster effective relationships between different levels of authority.

The central terms can be defined as follows:

Governance: What multilevel governance (MLG) means depends on how governance is defined. MLG can have as many different meanings as there are definitions of the term governance. To tackle the various ways in which multilevel relations can materialize, governance has to be defined broadly. Fukuyama's (2013) definition of governance as a government's ability to make and enforce rules, and to deliver services, covers a lot already, but leaves it unclear whether under 'services' also the engagement with societal stakeholders should be understood. An even broader approach is that, if policy is about *what* and *when* (the goals, targets, milestones), then governance is about *how* (which tools, instruments, processes) and *who* (actors, stakeholders) (Meuleman, 2021). The rationale behind this is that if a policymaker defines governance only as involving stakeholders, or as promoting accountability, or as focusing on cost-efficiency, she or he would limit the range of potential policy options. Therefore, only if governance covers all these perspectives, well-reflected choices are possible and governance frameworks can be designed which are contextualized.

Case study research (Meuleman, 2008) revealed that policymakers often understand this intuitively. A broad definition of governance that covers all typical governance styles and the whole repertoire of institutional mechanisms, rules, tools, and forms of actor involvement is: "Governance is the totality of interactions in which government, other public bodies, private sector and civil society participate (in one way or another), aimed at solving public challenges or creating public opportunities" (Meuleman, 2008, p. 11).

Based on this definition of governance, three other concepts can now be defined: governance styles, governance frameworks, and metagovernance.

Governance styles: Many scholars distinguish three ideal-typical governance styles which have their own values, logics and tools (see e.g. Kooiman, 2003; Meuleman, 2008; Peters, 1998; Pollitt and Bouckaert, 2011). The styles are normative as they are carriers of values: network governance embraces consensus and empathy, market governance entrepreneurship and competition, and hierarchical governance authority and control (Meuleman, 2018). Hierarchical, network and market governance usually occur in combinations but some of their characteristics are incompatible and could undermine the effectiveness of a specific governance style combination. The three styles differ in at least fifty features (Meuleman, 2018). How effective they are in practice largely depends on the context. For multilevel governance, the following features listed in Table 5.1 below seem crucial, because they focus on relations, including conflicts, between actors.

The last example (suitability for problem types) is also illustrative: for crisis management a certain/high geographical level might be necessary, for dealing with very complex problems, being close to citizens might help understanding the challenge better, and certain routine issues should not be dealt with bureaucratically or in a long-lasting dialogue, but might benefit from outsourcing to an efficient operator.

There is no blueprint for successful multilevel governance of the transitions needed to implement the 17 SDGs. Rather than that, any combination of elements such as those shown in Table 5.1 is in theory possible. But when a central government relies on a hegemonic, top-down and power-based approach, one cannot expect subnational governments to risk stepping out of their (narrow) comfort zone and develop innovative approaches. At the same time, when a national government relies solely on informal arrangements with subnational authorities, reaching national policy targets might become very difficult. As Fleming and Rhodes (2005, p. 203) have stated

TABLE 5.1 Selected features of hierarchical, network and market governance relevant for multilevel governance (based on Meuleman, 2018)

Feature	*Hierarchical. Governance*	*Network governance*	*Market governance*
Relational values	Hegemonism	Tolerance, pluralism	Individualism
Relation types	Dependent	Interdependent	Independent
Roles of government	Government rules society	Government is a partner in a network society	Government delivers societal services
Orientation of organizations	Top-down, formal, Internal	Horizontal, informal, open-minded,	Bottom-up, Competitive, external
Public sector reform approach	Top-down	Inclusive	Outsourced expertise
Conflict resolution types	Classical negotiation, power-based (win-lose)	Mutual gains approach to negotiation (win-win); diplomacy	Classical negotiation, competition based (win-lose)
Suitability for problem types	Crises, disasters	Complex, multi-actor issues	Routine, non-sensitive issues

Source: Author's own elaboration

pointedly: "The future will not lie with either markets, or hierarchies or networks but with all three. The trick will not be to manage contracts or steer networks but to mix the three systems effectively when they conflict with and undermine one another".

Governance frameworks: A governance framework can be defined as "*the totality of instruments, procedures and processes designed to tackle a societal problem*", followed by a normative recommendation that "(t)hey should be adapted to legal, cultural, and physical conditions of the problem environment and internally consistent; the normative assumptions (values, hypotheses) should be clear" (Meuleman, 2014, p. 978). Governance frameworks are necessary to support implementation of a policy.

Multilevel governance: According to (Pierre and Peters, 2021), multilevel governance has long been thought of as "central, regional and local government neatly organized in a hierarchy", but there are many different forms of MLG – some indeed hierarchical, others more based on collaboration and/or more on an ad hoc basis. In the EU system, the European Commission is a powerful fourth level. The UN can be seen as a fifth level, which is more influential than powerful. Liesbet and Gary (2003) distinguished two types of multi-level governance. In one type, every citizen is "located in a Russian Doll set of nested jurisdictions, where there is one and only one relevant jurisdiction". The other type is fragmented into functionally specific pieces, for example selecting a particular software standard or monitoring water quality of a particular river. The EU has often been described and analyzed as a MLG system, with a combination of a classical hierarchical polity and other, more informal forms of governance (e.g., Kull, 2016). The balance between formal and informal MLG is tricky. Peters and Pierre (2004: 76) warned that "the absence of distinct legal frameworks and the reliance on sometimes quite informal negotiations between different institutional levels could well be a "Faustian bargain" where actors only see the attractions of the deal and choose to ignore the darker consequences of the arrangement".

Governance failure: In terms of the triptych policy – polity – politics, governance is about polity (structures) and politics (processes). In this view, governance and policy are two sides of the same coin namely of the functioning of public administration. This might be logical from a theoretical perspective, but it is not always clear in the often ambiguous (Noordegraaf, 2015), complex, dynamic and 'wicked' (Termeer, Dewulf, and Biesbroek, 2019) reality of public administration, at all levels, especially in a political environment such as a ministry. Policymakers might be so much driven by policy objectives and targets imposed by political leaders that they neglect the governance dimension. A case in point is perhaps the European Green Deal (European Commission, 2019). This is a comprehensive policy programme with a range of strategies and legislative proposals, which itself has no governance section. Some of the Green Deal deliverables such as the EU climate and energy package (*EU 2018 – Regulation (EU) 2018/1999) do include a paragraph or section that contains recommendations regarding the necessary tools and instruments to implement the deal but others lack such a dimension altogether. For example, the 2020 EU Biodiversity Strategy (European Commission, 2020) did not include a governance section, but announced it as separate deliverable.

It is important to distinguish policy failure and governance failure. When a policy is unsuccessful, the reason might seem policy failure, but the underlying cause could well be governance failure. Mark Bovens and 't Hart (2016) observe three types of policy failure: farce (weak results but political success), tragedy (strong results but no political acknowledgement) and fiasco (weak results and weak political credits). Governance failure, in turn, can be defined as *"The ineffectiveness of governance goals, a governance framework or the management thereof, to achieve policy goals"* (adapted from Mark Bovens, t'Hart, & Peters, 2001). We can distinguish three types of governance failure, with a different action perspective (Meuleman, 2018). First, governance design failure that results from a mismatch between problem context and governance style, for example when a governance style (or combination of different styles) is incapable to successfully address a specific problem type. Second, governance capacity failure that results from the mismatch between governance style and governance capacity. Third, governance management failure that results from ineffective management of governance frameworks.

The analysis of governance failure from a multilevel perspective might show that the bottleneck is mainly on a certain administrative level. For example, a national governance framework to support preservation of biodiversity might be ineffective, when subnational authorities do not have the capacity to implement the necessary rules, collaboration and incentives, and/or when local authorities lack mechanisms to collaborate effectively with neighbouring cities and communities. For another example, Srigiri and Scheumann in this volume show that the integrated governance of the water-land-food nexus in Ethiopia is severely hampered by financial, technical, and human capacity deficits at the district and local levels.

Metagovernance – The complexity and dynamics of the governance environment require permanent reflection and management of governance frameworks. This 'governance of governance' is called metagovernance (Jessop, 1997; Kooiman, 2003). It can be defined as:

> *"a means by which to produce some degree of coordinated governance, by designing and managing sound combinations of hierarchical, market and network governance, to achieve the best possible outcomes from the viewpoint of those responsible for the performance of public sector organizations: public managers as 'metagovernors'"*
>
> *(Meuleman, 2008, p. 68)*

Metagovernance thinking can be integrated in the standard policy cycle by adding particular emphasis on mapping the governance environment and having the capacity and skills to know which governance features might be synergetic when combined, and which could be undermining the effectiveness of a governance framework. Applying metagovernance in a methodical way could follow seven steps (Meuleman, 2018): (1) Map the governance environment; (2) Evaluate the current situation; (3) Define, reframe, refine the problem; (4) Formulate context-specific goals and options; (5) Design a governance framework; (6) Metagovern the governance framework; and (7) Review the effectiveness of the governance framework.

Many examples of metagovernance practice have been observed in the EU's regional development and environmental policy. EU laws and policies need to be designed as packages that contains various approaches, instruments and tools. In some EU countries, informal institutions and a relative norm-free approach might work better, while in others the governance mix could require a strong legal basis that prescribes what needs to be done and how. UN conventions are, unlike EU law, rather self-binding. Still, also UN member states try to negotiate sufficient room to navigate effectively within their own implementation systems and make it possible to use metagovernance. International organizations may also establish rules for their engagement in partnerships, such as conditions for the registration of partnerships, e.g. in UN databases, and the provision of associated benefits like material and non-material resources (Beisheim, Ellersiek, Goltermann, and Kiamba, 2018).

An example of a metagovernance intervention by an international organization which has more executive power than the UN, is the establishment of the European Commission's Environmental Implementation Review (EIR) (Meuleman, 2018). In 2016 the European Commission observed that its extensive European legal framework for environmental protection with dozens of directives and strategies was not implemented sufficiently in many member states, causing economic, social and environmental damage of around €55 billion per year (European Commission, 2016). The existing governance framework combined two approaches: legal procedures (infringements) against countries, which could result in large fines, and financial support in the form of the EU's structural funds for regional development. As a response, in 2016 an informal dialogue tool was introduced: the EIR. This is a two- or three-yearly cycle of analytical country reports specifying where countries are with regard to environmental implementation, accompanied by bilateral high-level dialogues between Commission and countries, and a peer-to-peer tool that finances exchange of experiences and mutual learning between Member States. The peer-to-peer mechanism[1] was picked up at all levels: workshops were organized between national ministries, regions, and cities, based on their own demand. Although the third cycle has now started, after first rounds in 2017 and 2019, it is still too early to conclude how much the revision of the governance framework has resulted in better implementation. But the framework has now more options, owing to a combination of hierarchical, market- and network-style governance tools. And the country reports and dialogues, as well as the peer-to-peer tool have opened new avenues for navigating multilevel governance. The EIR is an example of a governance mechanism to manage SDG interlinkages and addressing power asymmetries between stakeholders from different sectors and levels (second research question of this volume).

Multilevel metagovernance: Finally, metagovernance can be applied in a multilevel context, for example by managing the balance between the three different multilevel approaches which will be distinguished in section 4. Multilevel metagovernance could be defined as designing and managing actionable and situationally adapted mechanisms to foster effective relationships between different levels of authority.

If multilevel metagovernance is not well-embedded in policies of the involved government levels, it can result in fragmentation. An example is the EU Urban Agenda (European Union, 2016) with 12 informal, voluntary partnerships of EU, national, and city officials to assess the appropriateness of existing policies for urban areas. According to Pazos-Vidal (2019), this was about EU and Member States mobilizing "with" and "for" subnational government rather than "by" urban authorities. He argued that it was a case of policy fragmentation because this exercise of subnational better regulation was not structurally integrated in the wider Better Regulation process of the European Commission.

Although most of the examples in this section are from EU countries, the non-normative definitions of the analytical and design concepts of governance, governance frameworks and metagovernance make them a good basis for tailor-made application in many different national political-administrative cultures and traditions, as research has shown. Metagovernance has been analyzed in Australia (Eberhard, 2018), Canada (Doberstein, 2013), China (Li, Homburg, de Jong, and Koppenjan, 2016), Colombia (Bonivento, 2014), Kenya (Beisheim et al., 2018), Nigeria (Agu, Okeke, and Idike, 2014), Tanzania (Lauwo, Azure, and Hopper, 2022), and The Netherlands (Hooge, Waslander, and Theisens, 2021), for example. In addition, there are examples of comparative research using a metagovernance lens on Germany, the Netherlands, Australia, China, and South Africa (Pahl-Wostl, 2019), on Chile, China, Denmark, Netherlands, Portugal, and Vietnam (Monteiro, do Rosário Partidário and Meuleman, 2018), and England, The Netherlands and Germany (Meuleman, 2008). Another example is a research on the metagovernance of fair trade and sustainable forestry challenges (Murphy-Gregory and Gale, 2019).

There is not yet much research published the application of metagovernance on the multilevel dimension of SDG implementation, but there is no reason why that field of application would be an exception. Some first examples include research in Brazil (Kull, Pyysiäinen, Christo, and Christopoulos, 2018; Martin, Teles da Silva, Duarte dos Santos, and Dutra, 2022), Denmark (Engberg, 2018), Belgium (Temmerman, De Rynck, and Voets, 2015), and Norway (Tønnesen, Krogstad, Christiansen, and Isaksson, 2019).

Multilevel (Meta)governance and the SDGs: the Rise of the Local Level

Although the SDGs have been adopted by national governments, the goals and targets often mention that action is needed "at all levels". Since subnational governments have begun adopting Voluntary Local Reviews (VLRs) on their progress on implementing the SDGs, and national governments have committed to produce a Voluntary National Review (VNR)_ every several years, it would be useful to see whether and how the existence of such a two-level self-reporting mechanism is able to stimulate effective MLG.

In the 2030 Agenda agreement, national governments were invited to prepare VNRs of their national planning to implement the SDGs, as part of the global follow-up and review mechanism for the Agenda. These VNRs are presented during the UN's annual sessions of the High-level Political Forum on Sustainable Development (HLPF). The UN keeps track of progress on their webpages dedicated to the VNRs, issues guidance for its member states, and annually presents reviews of the VNRs presented that year. The regional UN offices organize workshops to help countries to improve their VNRs.

Subnational governments have become increasingly involved in the SDG discourse since the adoption of the SDGs in 2015. Cities have their own international networks, such as the Covenant of Mayors on climate action. Many cities have become frontrunners on addressing climate change, other environmental issues and social challenges. Front-running big cities often complain that national governments are frustrating innovation and blocking progress. This points at least at a lack of effective collaboration and communication in a multilevel setting.

In 2018 local and regional governments started to engage in sub-national reviews of SDG implementation, the VLRs. Four VLRs were launched during the July 2018 meeting of the HLPF, by Kitakyushu, Shimokawa, and Toyama in Japan, and New York City in the US (Koike, Ortiz-Moya, Fujino, and Kataoka, 2020). This practice is stimulated by the UN with guidance and a series of events. In 2021 the UN website showcased more than 60 examples of VLRs.[2] The European Commission's Joint Research Centre published a handbook to support local-level authorities in preparing reviews of VLRs, with examples from 14 reviews presented since 2018. This 'European Handbook for SDG Voluntary Local Reviews' provides examples of official and experimental indicators that municipalities can use to monitor local SDG implementation (Siragusa, Vizcaino, Proietti, and Lavalle, 2020). The indicators can help local-level authorities establish baselines for their communities, compare action with that of other cities, and monitor their progress on addressed specific challenges. Other guidance was issued by United Cities and Local Government and UN Habitat (Ciambra, 2020).

A comparative analysis of 16 VLRs (Ortiz-Moya, Koike, Ota, Kataoka, and Fujino, 2020) concluded that a VLR: i) allows the local government to listen to the needs of its people and reflect them into local policymaking; ii) invites self-reflection; iii) provides for a process that is data-driven and can be used to plan for action to achieve the future we want; and iv) gives a local take on the global conversation on sustainable development. The fourth point has a clear multilevel dimension. As the authors of this study argue, local and regional governments rarely have all necessary means to achieve the SDGs: "Fiscal transfer, energy source, setting up financial regulations, tax intervention, and many other policies require the national government to take action". Ortiz-Moya et al. (2020) recommend that all VLRs should include messages directly addressed to the national governments. Such messages should include policy demands, and good practices that could be scaled up and shared through the national government channels.

Besides that sub-national level often do not have the legal or political 'license' to apply policy and governance tools that could best tackle their sustainability challenges, sub-national authorities also often lack technical capacities and skills. The real possibilities and capacities of subnational governments are generally not at par with the complexity of the 2030 Agenda with its interlinked targets (Zarrouk and Rodas, 2022). This is an important part of the challenge identified in the theoretical introduction chapter in this book, about state capacity as an influencing factor for integrated implementation of the SDGs, and is connected with the fifth research question of this volume, on political-institutional preconditions.

Having Voluntary Local Review reports, and the (usually inclusive) preparation processes as empowering and mobilizing mechanisms, could improve the dialogue between the different levels, and at the same time bring the lack of resources, capacities and ability to use policy tools on the national strategic SDG agenda. Moreover, producing VNRs and VLRs in the same country and at the same time could be a good accelerator of effective multilevel (meta)governance for the SDGs. VLRs are a good means to reinforce vertical coherence and with this complement the VNR process (UN Department of Economic and Social Affairs, 2020).

In addition, because of its pluralistic view – the same policy challenge may be governed by a specific governance framework at each level – metagovernance might help making MLG effective (Kull, 2016). Each level of government should have the possibility to develop sustainability governance arrangements which are tailor-made to both the area and to the type of challenges. Metagovernance is able to provide coherence between the increasingly fragmented landscapes of governance, where new sub-national governance bodies are created at different scales and with different mixes of policy mandates (Christopoulos, Horvath, and Kull, 2012).

The multilevel dimension is relevant in all SDG implementation processes, but there are differences, not only as regards the specific SDGs, but also concerning the typical governance cultures in countries and for specific policy areas. SDG 11 on sustainable cities and communities is a key example. Cities are hotspots of innovation, wealth and at the same time extreme poverty. They are to some extent independent but need excellent collaboration with national governments to address many of their larger challenges. Cities have formed international networks and communities of practice that might make them slightly less dependent from national governments and their often primarily hierarchical governance approach to multilevel relations. Other SDGs such as SDG 4 (health), SDG 6 (education), and SDG 7 (energy) are characterized by challenges in many countries, owing to privatization of these formerly public services. Privatization implies that (national) governance have given up their governing capacity – and this will have an impact also at the subnational level. These issues are touched upon in Meuleman (2018) for example, but probably deserve more in-depth analysis.

A Typology of Multilevel Governance Aligned with the Three Governance Styles

Effective governance for the SDGs depends to a large extent on the quality of horizontal (multi-sector), inclusive (multi-actor) and vertical (multi-level) mechanisms and how they are functioning (Dewulf, Meijerink, and Runhaar, 2015). In addition, as Köhler et al. (2017) have shown, multilevel governance challenges can differ widely between policy sectors, because actor constellations and path dependencies, to name two factors, are not the same. With the three classical governance styles as point of departure, three different and partly incompatible types of multilevel governance can be distinguished (Figure 5.2).

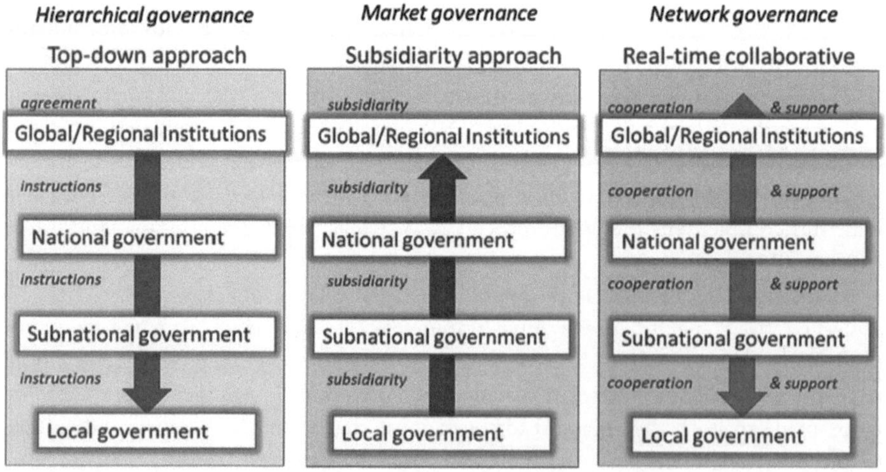

FIGURE 5.2 Three types of multilevel governance
Source: Author's own elaboration

1. The first style is most compatible with a hierarchical mindset and can be characterized as 'top-down'. This approach might be fast in times of crisis, as became clear during the eCOVID-19 pandemic, but is otherwise typically slow. National laws and policies might require years of preparation and even more time before they become local practice. For example, the implementation of new EU legislation by local authorities can take up to six or seven years after the initiative was taken by the European Commission, among others because the rules first need to be translated into national legislation and related tasks then delegated across the different government levels.
2. The second type is based on the principle of subsidiarity: decisions should be taken at the lowest level that can handle them. This principle has led to many innovations at the local level, but scaling up successful local sustainability practices is generally a challenge, for which some have suggested that national sustainable development councils could become catalysers (Cornforth, Niestroy, and Osborn, 2013). In any case, also the second type can be characterized as slow.
3. A third approach has emerged in some countries, referred to as 'real-time collaborative multilevel governance' approach (Meuleman, 2019). The Netherlands with its centuries-old network governance culture (Kickert, 2003) has a general multi-level governance mechanism that is also used for collaboration on SDG implementation (Meuleman, 2021). For strategic policy issues, so-called 'Inter-governmental Dossier teams' are established to discuss what each of the three tiers in the Dutch administrative system (national, provincial, local) can contribute. This means that in real-time the different governance levels get together to discuss about how to tackle a specific pressing problem. It is an addition, not replacement of the bottom-up subsidiarity style or the classical top-down hierarchical style of governance. In other countries, the approach could be different: comparative research on urban sustainability transitions has shown that multilevel relations can differ, according to national governance cultures (Ehnert et al., 2018). This third MLG approach is an example of the governance mechanisms mentioned in the 2nd research question of this volume.

A crisis is a good moment to observe "real-time" collaborative multilevel governance. During the COVID-19 pandemic, the German federal government used a two-level pandemic crisis management mechanism: the Conference of Premiers of the German federal states (Ministerpräsidentenkonferenz, MPK) with participation of the German Federal Chancellor (Bundeskanzler). The MPK took a leading role during the pandemic, a role that was unusual when compared to normal times (see e.g. Schnabel, Freiburghaus, and Hegele, 2022), holding frequent meetings and taking decisions, some of which were implemented successfully but others were not. As a result, citizens felt at times confused by the complex results, and the outcomes were mixed.

Each of the three types of multilevel governance represents a dominant governance style. Top-down multilevel governance represents hierarchical values. The bottom-up subsidiarity approach relates to the principles of devolvement and empowerment of market governance. The real-time approach requires mechanisms from network governance. The network style is underused and adding the 'real-time collaborative' approach is an example of a 'metagovernance' response to close the gap.

However, metagovernance implies a situational approach. Scaling up local innovation does not have to be (only) based on the collaborative style. Governance interventions for scaling up local innovation, can have three different forms, congruent with the three typical styles: Coercive, rule-based mechanisms such as impact assessment mechanisms, usually required by law and containing legal guarantees with regards to transparency and consultation; collaborative mechanisms such as peer reviews and (multi-level) stakeholder participation; and voluntary and market-based mechanisms to induce decision/behaviour change (Cornforth et al., 2013).

Thus, three specific challenges of multilevel (meta)governance are:

- How effective (and how fast) are national objectives and instructions being translated into subnational responses? This is a question in the context of classical top-down multilevel relations.
- How effective (and how fast) are subnational observations, solutions and needs landing at the relevant desks in a national administration? This is about the effectiveness of the subsidiarity role of subnational governments.
- Are there effective mechanisms to bridge the levels effectively and fast enough for important and urgent challenges which regard more than one level? This links to 'real-time participatory multilevel governance'.

A metagovernance perspective to tackle these questions is among others about whether and how each of the three above-mentioned questions could be answered better when ideas or tools from the other styles are integrated. The compatibility between the three MLG types and the three ideal-typical governance styles which are the material with which metagovernance works, suggests that metagovernance could help making MLG more reflexive, diverse, flexible, adaptive, and pluralist.

Multilevel (Meta)governance and Policy Coherence for Sustainable Development

Multilevel governance or 'vertical coherence' is not an island, isolated from other governance challenges. It is one of the eight dimensions (sub-indicators) of the SDG indicator 17.14.1 on policy coherence for sustainable development (PCSD) (UN Environment Programme, 2020). This is formulated in the indicator as follows: "The country has mechanisms in place for aligning priorities, policies and plans between various levels of government." Responding to a growing demand of both member and non-members about guidance on how to deal with the "how" of coherent 2030 Agenda implementation, in 2019 the OECD adopted eight principles of PCSD. The need for increased vertical coherence is addressed by the principle of "*Engaging appropriately sub-national levels of government in areas where they have a role in policy coordination* " (Organisation for Economic Co-operation and Development, 2019, lll.2), under which the OECD countries have committed to "Promote synergies among national, regional and local policies to better align with and contribute to relevant economic, social and environmental goals, including international commitments (...)".

Federal states often have a powerful second government level. The three Belgian Regions are a case in point, but also Germany's and Austria's Laender, and the Spanish Regions, are part of a MLG system that is not primarily hierarchical, as important responsibilities in relation to the SDGs have been put at the second, regional level. This illustrates how important effective multilevel metagovernance is to attain the whole SDG agenda.

In fact, the challenges posed by the 2030 Agenda have inspired several countries to improve their mechanisms for multilevel governance. Belgium, for example, revitalized its Inter-Ministerial Conference for Sustainable Development (IMCSD), which gathers ministers in charge of sustainable development and development co-operation at different levels (Federal, Communities, and Regions) as the central co-ordination mechanism for SDG implementation. The IMCSD is used as a central coordination mechanism for PCSD at all levels. In addition, all SD actions undertaken at a local level are collected by the Advisory Council for Policy Coherence for SD and publicly displayed online.[3] (Organisation for Economic Co-operation and Development, 2018).

However, as shown in the comparative analysis of 137 national government SDG bodies by Breuer, Leininger and Malerba in this volume, the creation of such mechanisms to ensure multi-level PCSD SDG implementation still constitutes the exception rather than the norm as the majority of countries has chosen an institutional design whereby sub-national governments are not represented in national SDG government bodies.

Another important principle formulated by the OECD refers to "Defining, implementing and communicating a strategic long-term vision that supports policy coherence (Organisation for Economic Co-operation and Development, 2019, ll.2). Having a long-term strategy or plan for sustainable development at the national level is one thing, but mobilizing subnational governments to become part of the action is also important. In 2016 the Republic of Korea, for example, established a mechanism to implement the alignment of national sustainable development policies at subnational and local levels. Local governments voluntarily established their own implementation strategies for sustainable development and drafted evaluation reports to measure their progress at the local level. This provided a strong basis for the local implementation of the SDGs (Government of the Republic of Korea, 2016).

Another example of a strategic long-term vision suited to foster multi-level PCSD is Colombia's National Development Plan, which includes a territorialization approach. Regional pacts for productivity and equity define development visions and strategic projects prioritized by the regions themselves to boost their economies and take advantage of their capabilities. Each regional pact presents a roadmap for a coordinated investment approach in the territory, as well as the articulation of efforts between levels of government. All indicators and targets of the regional pacts are linked to one or more SDG targets; this enables alignment of national and subnational plans (Alvarez, 2020).

These examples illustrate that how MLG can support, or slow down SDG implementation depends on a multitude of factors. It is no exception to the global observation that national administrative cultures and traditions might lead to very different institutional solutions to the same challenge: governance styles are normative – they are characterized by specific sets of values. Effective metagovernance of MLG therefore requires a minimum amount of cultural sensitivity (Meuleman, 2013).

Another category of states with specific challenges for effective MLG is formed by fragile states. A generally weak rule of law, the aftermath of a violent conflict, and the results of natural disasters might all result in, or exacerbate the fragility of the state at all levels. Further research could bring together examples of re-establishing the state at the different levels.

Conclusion and Recommendations for Further Research

This chapter explored how multilevel governance and -metagovernance can contribute to effective public governance of implementing the SDGs.

Concerning the second research question of this volume on governance mechanisms to manage SDG interlinkages and power asymmetries between stakeholders from different levels, it can be concluded that new governance mechanisms for MLG are indeed needed. One example mentioned is the European Commission's Environmental Implementation Review which aims at improving environmental multilevel governance.

An emerging mechanism is the 'real-time collaborative multilevel governance', with specific institutional arrangements to bring all levels together on important and urgent issues. This approach does not replace but complements the traditional top-down and bottom-up mechanisms of MLG.

The real-time approach reflects a network governance style, while the other MLG approaches can be linked to hierarchical and market governance. Combining the three approaches is an example of multilevel *meta*governance.

The need to have more effective mechanisms for MLG could trigger the beginning of a much wider use of the hybrid concept of multilevel metagovernance (MLMG), almost 20 years after it was coined (Jessop, 2004). Research projects on MLMG should, however, take into account that the term MLMG is not often used, which is not an indication that the practice is not there. Long before the term metagovernance was coined, it was already practiced by public managers. The same applies to multilevel governance: the term was coined after the relevant practice was 'discovered'.

Because metagovernance is a concept 'above and beyond' governance, it can help setting up the conditions and rules that can foster MLG systems with 'good governance' characteristics (Daniell, Hogan, and Cleary, 2017), and with the CEPA principles of effective governance for sustainable development (UN Committee of Experts on Public Administration, 2018).

The pandemic as well as other recent crises have made clear that different levels of government cannot provide the necessary services and protection to citizens if they work in 'silos'. As part of their crisis management, many countries have created ad hoc multi-level committees or other forms of collaboration, between national government and the tiers of subnational government. In global regions where supranational rules or agreements apply, that regional level is part of the multilevel governance architecture. These new institutional mechanisms for fast and effective collaboration between government levels should not be completely abolished after a crisis but remain at least in a kind of 'stand by' mode, ready to be used when again necessary. This would add to the institutional resilience of governments and societies as a whole.

A structured approach to improve multilevel governance for the SDGs should not only follow the fifth OECD principle of policy coherence for sustainable development on 'subnational engagement', but also the other seven principles. For example, the silos between policy sectors and government departments hamper multilevel (meta)governance in many ways. Improving working across silos is the aim of the fourth principle.

As regards the fifth research question of the volume on political-institutional preconditions to manage SDG interactions, it is clear that sub-national level often lacks the legal or political 'license' to apply the necessary policy and governance tools, and the technical capacities and skills. The real possibilities and capacities of subnational governments are generally not at par with the complexity of the 2030 Agenda with its interlinked targets.

Notes

1 https://ec.europa.eu/environment/eir/p2p/index_en.htm.
2 https://sdgs.un.org/topics/voluntary-local-reviews.
3 www.SDGs.be.

References

Agu, S.U., Okeke, R.C., and Idike, A.N. (2014). Meta-Governance and the Metaphysics of Political Leadership in 21st Century Africa: A Focus on Election Administration in Nigeria. *Mediterranean Journal of Social Sciences*, 5(9), 177.

Alvarez, G.R. (2020). The Battle for Guatemala: Multilevel Governance and the Nation-state. *Leadership and Developing Societies*, 5, 55–68. doi:10.47697/lds.34348004.

Beisheim, M., Ellersiek, A., Goltermann, L., and Kiamba, P. (2018). Meta-Governance of Partnerships for Sustainable Development: Actors' Perspectives from Kenya. *Public Administration and Development*, 38(3), 105–119. doi:10.1002/pad.1810.

Bonivento, J.H. (2014). *Del Gobierno a La Gobernanza Local. Capacidades, Instituciones y Visiones de Lo Público En El Proceso de Descentralización En Colombia: Estudio de Casos.* Universidad Complutense de Madrid.

Bovens, M. and 't Hart, P. (2016). Revisiting the study of policy failures. *Journal of European Public Policy*, 23, 653–666. doi:10.1080/13501763.2015.1127273.

Bovens, M., t'Hart, P., and Peters, B.G. (2001). Analysing governance success and failure in six European states. In *Success and failure in public governance: A comparative analysis*. Cheltenham: Edward Elgar Publishing.

Christopoulos, S., Horvath, B., and Kull, M. (2012). Advancing the governance of cross-sectoral policies for sustainable development: a metagovernance perspective. *Public Administration and Development*, 32, 305–323. doi:10.1002/pad.1629.

Ciambra, A. (2020). Guidelines for Voluntary Local Reviews. In *A Comparative Analysis of Existing VLRs* (Vol. 1). UGLG.

Cornforth, J., Niestroy, I., and Osborn, D. (2013). The Governance of Scaling up Successful Sustainability Practices: How Can National Councils for Sustainable Development Organise the Wider Use of National and Regional Examples?. In *Stakeholder Forum*.

Daniell, K.A., Hogan, A., and Cleary, J. (2017). *Breaking Down the@ One-Size-Fits-AlŒ Approach to Rural and Regional Policy: Enhancing Policy Initiatives through Multi-level Governance.*

Dewulf, A., Meijerink, S., and Runhaar, H. (2015). Editorial: The governance of adaptation to climate change as a multi-level, multi-sector and multi-actor challenge: a European comparative perspective. *Journal of Water and Climate Change*, 6(1), 1–8. doi:10.2166/wcc.2014.000.

Doberstein, C. (2013). Metagovernance of urban governance networks in Canada: In pursuit of legitimacy and accountability. *Canadian Public Administration*, 56(4), 584–609. doi:10.1111/capa.12041.

Eberhard, R. (2018). The Metagovernance of Australian Water Policy: Practices, Rationales and Outcomes. (PhD thesis). Queensland University of Technology.

Ehnert, F., Kern, F., Borgström, S., Gorissen, L., Maschmeyer, S., and Egermann, M. (2018). Urban sustainability transitions in a context of multi-level governance: A comparison of four European states. *Environmental Innovation and Societal Transitions*, 26, 101–116. doi:10.1016/j.eist.2017.05.002.

Engberg, L.A. (2018). Climate Adaptation and Citizens' Participation in Denmark: Experiences from Copenhagen. In S. Hughes, E.K. Chu, and S.G. Mason (Eds), *Climate Change in Cities: Innovations in Multi-Level Governance* (pp. 139–161). Cham: Springer International Publishing.

European Commission. (2019). The European Green Deal. In *COM(2019) 640 Final, 11.12.2019.*

European Union. (2016). *Urban Agenda for the EU: Pact of Amsterdam*. Retrieved from https://ec.europa.eu/regional_policy/sources/policy/themes/urban-development/agenda/pact-of-amsterdam.pdf.

Fleming, J. and Rhodes, R.A.W. (2005). Bureaucracy, Contracts and Networks: The Unholy Trinity and the Police. *Australian and New Zealand Journal of Criminology*, 38, 192–215. doi:10.1375/acri.38.2.192.

Fukuyama, F. (2013). What Is Governance? *Governance: An International Journal of Policy, Administration, and Institutions*, 26(3), 347–368. doi:10.1111/gove.12035.

Government. of the Republic of Korea. (2016). Voluntary National Review 2016. Retrieved from https://sustainabledevelopment.un.org/content/documents/10446Executive%20Summary%20Review_ROK.pdf.

Hooge, E.H., Waslander, S., and Theisens, H.C. (2021). The many shapes and sizes of meta-governance. An empirical study of strategies applied by a well-advanced meta-governor: the case of Dutch central government in education. *Public Management Review*, 1–19. doi:10.1080/14719037.2021.1916063.

Jessop, B. (1997). The Governance of Complexity and the Complexity of Governance: Preliminary Remarks on Some Problems and Limits of Economic Guidance. In *Beyond market and hierarchy: interactive governance and social complexity* (pp. 95–128).

Jessop, B. (2004). Multilevel Governance and Multilevel Metagovernance: Changes in the EU as Integral Moments in the Transformation and Reorientation of Contemporary Statehood. In I. Bache and M. Flinders (Eds), *Multi-level Governance* (Vol. 2, pp. 49–74). Oxford: Oxford University Press.

Kamau, A.M., Chasek, P.S., and O'Connor, D. (2018). *Transforming multilateral diplomacy: the inside story of the sustainable development goals*: Routledge.

Kickert, W.J.M. (2003). Beneath consensual corporatism: traditions of governance in the Netherlands. *Public Administration*, 81, 119–140. doi:10.1111/1467-9299.00339.

Köhler, J. et al. (2017). Anwendung Der Mehr-Ebenen-Perspektive Auf Transitionen: Initiativen in Den Kommunal Geprägten Handlungsfeldern Energie, Wasser, Bauen & Wohnen. In *Working Paper Sustainability and Innovation*.

Koike, H., Ortiz-Moya, F., Fujino, J., and Kataoka, J. (2020). *How Can Voluntary Local Reviews Contribute to the SDG Decade of Action? An Assessment of VLRs to Date*. Retrieved from http://sdg.iisd.org/commentary/guest-articles/how-can-voluntary-local-reviews-contribute-to-the-sdg-decade-of-action-an-assessment-of-vlrs-to-date.

Kooiman, J. (2003). *Governing as Governance*: Sage Publications.

Kull, M. (2016). *European Integration and Rural Development*. Routledge.

Kull, M., Pyysiäinen, J., Christo, G., and Christopoulos, S. (2018). Making sense of multilevel governance and governance coordination in Brazil: The case of the Bolsa Verde Programme. *Regional & Federal Studies*, 28(1), 47–78. doi:10.1080/13597566.2017.1355788.

Lauwo, S.G., Azure, J.D.-C., and Hopper, T. (2022). Accountability and governance in implementing the Sustainable Development Goals in a developing country context: evidence from Tanzania. *Accounting, Auditing & Accountability Journal*. doi:10.1108/AAAJ-10-2019-4220.

Li, Y., Homburg, V., de Jong, M., and Koppenjan, J. (2016). Government responses to environmental conflicts in urban China: the case of the Panyu waste incineration power plant in Guangzhou. *Journal of Cleaner Production*, 134, 354–361. doi:10.1016/j.jclepro.2015.10.123.

Liesbet, H. and Gary, M. (2003). Unraveling the Central State, but How? Types of Multi-level Governance. *American Political Science Review*, 97. doi:10.1017/s0003055403000649.

Martin, P., Teles da Silva, S., Duarte dos Santos, M., and Dutra, C. (2022). Governance and metagovernance systems for the Amazon. *Review of European, Comparative & International Environmental Law*, 31(1), 126–139. doi:10.1111/reel.12425.

Meadowcroft, J. (2011). Sustainable Development. In M. Bevir (Ed.), *The SAGE Handbook of Governance*. doi:10.4135/9781446200964.

Meuleman, L. (2008). *Public Management and the Metagovernance of Hierarchies, Networks and Markets: the Feasibility of Designing and Managing Governance Style Combinations*: Springer Science & Business Media.

Meuleman, L. (2013). Cultural Diversity and Sustainability Metagovernance. In L. Meuleman (Ed.), *Transgovernance: Advancing Sustainability Governance* (pp. 37–81). Berlin, Heidelberg: Springer.

Meuleman, L. (2014). *Global environmental change Governance Frameworks*. Dordrecht: Springer.

Meuleman, L. (2018). *Metagovernance for Sustainability: A Framework for Implementing the Sustainable Development Goals*. Routledge.

Meuleman, L. (2019). Why We Need 'Real-Time' Multi-Level Governance for the SDGs. In *ISD SDG Knowledge Hub 13 June 2019*.

Meuleman, L. (2021). Public Administration and Governance for the SDGs: Navigating between Change and Stability. *Sustainability*, 13, 5914. doi:10.3390/su13115914.

Monteiro, M.B., do Rosário Partidário, M., and Meuleman, L. (2018). A comparative analysis on how different governance contexts may influence Strategic Environmental Assessment. *Environmental Impact Assessment Review*, 72, 79–87. doi:10.1016/j.eiar.2018.05.010.

Murphy-Gregory, H. and Gale, F. (2019). Governing the Governors: The Global Metagovernance of Fair Trade and Sustainable Forestry Production. *Politics & Policy*, 47(3), 569–597. doi:10.1111/polp.12300.

Noordegraaf, M. (2015). Hybrid professionalism and beyond: (New) Forms of public professionalism in changing organizational and societal contexts. *Journal of Professions and Organization*, 2, 187–206. doi:10.1093/jpo/jov002.

Organisation for Economic Co-operation and Development. (2018). *Policy Coherence for Sustainable Development 2018: Towards Sustainable and Resilient Societies*. Retrieved from: www.oecd-ilibrary.org/development/policy-coherence-for-sustainable-development-2018_9789264301061-en.

Organisation for Economic Co-operation and Development. (2019). Recommendation on Policy Coherence for Sustainable Development. Retrieved from www.oecd.org/gov/pcsd/oecd-recommendation-on-policy-coherence-for-sustainable-development.htm.

Ortiz-Moya, F., Koike, H., Ota, J., Kataoka, Y., and Fujino, J. (2020). State of the Voluntary Local Reviews 2020: Local Action for Global Impact in Achieving the SDGs. Retrieved from www.iges.or.jp/en/pub/vlrs-2020/en.

Pahl-Wostl, C. (2019). The role of governance modes and meta-governance in the transformation towards sustainable water governance. *Environmental Science & Policy*, 91, 6–16. doi:10.1016/j.envsci.2018.10.008.

Pazos-Vidal, S. (2019). *Subnational Actors and EU Multilevel Governance: Between the Demise of "Europe of the Regions" and the Emergence of Active Subsidiarity*. Paper presented at the 19th ECPR General Conference, Wroclaw.

Peters, B.G. (1998). *Managing horizontal government: the politics of coordination*. Canadian Centre For Management Development.

Pierre, J. and Peters, B.G. (2021). *Advanced Introduction to governance*. Cheltenham: Edward Elgar Publishing.

Pollitt, C. and Bouckaert, G. (2011). *Public Management Reform: A Comparative Analysis of NPM, the Neo-Weberian State, and New Public Governance*. Oxford: Oxford University Press.

Schnabel, J., Freiburghaus, R., and Hegele, Y. (2022). Crisis management in federal states: the role of peak intergovernmental councils in Germany and Switzerland during the COVID-19 Pandemic. *dms – der moderne staat – Zeitschrift für Public Policy, Recht und Management*, 15(1), 1–20. doi:10.3224/dms.v15i1.10.

Siragusa, A., Vizcaino, M., Proietti, P., and Lavalle, C. (2020). *European Handbook for SDG Voluntary Local Reviews*. Retrieved from https://publications.jrc.ec.europa.eu/repository/handle/JRC118682.

Temmerman, C., De Rynck, F., and Voets, J. (2015). Opening the black box of metagovernance: The roles of central government in local multilevel networks – the case of the local job centers in Flanders. *International Review of Public Administration*, 20(3), 227–242. doi:10.1080/12294659.2015.1039777.

Termeer, C.J.A.M., Dewulf, A., and Biesbroek, R. (2019). A critical assessment of the wicked problem concept: relevance and usefulness for policy science and practice. *Policy and Society*, 38, 167–179. doi:10.1080/14494035.2019.1617971.

Tønnesen, A., Krogstad, J.R., Christiansen, P., and Isaksson, K. (2019). National goals and tools to fulfil them: A study of opportunities and pitfalls in Norwegian metagovernance of urban mobility. *Transport Policy*, 81, 35–44. doi:10.1016/j.tranpol.2019.05.018.

UN Committee of Experts on Public Administration. (2018). *Principles of Effective Governance for Sustainable Development*. New York.

UN Department of Economic and Social Affairs. (2020). Multilevel Governance and Subnational Reporting on VNRs and VLRs. In *Concept Note*.

UN Environment Programme. (2020). Methodology for SDG-Indicator 17.14.1: Mechanisms in Place to Enhance Policy Coherence for Sustainable Development. Retrieved from https://city2city.network/file/5237/download?token=ejTPUj76.

Zarrouk, N. and Rodas, M. (2022). Application of the Principles of Effective Governance for Sustainable Development at the Subnational Level. Retrieved from: https://documents-dds-ny.un.org/doc/UNDOC/GEN/N22/237/51/PDF/N2223751.pdf?OpenElement.

6

TRADE-OFFS AND SDG POLITICS IN SOUTH AMERICAN AGRIFOOD GOVERNANCE

The risks from cherry-picking

Karen M. Siegel[1] and Mairon G. Bastos Lima

Introduction

Agrifood governance is one of the most important sectors for a potential sustainability transition, and recognition that the sector requires changes to become more sustainable is on the rise. However, general acknowledgement of the need for more sustainable agrifood systems does not occlude the fact that this is one of the most contested policy arenas. Opposing views have long been advanced as to what constitutes sustainability in agrifood systems. On the one hand, there are actors (notably some scientists and civil society organizations, and many rural social movements) who advocate for agroecology and smaller-scale, more diverse and perhaps organic production for local consumers (Altieri and Toledo, 2011; IPES-Food and ETC Group, 2021). On the other hand, there are those (other scientists and the private sector, among others) who favour the dominant large-scale industrial agriculture, often for export and based on monocultures but using new technologies and approaches (Finger, Swinton, El Benni, and Walter, 2019). The latter group champions mostly technical innovations (such as precision farming) to reduce chemical inputs and mitigate adverse environmental impacts, sometimes under the banner of "sustainable intensification". This concept is frequently taken up in policy documents, but it has received much criticism from environmentalists and civil society organizations who argue that it constitutes a contradiction in terms (Mahon, Crute, Simmons, and Islam, 2017; Siegel, Deciancio, Kefeli, Queiroz-Stein, and Dietz, 2022, p. 2). Although the notion of "sustainable intensification" remains quite vague and not well defined, an important concern is the emphasis on increasing agricultural production while social and ecological issues do not receive sufficient attention (Garnett et al., 2013).

These diverging approaches to agrifood governance were also evident in the process of formulating SDG 2, which aims to "End hunger, achieve food security

DOI: 10.4324/9781003254683-6

and improved nutrition and promote sustainable agriculture". In the end, both approaches can be found in the targets under SDG 2 and this circumstance obscures the existence of substantial disagreements as to what exactly the problem is and how it should be addressed (McNeill, 2019, p. 17). While the text of SDG 2 acknowledges that agriculture and food production need to become more sustainable, it does not help to promote consensus on what exactly that means or how it should be done. Against this background, it is particularly important to examine interlinkages and trade-offs between different dimensions of sustainability and different SDGs in agrifood governance.

To do so, in this chapter, we look at the Southern Cone of South America. As one of the most important producers and exporters of agricultural commodities, notably soybean used as the world's chief animal feed protein, this region holds a central position in the global agrifood system. Since the 1990s, the expansion of large-scale industrial agriculture has played a pivotal role for economic growth in the region, but it has also been heavily criticized for its negative social and environmental impacts (Bastos Lima and Kmoch, 2021; Garrett and Rausch, 2016; Russo Lopes, Bastos Lima, and Reis, 2021). Environmental impacts include biodiversity loss, deterioration of soil and water quality, desertification, and climate change – both at regional and global levels (Escobar et al., 2020; Green Jonathan et al., 2019). Land-use change and agriculture are the primary sources of greenhouse gas emissions as well as of landscape transformations that already entail worrisome environmental changes in South America, such as droughts and reduced rainfall, with an already noticeable impact on rain-fed agriculture (ECLAC, 2019, pp. 155–157; Edwards and Roberts, 2015, p. 9; Flach et al., 2021; Leite-Filho, Soares-Filho, Davis, Abrahão, and Börner, 2021). Social impacts, in turn, include the displacement of small-scale farmers or indigenous communities while offering few if any employment opportunities, eroding rural communities, and contributing to urban sprawl (Bastos Lima, and Kmoch, 2021). Because of the intensive use and aerial spraying of pesticides across large areas, the expansion of soy and other industrial monocultures has also been associated with severe health impacts (Oliveira and Hecht, 2016; Russo Lopes et al., 2021).

In the following sections, we first examine how the changes in agricultural production and land use that the Southern Cone has witnessed in the last few decades interlink with the SDGs. We then show how the cherry-picking of selected SDGs (while neglecting others) and selective reporting obscures trade-offs and risks thus entrenching existing marginalization patterns despite the Agenda's commitment to inclusiveness and leaving-no-one-behind. In doing so, we build on our previous in-depth qualitative research on agrifood governance, inclusiveness, and the politics of the SDGs in the Southern Cone (Siegel and Bastos Lima, 2020). In this chapter, we examine the SDGs more closely to map to what extent current developments in agrifood governance in the Southern Cone have supported the achievement of particular SDGs and their targets, while identifying the main trade-offs and adverse effects. Based on an analysis of the first Voluntary National Reviews (VNRs) on the SDGs submitted by Brazil, Uruguay and Paraguay, we then look at agenda-setting processes and highlight some of the risks of relying on voluntary targets and reporting.

In response to RQ 1 of this volume (What do we know about the most important interlinkages between SDGs?), our analysis demonstrates that the expansion of large-scale intensive agricultural production in the Southern Cone of South America has helped achieve some of the targets on economic growth (SDG 8) and industrialization (SDG 9), but for other SDGs and targets there are clear adverse effects, creating evident – if often unacknowledged – trade-offs. It is concerning that such trade-offs are rarely considered in SDG politics. Instead, cherry-picking and selective reporting is commonplace as a way to focus exclusively on particular SDGs of choice while ignoring adverse effects on others. That, however, poses a substantial risk of entrenching – and possibly augmenting – existing patterns of marginalization. We expose how structural power asymmetries in SDG politics constitute a significant barrier for the 2030 Agenda's potential for transformation while highlighting important differences between countries (e.g., more inclusive governance in the case of Uruguay).

These findings are relevant also to RQ 2 (What governance mechanisms are needed to manage SDG interlinkages and address power asymmetries between different stakeholders and sectors?). It is crucial that governance mechanisms are put in place to examine which trade-offs exist and seek to address them, taking into account the views of different actors and with particular attention to those actors that are frequently marginalized in policy processes. Facilitating ways to raise complaints at different levels from the local to the national and a functioning judiciary to examine such complaints are crucial elements. Regarding RQ 5 (What political-institutional preconditions are conducive to the establishment of effective governance mechanisms to manage SDG interactions?) our comparison of the three countries suggests that, at least in the South American context, a higher quality of democracy and stronger state capacity are conducive to more inclusive governance.

The Southern Cone region of South America: Trade-offs in Agricultural Production

South America has recently emerged as the world's protein breadbasket, notably as an exporter of soybean, the main crop used for animal feed worldwide. As much as 60% of global soy exports are expected to come from South America in 2022 (APK Inform, 2021). That production stems from the continent's Southern Cone region, its southern half, including Argentina, Uruguay, Paraguay, part of Brazil, and the south of Bolivia. Since at least the 1970s, driven in part by US economic and (geo)political interests, the region started to significantly embrace "green revolution" technologies linked to industrial monocultures, heavy use of agro-chemicals, and – eventually – genetically modified (GM) seeds (Nehring, 2022). The entire region then adopted neoliberal reform agendas in the 1990s to promote foreign investment and free trade, which added to South America's historical devotion to agricultural exports as its chief economic activity (Otero, 2012).

Finally, since the early 2000s, rapidly growing demand from China as well as a growing presence of Chinese capital have further boosted those countries' agro-exporting sectors (Oliveira, 2019).

Agribusiness has thus become an increasingly significant economic sector in the Southern Cone. The expansion of large-scale industrial agriculture has contributed towards the targets of some SDGs, notably target 1 of SDG 8 (Decent work and economic growth), which aims to sustain or increase per capita economic growth. However, results are mixed for other targets of the same SDG, such as decent work (target 8.5), labour rights (8.6), or even youth employment (8.6). Evidence from Brazil and other emerging economies has shown that the expansion of large-scale intensive agriculture has led to the precarization of rural labour and the loss of labour rights (Bastos Lima, 2021, p. 184). As the region's dominant cash crops (soy and maize) are highly mechanized and often expand over traditional family-farming areas, displaced rural youth notoriously migrate to the cities and little workforce is absorbed (Russo Lopes et al., 2021). The promotion of such consolidated large-scale agriculture has often come as a government-sponsored development project and, therefore, hand-in-hand with neglect for smaller-scale alternatives that usually employ more people (Bastos Lima and Kmoch, 2021; Henderson, Godar, Frey, Börner, and Gardner, 2021; Nehring, 2022).

The dominant agrifood system has contributed to progress on some aspects of SDG 9 (Industry, innovation and infrastructure) in terms of promoting industrialization and infrastructure development (see Garrett and Rausch, 2016). However, targets 9.1 and 9.2 clearly set out that such industrialization and infrastructure development should be inclusive and provide equitable access to all. Our research demonstrates that critical questions remain as to how inclusive and sustainable the industrialization of agricultural production in the Southern Cone has been.

On SDG 2 (Zero hunger), there is a critical decoupling between food production and food security in the region, both of which are aspects captured by targets 2.1 and 2.2. In Brazil, for example, soybean outputs and the production of other major agricultural commodities have continued to increase, yet over 40 million people were food-insecure in 2021, and as many as 19 million people (twice as many as two years before) were experiencing hunger (Rede Penssan, 2021). Accompanying this boom in commodity production, the UN Food and Agriculture Organization (FAO) notes in its regional overview of food security and nutrition for Latin America and the Caribbean that hunger has been steadily growing since 2014, increasing by 30% just between 2019 and 2020, with now four in every ten people in the region being food-insecure. Argentina, known for its export-oriented agroindustry, is of all Latin American countries the one where food insecurity has worsened the most since 2014 (FAO, IFAD, PAHO, WFP, and UNICEF, 2021). These apparent contradictions expose the exclusion of much of the population from the bounty of large-scale agriculture as well as their vulnerability to shocks. The Covid-19 pandemic, in particular, has exposed the fragility of such a dominant agrifood system (Clapp and Moseley, 2020). Its exclusionary nature has thus left target 2.3 – on prioritized improvements for smallholder agriculture, women, family farmers and indigenous peoples – vastly unmet. The dominant GM-based agricultural production has also hindered the achievement of target 2.4 (fostering sustainable and resilient food production systems) and – through its patented-seed dominance – target 2.5 (maintaining the genetic diversity of seeds and cultivated plants). All in all, only total food production has been increasing, but alongside the deterioration of food security in the Southern Cone and growing problems with respect to all other SDG 2 targets.

Broadly, there had been progress towards achieving SDG 1 (No poverty) in the region, notably as many South American governments used revenues from commodity exports to fund social programmes (Vergara-Camus and Kay, 2017). Conditional cash-transfer programmes such as Brazil's Bolsa Família in effect helped combat poverty and extreme poverty, but results started to be reversed as many such policies were downgraded or dismantled after 2015 (ECLAC, 2019, pp. 115–117). In the Brazilian case, a right-wing political turn through the impeachment of President Dilma Rousseff in 2016 ended 13 years of Workers' Party rule, paving the way for the election of the far-right presidential candidate Jair Bolsonaro in 2018. Bolsonaro's electoral success drew on substantive resentment from more conservative segments of the population and business elites (including those in the agri-food sector) against what they saw as excessive government regulation with socialist hues (Hatzikidi and Dullo, 2021; van Dijk, 2017). Across the region, poverty rates are rising again, raising questions about the ability of the current export-oriented, large-scale agribusiness-based model of economic development to generate durable improvements.

Finally, the agricultural sector in the Southern Cone has also embraced the broader agenda of promoting a bio-economy, understood as increasingly varied economic sectors for bio-based goods and services, many of which can replace fossil-based fuels and products (Bastos Lima and Siegel, 2020; Siegel et al., 2022). With the expansion of soybean and sugarcane, respectively used for producing biodiesel and ethanol, there are benefits to SDG 7 (Affordable and clean energy), as there is towards SDG 13 (Climate action) for reducing greenhouse gas emissions and helping mitigate climate change. Brazil, in particular, has long had a sizeable biofuels industry, and crop-based bioenergy accounts for as much as 19% of the country's total energy consumption (Empresa de Pesquisa Energética, 2020, p. 27). Argentina, too, has a growing soy-based biodiesel industry. It has replaced part of the country's fossil-diesel consumption and made it one of the world's largest biodiesel exporters (Baraibar Norberg, 2020, p. 217). Overall, there is a growing appetite for agribusiness in the Southern Cone countries to develop "value webs" based on their crops and diversify downstream markets (Bastos Lima, 2021). Yet, it is also clear that the bio-economy is not inherently more sustainable. In the Southern Cone, strategies to promote the agricultural sector as part of the bio-economy so far mostly expand on dominant practices, without sufficient attention to the socio-environmental concerns frequently raised by citizens and civil society (Deciancio, Siegel, Kefeli, Queiroz-Stein, and Dietz, 2022; Tittor, 2021).

SDG setbacks caused by the expansion of large-scale agriculture

As discussed above, SDGs 1, 2, 7, 8, 9 and 13 have all experienced either positive or mixed – sometimes contradictory – impacts from the expansion of large-scale agriculture in the Southern Cone. However, with respect to some other SDGs, the impacts have been unequivocally negative. We address two areas that civil society actors frequently report on: human health (SDG 3) and inclusiveness. The latter is reflected in SDG 16, but it is also an issue that cuts across several SDGs and is enshrined in the 2030 Agenda's overarching principle to leave no one behind. Health and inclusiveness have been key in the critiques against the unfettered expansion of industrial agriculture across the region.

Health issues have increasingly emerged, owing to the input-intensiveness of large-scale agriculture in South America. The aerial spraying of agrochemicals like glyphosate and others commonly used in soybean production has frequently resulted in air pollution and the contamination of food crops and water sources in rural communities. Rural communities have often reported symptoms such as headaches and nausea as direct effects and concerns about cancer and congenital malformations later on. In Paraguay and Brazil, aerial spraying has also been an important factor driving rural communities from the land – making it a displacement mechanism (Ezquerro-Cañete, 2016, p. 706; Hetherington, 2011, p. 63; Russo Lopes et al., 2021). Civil society organizations in those countries have long reported that aerial spraying takes place without respecting environmental regulations for keeping minimum distances to inhabited areas, as noted also by a World Bank representative visiting Paraguay in March 2019 (*Ultima Hora*, 2019). In Argentina, similar problems have been highlighted by the social movement 'Madres de Ituzaingó' in their campaign to change the regulatory frameworks for the use of pesticides for agriculture (Arancibia, 2013). In Uruguay, conflicts in relation to the expansion of industrial agriculture are less severe, but there are also health concerns in relation to the use of agrochemicals (Chiappe, 2020). These problems stemming from the expansion of intensive GM-based agriculture undermines the achievement of SDG 3 (Good health and well-being) and of target 3.9 in particular, which aims to 'reduce deaths and illnesses from chemicals and pollution'. In tandem, progress on SDG 12 (Responsible consumption and production), notably target 12.4 to 'achieve environmentally sound management of chemicals', becomes hampered.

Although such health issues linked to the spread of industrial agriculture exist across the region, it is important to note that the level of civil society contestation varies between countries. A comparison between Paraguay and Uruguay can be illustrative of how political-institutional preconditions are crucial for understanding such differences. Paraguay has high levels of inequality and corruption compared to other countries in South America (Siegel and Bastos Lima, 2020, p. 4). It is rated as a highly defective democracy on the Bertelsman Index of political transformation (BTI, 2021) and ranks 128 out of 180 countries on Transparency International's (TI) Global Corruption Perception Index (Transparency International, 2021). The country has been described as a 'predatory state' (Ezquerro-Cañete and Fogel,

2017, p. 282) that acts in the interest of its elites rather than to pursue a coherent development strategy for the country and the broader public good. Its political and economic elites represent a ruling class of landowners gaining large profits from intensive agricultural production. At the same time, both state capacity and state autonomy are extremely weak, with no state presence at all in large parts of the country (Baraibar Norberg, 2020, pp. 362–363). In this context, the health impacts from agrochemical runoffs have long been ignored by policy-makers. In the absence of functioning institutional channels, protests and conflicts over land have also continued and, in some cases, escalated violently. That constitutes a marked difference to Uruguay as a consolidated democracy (BTI, 2021) where the state is considerably stronger and levels of corruption significantly lower, ranking 18 out of 180 countries on TI's corruption index in 2021. There, it has been possible to use changes in agricultural production for the benefit of a broader national development agenda (Baraibar Norberg, 2020, pp. 364–365) while also paying much more attention to negative impacts. In Uruguay, too, the expansion of soybean has led to an increase in health problems related to agrochemicals. Complaints increased, but they took place via mechanisms that Uruguayan government agencies purposefully created for this purpose. This includes a range of different channels, from local authorities to the Ministry of Livestock, Agriculture and Fisheries, the National Institution for Human Rights and the civil justice system. Moreover, on several occasions local and national newspapers as well as environmental non-governmental organizations (NGO) have covered such complaints and thus increased the visibility of the problems. Over the years this has resulted in several fines and legal proceedings, but also an increased knowledge of where and how to complain and in some cases also changes in agricultural practices with less aerial spraying and more rotation of crops (Chiappe, 2020). Grievances over health impacts are, therefore, taken up to a large extent through institutional channels put in place by the government for reporting problems rather than through protests by excluded actors outside the institutional framework.

Another area with important SDG trade-offs to account for is inclusiveness. Inclusiveness underscores the entire 2030 Agenda, being enshrined in its overarching aim to 'leave no one behind' as well as in other commitments such as the one to 'reach the furthest behind first' (UN General Assembly, 2015, p. 3). Such an overarching commitment to improving social equity has also faced setbacks in South America. The dominant agricultural production of Southern Cone countries has frequently trampled on land rights and disregarded indigenous or other customary owners as monoculture expands (Ezquerro-Cañete, 2016; Russo Lopes et al., 2021). That runs counter SDG 1 (No poverty), notably target 1.4, which seeks to protect the poor and vulnerable in their ownership and control over land and natural resources. Owing in part to the highly mechanized nature of soybean cultivation, its production has generally entailed the economic and political marginalization of rural labourers and communities (Russo Lopes et al., 2021).

Meanwhile, those protesting against the negative impacts of large-scale industrial agriculture often fail to get access to justice, contravening SDG 16 (Peace, justice and strong institutions), especially in its mission to promote equal access to justice for all (target 16.3). Even multistakeholder sustainability initiatives – often aimed at stopping deforestation – have usually neglected local stakeholder concerns. For example, in Brazil, landscape sustainability initiatives coalescing around the tropical savanna ecoregion *the Cerrado* have narrowed broad sustainability concerns (related to water access, inclusiveness, land rights, etc.) to merely one of conversion-free soy supply chains, involving international environmental NGOs and commodity traders but none of the region's grassroots actors (Bastos Lima and Persson, 2020). Such tensions produce trade-offs that cut across multiple SDGs but in particular SDG 10 (Reduced inequalities) and SDG 16 (Peace, justice and strong institutions).

It is essential to draw attention to these setbacks not only for their practical relevance but also because they limit the 2030 Agenda's potential for transformation. As noted, inclusiveness is central to the Agenda and cuts across all SDGs. In particular, grassroots movements and civil society organizations have frequently and long drawn attention to such health and exclusion issues, but now they more clearly stand out also as trade-offs in the implementation of the 2030 Agenda – with important setbacks on its inclusiveness core. Yet, as we shall see, such trade-offs have been frequently ignored in the SDG politics of Southern Cone countries.

Cherry-picking and SDG Politics in South America

In many ways, the 2030 Agenda presents a novel and ambitious approach to global development. It applies to both high-income and low-income countries, and therefore recognizes that sustainable development is a universal challenge to be addressed by all countries. The adoption of SDG 16, in particular, was also a major difference and step forward on building fair institutions when compared to the MDGs (see introduction chapter of this volume). Yet, the implementation of the SDGs relies on what has sometimes been called "governing through goals" (Kanie, Bernstein, Biermann, and Haas, 2017). That means countries outline how they will pursue each SDG and document it in voluntary reports. Such an approach without legally binding obligations, however, leaves significant room for governments to interpret, prioritize, and implement the SDGs in different ways and with varying levels of commitment.

The nature of that arrangement gives particular salience to what we have elsewhere called 'SDG politics' in domestic contexts. It refers to the processes of contestation over how the SDGs and their associated targets are interpreted, institutionalized, and implemented (or not) (Siegel and Bastos Lima, 2020, p. 2). SDG politics is a crucial element in any analysis of SDG interlinkages and trade-offs, for it determines whether such interlinkages and trade-offs are recognized in the first place and how they are eventually addressed. SDG politics also influences which aspects are given priority or, possibly, neglected. This means that there is ample room for cherry-picking among the vast set of 17 SDGs and 169 targets (Forestier and Kim, 2020).

In that regard, our analysis of agrifood governance in three South American countries has revealed how cherry-picking and selective reporting are not only commonplace but also problematic. It means that certain influential actors have focussed on just a few SDGs and narrowly defined elements of sustainability, while official government reporting on SDG progress disregards the concerns raised by civil society. While it might be understandable that it is not always possible to address all 17 goals simultaneously, it is problematic if there is no regard for the remainder of SDGs. The Agenda is, after all, to be "indivisible" (UN General Assembly, 2015, paragraph 55).

Actors in agrifood governance have frequently put forth what has been described as a form of sustainability metonymy, that is, claiming adherence to the whole 2030 Agenda while addressing only selected parts of it (Siegel and Bastos Lima, 2020, p. 10). Consequently, interlinkages or trade-offs between different SDGs – particularly with those that are ignored – become obscured. Consequently, this practice gives tacit priority to certain SDGs and dimensions of sustainability without an open and transparent discussion. Significantly, cherry-picking works as a way for dominant actors to push through their agendas and priorities. It allows them to claim adherence to the 2030 Agenda as a whole even while disregarding key SDGs, targets and Agenda principles such as those identified above on health and exclusion.

From the three countries we have examined in South America (Brazil, Paraguay, and Uruguay), only Uruguay openly acknowledged the existence of trade-offs in the VNRs on the SDGs (Siegel and Bastos Lima, 2020, p. 4). The Uruguayan VNR from 2018, for example, states that 'as a result of the strong growth process that the country has undergone in recent years, there are great challenges in achieving sustainable production, protecting the environment and conserving the resources that are part of the food production systems' (Presidencia de la República Oriental del Uruguay, 2018, p. 13)[2]. There are other references to trade-offs and challenges throughout the document, and the report also outlines public policies and strategies that aim to address those challenges. In contrast, Paraguay's first VNR presented in 2018 is much less detailed. There are references to challenges, but specific negative impacts of the dominant model of soybean production are not acknowledged at all – notwithstanding the many civil society critiques that have documented such impacts. For example, the VNR highlights the 'good performance of the agricultural sector' (Comisión ODS Paraguay 2030, 2018, p. 52)[3] that has led to strong economic growth, but with no mention of the social or environmental costs. Finally, in Brazil, the country's only VNR acknowledged that 'poverty in Brazil has notably rural features' and reiterated the value of existing national policies that 'incentivise family agriculture, promoting its economic and social inclusion, with stimulus for sustainable production (Presidência da República, 2017, p. 55)[4]. Yet, by then such smallholder-oriented programs were already facing substantive budgetary cuts alongside diminishing political support from increasingly agribusiness-oriented governments. President Bolsonaro's election in 2018 would only deepen such disregard for sustainability issues (Bastos Lima and Da Costa, 2021), and the publication of VNRs was discontinued altogether. Meanwhile, agricultural commodity traders in the country have claimed adherence to the SDGs by contributing to increased food production and what they frame as 'global nutrition' (Siegel and Bastos Lima, 2020), regardless of the region's rapidly deteriorating state of food insecurity.

Power, Participation, and Agenda-setting

The adoption of the SDGs on its own clearly is insufficient to overcome inequity in domestic politics related to pre-existing power imbalances. Operating already at the agenda-setting stage, particularly in contexts of high inequality, as is the case of much of South America, establishment actors (such as large landowners, agrifood businesses and strong political groups associated with them) have frequently sought to muffle out civil society contestations present in agrifood governance. By often controlling the agenda-setting of SDG interpretation and implementation, those actors manifest structural power that can constrain efforts towards inclusiveness. Agenda-setting has to do, first, with 'what to think about', and then 'how to think about' issues or actors (Balmas and Sheafer, 2010). As McCombs (2014, p. 73) explains, it has to do with building a 'compelling argument' while highlighting 'certain characteristics [that] might resonate with the public in such a way' that builds support and legitimacy for one's agenda. In the case of such a broad framework as the SDGs, that exercise is a crucial part of the process. Moreover, the extent to which interlinkages and trade-offs are recognized often becomes evident already at that agenda-setting stage.

The first VNRs that we analysed mostly set out how the SDGs align with existing public policies and commitments made through other international treaties and outline the institutional set-up for the implementation of the SDGs. Paraguay created a new institution for SDG implementation, but this operates alongside older institutional mechanisms in similar areas, most notably the institutions put in place to implement the country's National Development Plan. The lack of coordination capacity together with institutional weaknesses and insufficient financing and social accountability mechanisms remain significant problems. Most of the information regarding stakeholder participation provided in its first VNR is relatively vague. A more detailed reading reveals, however, that the only concrete indication of collaboration between the government and non-state actors in relation to the agrifood sector has been a memorandum of understanding with a commercial chamber representing large agribusiness interests. This suggests unequal access to government and limited institutional space for the participation of civil society, particularly for organizations that are critical of the country's dominant agricultural production model. Instead of helping address Paraguay's skewed agrifood governance that favours large-scale soybean growers and other agribusiness elites at significant social and environmental costs, this form of SDG politics might have strengthened their hand while excluding most Paraguayan civil society organizations even further. Brazil, in turn, used to have significant space for civil society participation but has clamped down on those. In 2019, the Bolsonaro administration abolished the

National Commission for the SDGs, which was considered a role model to other Commissions around the world, alongside many other venues and multi-stakeholder fora that previous governments had created (see e.g. GTSC-A2030, 2020). Those were institutional spaces where actors with diverse perspectives could broaden the debate and eventually discuss SDGs trade-offs, for example in relation to health or inequality. Currently, the SDGs are a moot-point for the government while, as seen, large agricultural commodity traders cherry-pick goals of choice to frame business as usual as fitting to the 2030 Agenda. Finally, in Uruguay, implementation of the SDGs builds on previously created institutions, notably a large-scale multi-stakeholder dialogue called the 'Social Dialogue'. It included over 665 institutions from the public sector, international agencies, civil society organizations, trade and business representatives, academia and religious groups in a series of events to discuss the future direction of the country. With the adoption of the 2030 Agenda, that pre-existing institutional process then became linked to the SDGs (Siegel and Bastos Lima, 2020, pp. 4–8). These experiences in South America show that more inclusive institutional setups such as Brazil's National Commission for the SDGs or Uruguay's 'Social Dialogue' can help promote greater participation and discussion of trade-offs, either building on pre-existing institutions promoting inclusiveness (as in Uruguay) or trying to address widespread exclusion (Brazil). Conversely, the absence of such spaces can perpetuate – and even expand on – skewed agrifood governance, as the case of Paraguay demonstrates. Here, there is a strong risk of not only sidelining but also entrenching existing patterns of marginalization under the name of the SDGs.

SDGs politics and agenda-setting are part of broader domestic political dynamics. Our findings suggest they particularly relate to underlying political features such as quality of democracy[5] and participation in governance (Siegel and Bastos Lima, 2020). They suggest that certain political-institutional preconditions such as regime type, state capacity, the rule of law, good governance, quality and stability of institutions might be key to provide for more inclusive processes in agrifood governance and SDG implementation.

Conclusion

This assessment of agrifood governance in the Southern Cone provides at least three important lessons for governing the interlinkages between the SDGs. First, cherry-picking and selective reporting poses serious risks. It has gone beyond just focussing on the most relevant SDGs for a given topic or policy area. Instead, it has become a method for obscuring contradictions and limiting any discussion of trade-offs. Through metonymy, dominant actors linked to large-scale agribusiness (e.g., large soy farmers, agricultural commodity traders) have claimed conformity to the entire 2030 Agenda by simply doing their business as usual – pathways that demonstrably increase and entrench marginalization in the region.

Second, this assessment reveals that focussing predominantly on synergies and win-win scenarios is problematic. Rather, it is crucial to look for trade-offs and, eventually, address them. Some of those might have been obscured already at the agenda-setting stage and need to be acknowledged. Among our case studies, Uruguay's efforts to build on open processes that take different stakeholders and perspectives into account shows that it is possible to fruitfully acknowledge different views and discuss trade-offs between the SDGs. There are some similarities here with the process that Bennich, Weitz, and Carlsen (in Chapter 2 of this volume) identify with respect to the drafting of Sweden's 2021 VNR. That country explicitly considered all SDGs, their targets, and with the involvement of different stakeholders it uncovered trade-offs.

Inclusive, participatory and transparent VNR drafting processes thus play an important role in increasing accountability and lowering the risks of cherry-picking and selective reporting. However, it is important to realize that this in itself requires resources and capacities that are unevenly distributed between countries. Moreover, that depends on the presence of domestic political commitment to such inclusiveness – as we have seen, contexts such as Brazil under the Bolsonaro administration have gone in the opposite direction. Here, shadow reporting by civil society organizations plays an important role and needs to be supported internationally. The Brazilian experience shows the vital importance of enshrining sustainability and inclusiveness commitments in law (potentially even in countries' Constitutions) to safeguard advancements from being easily dismantled by any political volatility.

That brings us to our third lesson: that pre-existing institutions and power relations matter significantly. It is unlikely that the adoption of the 2030 Agenda on its own can be sufficient to ensure that no one will be left behind, particularly given the voluntary nature of governing through goals. Without clear implementation guidelines, policy instruments or accountability mechanisms, such an approach has significant limitations regarding inclusiveness and inequality, owing to elite capture. As such, it is crucial to account for the political circumstances in which the SDGs are taken up and implemented domestically. Likewise, it is important to recognize internationally the constraints stemming from very different, and unequal, positions in the global economy. Domestic actions on sustainability issues that are deeply embedded in international commodity supply chains, as is the case for agrifood governance in South America, are inevitably limited and even more so in contexts of weak state institutions. In this respect, SDG 17 (Partnerships for the Goals) becomes a crucial objective waiting for more action.

Further research is therefore needed on the international dimensions of SDG implementation. Moreover, how to ensure effectiveness and accountability in the context of voluntary national implementation remains a challenge – linked not only to the 2030 Agenda but also to other governance domains such as climate change. Finally, the link between the quality of democracy and inclusive sustainable development that seeks to address trade-offs also merits further study. Our comparison suggests that a stronger democracy increases the chances that trade-offs are recognized, but that relationship would benefit from more systematic exploration, potentially through a larger, cross-national comparative analysis.

Notes

1 I gratefully acknowledge funding by the German Federal Ministry of Food and Agriculture (grant number 2219NR291) and the University of Münster through the research project 'Transformation and Sustainability Governance in South American Bioeconomies'.
2 Author's translation, Spanish original 'Dado el proceso de fuerte crecimiento que ha tenido el país en los últimos anos, esto implica grandes desafios en el logro de producción sostenible, cuidando el medio ambiente y conservando los recursos que forman parte de los sistemas de producción de alimentos'.
3 Authors' translation, Spanish original 'el buen comportamiento del sector agrícola'.
4 Authors' translation, Portuguese original 'A pobreza no Brasil tem traços acentuadamente rurais' and 'incentiva a agricultura familiar, promovendo sua inclusão econômica e social, com fomento à produção sustentável'.
5 The BTI Transformation Index classifies Uruguay as a 'democracy in consolidation', while Brazil and Paraguay are considered 'defective democracies' (bti-project.org). The Economist Intelligence Unit Democracy Index 2020 ranks Uruguay as a 'full democracy', comparable to Germany or the UK, and Brazil and Paraguay as 'flawed democracies' in a similar range as many Eastern European states (www.eiu.com/n/campaigns/democracy-index-2020).

Bibliography

Altieri, M.A. and Toledo, V.M. (2011). The agroecological revolution in Latin America: rescuing nature, ensuring food sovereignty and empowering peasants. *The Journal of Peasant Studies*, 38(3), 587–612. doi:10.1080/03066150.2011.582947.

APK Inform. (2021). South America to keep expansion on global soybean market. Retrieved from www.apk-inform.com/en/news/1521790.

Arancibia, F. (2013). Challenging the bioeconomy: The dynamics of collective action in Argentina. *Technology in Society*, 35(2), 79–92. doi:10.1016/j.techsoc.2013.01.008.

Balmas, M. and Sheafer, T. (2010). Candidate Image in Election Campaigns: Attribute Agenda Setting, Affective Priming, and Voting Intentions. *International Journal of Public Opinion Research*, 22(2), 204–229. doi:10.1093/ijpor/edq009.

Baraibar Norberg, M. (2020). *The Political Economy of Agrarian Change in Latin America - Argentina, Paraguay and Uruguay*. Cham: Palgrave Macmillan.

Bastos Lima, M.G. (2021). Corporate Power in the Bioeconomy Transition: The Policies and Politics of Conservative Ecological Modernization in Brazil. *Sustainability*, 13(12), 6952. Retrieved from: www.mdpi.com/2071-1050/13/12/6952.

Bastos Lima, M.G. and Da Costa, K. (2021). Quo vadis, Brazil? Environmental Malgovernance under Bolsonaro and the Ambiguous Role of the Sustainable Development Goals. *Bulletin of Latin American Research*. doi:10.1111/blar.13336.

Bastos Lima, M.G. and Kmoch, L. (2021). Neglect paves the way for dispossession: The politics of "last frontiers" in Brazil and Myanmar. *World Development*, 148, 105681. doi: https://doi.org/10.1016/j.worlddev.2021.105681.

Bastos Lima, M.G. and Persson, U.M. (2020). Commodity-Centric Landscape Governance as a Double-Edged Sword: The Case of Soy and the Cerrado Working Group in Brazil. *Frontiers in Forests and Global Change*, 3(27). doi:10.3389/ffgc.2020.00027.

Bastos Lima, M.G. and Siegel, K.M. (2020). Promoting inclusive bioeconomies? Lessons from agri-food governance and the politics of the SDGs in South America. *ZEF Policy Brief* (Vol. 37, pp. 1–4).

BTI. (2021). The BTI Transformation Index. Retrieved from https://bti-project.org/en/?&d=D&cb=00000.

Chiappe, M. (2020). Conflictos por uso de agroquímicos: el papel de las mujeres rurales en Uruguay. *Agrociencia Uruguay*, 24(352), 1–15. doi:10.31285/agro.24.352.

Clapp, J. and Moseley, W.G. (2020). This food crisis is different: COVID-19 and the fragility of the neoliberal food security order. *The Journal of Peasant Studies*, 47(7), 1393–1417. doi:10.1080/03066150.2020.1823838.

Comisión ODS Paraguay 2030. (2018). Informe Nacional Voluntario Paraguay 2018.

Deciancio, M., Siegel, K.M., Kefeli, D., Queiroz-Stein, G., and Dietz, T. (2022). Bioeconomy governance and (sustainable) development. In M. Deciancio, P. Nemiña, and D. Tussie (Eds), *Handbook on the Politics of International Development*: Cheltenham: Edward Elgar Publishing.

ECLAC. (2019). Quadrennial report on regional progress and challenges in relation to the 2030 Agenda for Sustainable Development in Latin America and the Caribbean. Retrieved from: www.cepal.org/en/publications/44552-quadrennial-report-regional-progress-and-challenges-relation-2030-agenda.

Edwards, G. and Roberts, J.T. (2015). *A Fragmented Continent - Latin America and the Global Politics of Climate Change*. Cambridge, Massachusetts: MIT Press.

Empresa de Pesquisa Energética. (2020). *Balanço Energético Nacional: Ano Base 2019*. Rio de Janeiro, EPE.

Escobar, N., Tizado, E.J., zu Ermgassen, E. K.H.J., Löfgren, P., Börner, J., and Godar, J. (2020). Spatially-explicit footprints of agricultural commodities: Mapping carbon emissions embodied in Brazil's soy exports. *Global Environmental Change*, 62, 102067. doi:10.1016/j.gloenvcha.2020.102067.

Ezquerro-Cañete, A. (2016). Poisoned, Dispossessed and Excluded: A Critique of the Neoliberal Soy Regime in Paraguay. *Journal of Agrarian Change*, 16(4), 702–710. doi:10.1111/joac.12164.

Ezquerro-Cañete, A. and Fogel, R. (2017). A coup foretold: Fernando Lugo and the lost promise of agrarian reform in Paraguay. *Journal of Agrarian Change*, 17(2), 279–295. doi:10.1111/joac.12211.

FAO, IFAD, PAHO, WFP, and UNICEF. (2021). *Latin America and the Caribbean – Regional Overview of Food Security and Nutrition*. Santiago, FAO. https://doi.org/10.4060/cb7497en

Finger, R., Swinton, S.M., El Benni, N., and Walter, A. (2019). Precision Farming at the Nexus of Agricultural Production and the Environment. *Annual Review of Resource Economics*, 11(1), 313–335. doi:10.1146/annurev-resource-100518-093929.

Flach, R. et al. (2021). Conserving the Cerrado and Amazon biomes of Brazil protects the soy economy from damaging warming. *World Development*, 146, 105582. doi:10.1016/j.worlddev.2021.105582.

Forestier, O. and Kim, R.E. (2020). Cherry-picking the Sustainable Development Goals: Goal prioritization by national governments and implications for global governance. *Sustainable Development*, 28(5), 1269–1278. doi:10.1002/sd.2082.

Garnett, T. et al. (2013). Sustainable Intensification in Agriculture: Premises and Policies. *Science*, 341(6141), 33–34. doi:10.1126/science.1234485.

Garrett, R.D. and Rausch, L.L. (2016). Green for gold: social and ecological tradeoffs influencing the sustainability of the Brazilian soy industry. *The Journal of Peasant Studies*, 43(2), 461–493. doi:10.1080/03066150.2015.1010077.

Green Jonathan et al. (2019). Linking global drivers of agricultural trade to on-the-ground impacts on biodiversity. *Proceedings of the National Academy of Sciences*, 116(46), 23202–23208. doi:10.1073/pnas.1905618116.

GTSC-A2030. (2020). 2030 Agenda for Sustainable Development Spotlight Report Synthesis IV Brazil.

Hatzikidi, K. and Dullo, E. (Eds). (2021). *A Horizon of (Im)possibilities: A Chronicle of Brazil's Conservative Turn*. London: University of London Press.

Henderson, J., Godar, J., Frey, G.P., Börner, J., and Gardner, T. (2021). The Paraguayan Chaco at a crossroads: drivers of an emerging soybean frontier. *Regional Environmental Change*, 21(3), 72. doi:10.1007/s10113-021-01804-z.

Hetherington, K. (2011). *Guerrilla Auditors - The Politics of Transparency in Neoliberal Paraguay*. Durham, NC and London: Duke University Press.

Ultima Hora (2019, March 11). *Alertan sobre plantaciones de soja sin barreras verdes de protección*. Retrieved from: www.ultimahora.com/alertan-plantaciones-soja-barreras-verdes-proteccion-n2806224.html.

IPES-Food and ETC Group. (2021). *A Long Food Movement: Transforming Food Systems by 2045*. Retrieved from: http://www.ipes-food.org/pages/LongFoodMovement

Kanie, N., Bernstein, S., Biermann, F., and Haas, P.M. (2017). Introduction: Global Governance through Goal Setting. In N. Kanie and F. Biermann (Eds), *Governing through Goals: Sustainable Development Goals as Governance Innovation* (pp. 1–28). Cambridge, MA: MIT Press.

Leite-Filho, A.T., Soares-Filho, B.S., Davis, J.L., Abrahão, G.M., and Börner, J. (2021). Deforestation reduces rainfall and agricultural revenues in the Brazilian Amazon. *Nature Communications*, 12(1), 2591. doi:10.1038/s41467-021-22840-7.

Mahon, N., Crute, I., Simmons, E., and Islam, M.M. (2017). Sustainable intensification – "oxymoron" or "third-way"? A systematic review. *Ecological Indicators*, 74, 73–97. doi:10.1016/j.ecolind.2016.11.001.

McCombs, M. (2014). *Setting the Agenda*. Cambridge: Polity Press.

McNeill, D. (2019). The Contested Discourse of Sustainable Agriculture. *Global Policy*, 10 (S1), 16–27. doi:10.1111/1758-5899.12603.

Nehring, R. (2022). The Brazilian Green Revolution. *Political Geography*, 95, 102574. doi:10.1016/j.polgeo.2021.102574.

Oliveira, G. and Hecht, S. (2016). Sacred groves, sacrifice zones and soy production: globalization, intensification and neo-nature in South America. *The Journal of Peasant Studies*, 43(2), 251–285. doi:10.1080/03066150.2016.1146705.

Oliveira, G.d.L.T. (2019). Boosters, brokers, bureaucrats and businessmen: assembling Chinese capital with Brazilian agribusiness. *Territory, Politics, Governance*, 7(1), 22–41. doi:10.1080/21622671.2017.1374205.

Otero, G. (2012). The neoliberal food regime in Latin America: state, agribusiness transnational corporations and biotechnology. *Canadian Journal of Development Studies / Revue canadienne d'études du développement*, 33(3), 282–294. doi:10.1080/02255189.2012.711747.

Rede Penssan (2021). *Insegurança Alimentar e Covid-19 no Brasil* 1. Retrieved from: http://olheparaafome.com.br/VIGISAN_Inseguranca_alimentar.pdf

Presidência da República. (2017). *Relatório Nacional Voluntário sobre os Objetivos de Desenvolvimento Sustentável: Brasil 2017*. Secretaria de Governo da Presidência da República, Ministério do Planejamento, Desenvolvimento e Gestão.

Russo Lopes, G., Bastos Lima, M.G., and Reis, T.N.P.d. (2021). Maldevelopment revisited: Inclusiveness and social impacts of soy expansion over Brazil's Cerrado in Matopiba. *World Development*, 139, 105316. doi:10.1016/j.worlddev.2020.105316.

Siegel, K.M. and Bastos Lima, M.G. (2020). When international sustainability frameworks encounter domestic politics: The sustainable development goals and agri-food governance in South America. *World Development*, 135, 105053. doi:10.1016/j.worlddev.2020.105053.

Siegel, K.M., Deciancio, M., Kefeli, D., Queiroz-Stein, G., and Dietz, T. (2022). Fostering Transitions Towards Sustainability? The Politics of Bioeconomy Development in Argentina, Uruguay and Brazil. *Bulletin of Latin American Research*. doi:10.1111/blar.13353.

Tittor, A. (2021). The key role of the agribusiness and biotechnology sectors in constructing the economic imaginary of the bioeconomy in Argentina. *Journal of Environmental Policy & Planning*, 23(2), 213–226. doi:10.1080/1523908X.2021.1893162.

Transparency International. (2021). Corruption Perceptions Index.

UN General Assembly. (2015). Transforming our world: the 2030 Agenda for Sustainable Development. Retrieved from: https://sustainabledevelopment.un.org/post2015/transformingourworld.

van Dijk, T.A. (2017). How Globo media manipulated the impeachment of Brazilian President Dilma Rousseff. *Discourse & Communication*, 11(2), 199–229. doi:10.1177/1750481317691838.

Vergara-Camus, L. and Kay, C. (2017). Agribusiness, peasants, left-wing governments, and the state in Latin America: An overview and theoretical reflections. *Journal of Agrarian Change*, 17(2), 239–257. doi:10.1111/joac.12215.

7

GOVERNANCE OF THE WATER-LAND-FOOD NEXUS FOR INTEGRATED ACHIEVEMENT OF THE 2030 AGENDA

The case of Lower Awash River Basin, Ethiopia

Srinivasa Srigiri and Waltina Scheumann

Introduction

The 2030 Agenda is unique in emphasizing the indivisibility of social, economic, and environmental dimensions of sustainable development and the need for governance innovation and increased policy coherence to manage the interdependencies among the sustainable development goals (SDGs) and targets (Le Blanc, 2015; Pahl-Wostl, 2019). Further, the 2030 Agenda is innovative in adopting five core principles to guide the implementation of the SDGs namely, universality; leaving no one behind (LNOB); interconnectedness and indivisibility; inclusiveness; and participatory decision-making (UN System Staff College, 2021).

Several approaches to study the interlinkages among goals and targets have identified both synergies and conflicts at global, regional or national levels. The water–energy–food (WEF) nexus is one such approach, which primarily focuses on the interlinkages among the SDGs 2 (zero hunger), 6 (access to water and sanitation), and 7 (access to energy). However, these SDGs are further found to have strong interlinkages with most other SDGs (International Council for Science and International Social Science Council, 2015). Applying the WEF nexus approach in a selected context might offer an "issue-based entry point" to unravelling the nature of the complex interdependencies and the factors determining those (Breuer, Janetschek, and Malerba, 2019). As an analytical and governance approach, the WEF nexus concept seeks to overcome policy incoherence that might result from fragmented policies across the water, energy and food sectors, and centralized hierarchic political regimes and promote policy coherence through identifying optimal policy mixes and governance arrangements (Páez-Curtidor, Keilmann-Gondhalekar, and Drewes, 2021; Weitz, Strambo, Kemp-Benedict, and Nilsson, 2017, p. 165). Coordination across sectors (horizontal) and across levels (vertical) is crucial to mitigate trade-offs and enhance synergies among the SDGs associated with the WEF nexus. However, there is

DOI: 10.4324/9781003254683-7

insufficient understanding of the factors that determine the effectiveness of institutions and governance mechanisms that achieve coherent policy design and implementation.

In this chapter, we aim to understand the social, political and institutional conditions that determine the effectiveness of institutional arrangements for water and land governance in the lower Awash River Basin in Ethiopia in achieving an integrated implementation of water and land dependent SDGs, mainly SDG 2 (food security), SDG 6 (water security), SDG 8 (economic growth), and SDG 15 (sustainable ecosystems). As we identify and explain the interlinkages between water, land and food security, we focus on the governance of the water–land–food (WLF) nexus and the related SDGs. In doing so, this chapter contributes to answering three research questions set forth by the book: one, by exploring the key interlinkages among water and land related SDGs in the study region; two, by throwing light on the effectiveness of existing governance mechanisms in overcoming the power asymmetries and minimizing trade-offs among interlinked SDGs; and three, by identifying the political-institutional conditions that influence the emergence and effectiveness of suitable governance mechanisms.

Ethiopia faces important challenges with regard to democratization and governance. It is classified as a "hard line autocracy" (3.02) on the Bertelsmann Index of political transformation (BTI, 2021). Further, Ethiopia also has low state capacity (-0.64) as measured by the World Governance Indicators sub-indicator of Government Effectiveness (World Bank, 2016). Therefore, we consider the country as a typical case to study the influence of political-institutional conditions and the capacity of the state on effectiveness of governance mechanisms to manage the SDG interlinkages. Further, the lower Awash River Basin shows low population density, arid and semi-arid climatic conditions and, above all, predominant pastoralism, which is a mode of production and a cultural way of life. Conversion into cropland of prime dry-season grazing areas has restricted pastoralists' access to riverine resources, and governments have favoured the sedentarization of pastoralists. Competition exists over land between commercial state and private farms on the one hand, and pastoralists and smallholder farmers on the other. Therefore, the lower Awash River Basin makes an ideal case to study the interlinkages between different SDGs and targets pursued that are based on land and water resources. Moreover, it enables to assess in how far policies and the existing governance mechanisms for land and water resources are inclusive to benefit marginalized (pastoralists) groups, in compliance with the LNOB core principle of the 2030 Agenda.

In the following section, we analyze the key interlinkages in pursuit of those SDGs (1, 2, 8, and 15) that relate to Water-Land-Food (WLF) nexus from a polycentric governance perspective. In doing so, we identify major trade-offs among goals and explain which institutional factors influence the governance mechanisms and their effectiveness in managing these trade-offs. We then assess the effectiveness of the existing coordination mechanisms in light of the 2030 Agenda principles of indivisibility, LNOB, and inclusiveness, followed by conclusions.

Interlinked Action Situations of Land and Water Governance in the Lower Awash River Basin

Polycentric understanding of WLF nexus governance

Systems for providing water, energy, food and other securities in the WLF nexus exhibit features of polycentric governance, which involves decision-making centers across different sectors and at various levels. While these centers might be formally independent, they are often functionally dependent but might sometimes overlap. Therefore, we apply polycentricity as an analytical approach to investigate the interactions and coordination among different decision-making centers in the Ethiopian context. In order to assess the outcomes produced by the existing governance structure of the WLF nexus (i.e. synergies and/or trade-offs among SDGs) and whether existing governance processes and mechanisms adhere to the 2030 Agenda principles (indivisibility of the SDGs, LNOB, participatory and inclusive decision-making), we adapted the Institutional Analysis and Development (IAD) framework put forth by Ostrom (1990) and the concept of the network of adjacent action situations (NAAS) by McGinnis (2011). Srigiri and Dombrowsky (2022) provide a more detailed adaptation of the analytical framework for studying the governance of interdependencies within the nexus. While the IAD framework is often applied to analyze the behaviour of actors in singular action situation, the NAAS concept highlights the importance of considering complex policy networks in which overlapping sets of actors in different action situations perform distinct governance functions such as production, provision[1], financing, coordination, monitoring and enforcement, and dispute resolution.

According to the IAD framework, an *action situation* is a situation in which two or more actors participate by taking specific positions and choosing from a set of possible actions that lead to outcomes, which in turn, have different payoffs for each participant (Ostrom, 2005). *Actors* might be individuals or an organized entity of individuals who participate in a given action situation. They act upon information available to them about the costs and benefits of actions, outcomes and their individual payoffs that depend on the rules for distribution of costs and benefits (Ostrom, 2005). McGinnis' NAAS concept extends the IAD framework by stating that various functions of polycentric governance occur in distinct action situations adjacent to each other. These action situations might be spread across different action arenas or conceptual levels of analysis (Ostrom, 2005), namely (i) the operational-choice level, wherein the outcomes of action situations are related to the wellbeing of actors involved and natural resource conditions; (ii) the collective-choice level, wherein the outcomes of action situations are institutions or rules that define the set of action choices at operational-choice level; and (iii) the constitutional-choice level, wherein the outcomes of action situations are in rules-in-use (North, 1993) that legitimize and/or impose constraints on the action and mutual interaction for actors at the collective and operational choice levels. They might include both formal rules (laws, regulations, statutes, and so on) and customary rules (for instance, societal norms, customs, values, beliefs) and their enforcement characteristics. It is important to understand both formal and customary rules-in-use to explain the behaviour of actors in different action situations and their outcomes.

The results presented in the following section are based on the analysis of qualitative data collected through 29 semi-structured interviews with key actors[2] in November 2018, as well as an extensive review of primary and secondary literature relating to natural resource governance and implementation of the 2030 Agenda in Ethiopia. We analyze multiple interlinked action situations that spread across operational, collective- and constitutional-choice levels. In the following, we describe these action situations and their outcomes, and explain the rules that determine the actions and outcomes in the three arenas.

National planning process – Constitutional choice level

The main national planning instruments, i.e., the Growth and Transformation Plans (GTPs) I (2010–2015) and II (2015–2020) focused on increasing agricultural productivity and accelerating growth through substantial public investment in infrastructure, with GTP II advancing the reform efforts launched under the GTP I. The preparation and ratification of the GTPs occurs in a sequence of action situations involving various actors influenced by several institutions that are briefly described below.

At the first stage, a Macro-Economy Team consisting of representatives from the Planning and Development Commission (PDC), the Ministry of Finance and Economic Development, the National Bank of Ethiopia, and the Ministry of Revenue develops a macro-economic framework. Essentially, the macro-economic framework is based on Ethiopia's long-term strategic vision to become a lower-middle-income country with a climate-resilient green economy by 2025, which

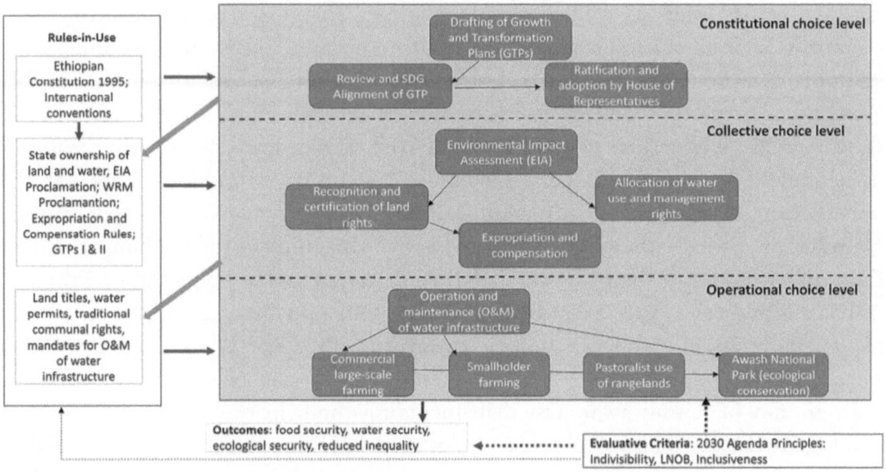

FIGURE 7.1 Interlinked action situations of the water-land-food nexus governance in Lower Awash River Basin

Source: Author's own elaboration

was first formulated in 2011 (Ayalew, 2013). Then, the macro-framework is transmitted to line ministries, together with guidelines and directions for the formulation of their respective sectoral plans and targets. Throughout the process of preparing sectoral plans, line ministries receive technical support and orientation from the PDC. The PDC then consolidates the individual, sectoral plans into a coordinated set of plans that constitutes the first draft version of the GTP.

Alignment of SDGs with GTP II fell under responsibility of the PDC, in which neither non-state stakeholders nor sub-national governments were represented. Further, the fact that the term "2030 Agenda" appears only three times in the 225-page GTP II document indicates weak degree of mainstreaming the SDGs into the national plans. In its Voluntary National Review (VNR) Report of 2017, Ethiopia presented a "mapping" of SDGs against the ten national development priorities defined in GTP II. However, the VNR does not provide information about the logic and methodology underlying this mapping exercise. Against this background, it is interesting to see that the top national development priority (agricultural sector development as a major source of economic growth) alone is deemed to contribute to the achievement of eleven SDGs (including, for example, SDG 10 "reduced inequalities", SDG 14 "life below water", and SDG 15 "life on land"). However, there is no explanation of the pathways by which industrialization of agricultural development would lead to achievement of the mapped SDGs. A detailed analysis of the national indicator framework of GTP II by Srigiri, Breuer, and Scheumann (2021) reveals that it is only weakly aligned with the Global indicator framework for the Sustainable Development Goals. These observations suggest that mainstreaming the SDGs into the national planning framework has, to some extent, been a cosmetic exercise of SDG labelling in the sense of adapting previously existing national priorities to the rhetoric of the 2030 Agenda.

At the second stage, the draft version of the GTP is submitted for review to the Council of Ministers as well as by the National Planning Council in which the country's regions are represented by their respective heads of government. In addition, a nationwide multi-stakeholder consultation process is conducted. The PDC then integrates feedback from the review into the draft. In the case of GTP II, the PDC was tasked with integrating the SDGs into the existing national development framework (Federal Democratic Republic of Ethiopia, 2017a). The presentation of the draft to city administrations and other stakeholders during the consultation process therefore officially served the purpose of enabling them to fully understand and own the GTP II (Federal Democratic Republic of Ethiopia, 2017b). However – and as has been the case with previous national development plans (Haile, 2015) – the stakeholder consultation for GTP II occurred in a rather hierarchical top-down mode at a rather late phase in the planning process, more with an aim of raising awareness than including them in the process of decision-making At the third and final stage, the revised GTP is presented to the House of People's Representatives for ratification, upon which it becomes a legally binding document for the next five years.

The Ethiopian Constitution of 1995 (Article 52, 2c) mandates regional states to formulate and execute their own policies, strategies and plans. Regional planning cycles are consistent with the national planning cycle. Regional state governments have their own planning units, which compile the sectoral plans formulated by the regional sector bureaus into regional five-year planning documents. While regional development plans attempt to reflect regional priorities and needs, they can be considered as sub-sets of the national GTP II given their strong alignment with the strategic directions provided by the national planning framework. Interview partners from donor organizations have stressed the high degree of bindingness of GTP II across all levels of government as a characteristic that sets Ethiopia apart from many other developing economies. Some interview partners attributed the high commitment to the national development plan to Ethiopia's tradition of a socialist planned economy. Officials at sub-national level consistently described GTP II as their main point of reference in interviews. By comparison, awareness and ownership of the 2030 Agenda was much lower at the sub-national level. The general misperception among interviewees appeared to be that the SDGs had been fully mainstreamed into GTP II and that, consequently, pursuing GTP II was equivalent to pursuing the 2030 Agenda.

Allocation of land and water rights – collective choice level

Land rights and land alienation

Land tenure in Ethiopia is governed by co-existing formal and informal rules, which are contradicting each other. The Constitution of Ethiopia adopted in 1995 vests "the right to ownership of rural land and urban land, as well as all natural resources [...] exclusively in the state and the peoples of Ethiopia. Land is the common property of the Nations, Nationalities and Peoples of Ethiopia" (Article 40). The Constitution covers individual as well as collective rights including those of traditional (pastoralist) communities, and recognizes their communal customary land holding systems (Abdulahi, 2007, p. 103). On the other hand, subsequent legislation on land administration and land use in 2005, stipulates that land is a public property owned by the state. However, de facto, a majority of the rural areas in the lowlands, continue to be controlled by traditional rules and local authorities (clan leaders), and not by the state (Ambaye, 2015; World Bank, 2012), except the large tracts of land that were appropriated to establish state farms (Abdulahi, 2007). Land in pastoral areas is accessed on the basis of clan, sub-clan, and lineage group membership (World Bank, 2012, p. 22).

The land administration and land use legislation of 2005 introduced Rural Land Holding Certificates to provide security to private landholders "while it said almost nothing about the security of communal landholding systems" (Abdulahi, 2007, p. 118). The legislation encourages "private investors in pastoral areas having tribe-based communal land holding systems" (Abdulahi, 2007, p. 119). It thus created the potential for the state to easily appropriate communal land in order to encourage investments and facilitate state-driven projects (World Bank, 2016, p. 12). Although the Constitution rules that "pastoralists have the right to free land for grazing as well as the right not to be displaced" (Article 40(5)), the existing legislation in use violates the constitutionally granted rights of the pastoralists, which is also in contradiction of the LNOB principle of the 2030 Agenda.

However, regional governments constitutionally hold the legitimate authority to administer land, enact legislation, and implement the respective laws (Nega, Adenew, and Gebre Sellasie, 2003; World Bank, 2012, p. 23). Relevant for the middle and lower Awash Basin, the regulatory frameworks for land use in the regional states of Oromia and Afar explicitly recognize the communal holding systems of pastoralists and communal rights to grazing land. Groups of people might hold a land title which is inseparable. Communities are granted usufruct rights, the right to inherit, transfer and lease land but not to sell it.

Both the Ethiopian Constitution and the Proclamation on the Expropriation of Landholdings (2005) allow expropriation of private and communal land for *public purposes*, which is a term that is not legally specified (Tamrat, 2010; World Bank, 2016, p. 12). A purpose should be beneficial to the public, but in practice land acquisitions for investors who might use land more productively, are also declared as being of public purpose (LANDac, 2018; Tura, 2018). In the absence of a clear legal definition of the term public purpose, the government might expropriate pastoral land in order to achieve economic growth through commercial agriculture at the expense of food and income security of smallholder farmers and pastoralists (Aneme, 2015; Anteneh, 2007, p. 20).

Only recently has the government taken an initiative to register and certify communal pastoral landholding systems (World Bank, 2016, p. 52). However, interviewees emphasized that the process of registration and certification of common titles for pastoral communities is a complex issue. The vast expanse of pastoral rangelands, the transection of migration routes across multiple state and private farms, *woredas* (districts) and *kebeles* (villages), and the internal hierarchical structure of pastoral clans and sub-clans are some of the main factors complicating the certification process for communal land rights (Kassa, 2001). A notable example is the certification of land titles for the Borana pastoralists, the largest Oromo tribe. Based on the Oromia land law, they received the first ever communal land holding title within the context of a USAID-financed Land Administration to Nurture Development (LAND) project (Haddis, 2018; Napier and Desta, 2011; Woldegiorgis, 2018). However, in the Awash Basin, communities have not succeeded so far in securing communal titles. This might be because of its proximity to markets and the very high demand for fertile land and water.

In the absence of certified communal and group land rights, which is common in the lowlands and pastoral areas (Rahmato, 2011), pastoralists are disadvantaged by the compensation framework and practices: subject to compensation payments is not land and not the resources which grow naturally on the land, but immovable property and the improvements one brings about on the land by one's labor or capital (Ayane, 2014). Only those land users are eligible to compensation payments who hold lawful possession of land registered in land holding books. Furthermore, there is no comprehensive directive to guide certified valuators to carry out valuations of property on which compensation payments are calculated. While complaints and appeals can be raised at local units, or at regular courts, complaints can be made only on the amount of compensation payments. Finally, the legislation does not specify which authority decides on compensation levels, and on the method by which payments are calculated. These regulatory deficits are leading to a further marginalization (against the LNOB principle) of pastoralist communities thus increasing their food insecurity.

Water rights and water allocation

Endeavours to develop a water allocation regime were already emphasized in the Ethiopian Constitution, which ruled that water resources are not subject to private but public ownership vested in the State. Furthermore, the Ethiopian Water Resources Management Proclamation No. 197/2000 forms the legal framework for a permit-based water allocation and utilization, while vesting the power and responsibility of planning, management, allocation and protection in the Ministry of Water, Irrigation and Electricity (MoWIE). This Ministry can further delegate power and responsibilities to regional governments, river basin councils and their executive arms, i.e., river basin authorities. Water User Associations (WUAs) – as the lowest level of irrigation scheme management – were formed only later following the Proclamation No. 841/2014.

Ethiopian governments have initiated a formal water permit system to regulate water allocation and use. Applicants of water permits are required to submit investment certificates, environmental clearance documents (see next sub-section), feasibility studies, documents clarifying land-use rights, no-objection letters from potentially affected communities or *woreda* administrations, specifications related to the source from which water will be abstracted, for what purpose, the irrigation technology used and on crops' monthly and annual water requirements.

However, water permits remain far from being an effective instrument to balance demands, since not all users in the basin hold a permit, and water abstractions are in most cases not measurable owing to the non-existent or dysfunctional gauging stations and measurement devices. Overall, permits are not used as an instrument to balance demands based on water availability and changing demand patterns, but as a tool to mobilize financial resources from larger water users. Rather, the yearly water allocation plans developed by the Awash Basin Authority (ABA) are the instrument to allocate water and balance demand.

Respondents from the WUA in the kebele Dire Sede, for example, stated that their water permit did not specify the amount of water it is entitled to receive. It only gets water that is in excess of the actual demand of the Nura Era estate (located upstream). Administrative guidelines are being developed that specify eligibility criteria, requirements for applications, issuance, duration, suspension and revocation of water permits. However, data on water availability, water permits already approved and water actually use by permit holders are not available in a systematic and reliable manner. The limited technical and financial capacity of public administrations to integrate robust scientific criteria, namely, hydrological water balance studies, water resources management and water resources allocation plans including allocations for smallholdings and wildlife, is leading to non-transparent decisions regarding the allocation of water-use rights. Moreover, while rural populations, smallholders and pastoralists are exempted from water permit applications, in the case of conflicts, these users have a weak legal standing compared to permitted users (van Koppen & Schreiner, 2018). Therefore, instead of easing the access of pastoralists to water points and granting rights-of-way, the exemption rule further excludes them from benefits deriving from the use of water resources, leaving them further behind.

Further constraints for an effective allocation regime also arise from coordination deficits between different levels of government, specifically between the federal ministry, the regional states and river basin authorities that all have the authority to issue water permits. Ethiopia's state apparatus for water allocation has been traditionally characterized by a two-tier organizational set-up, namely the Basin High Councils, the highest political, strategic decision-making body, and the basin authorities being the administrative, technical arm of the High Basin Councils. In the Awash High Basin Council, for instance, five regional states, two city administrations as well as representatives of federal ministries (Agriculture; Water, Irrigation and Energy; Environment, Forestry and Climate Commission; and Finance) were represented and headed by the Deputy to the Prime Minister. The High Basin Council was mandated to decide on master plans, water resources management and water allocation plans to balance sectoral demands, while the Awash Basin Authority prepared basin plans, issued permits, and monitored compliance. The high-level composition of the Awash High Basin Council and unclear competencies and roles of different members in the process of approving water permits hinder its effective functioning.

Furthermore, inter-regional rivers are under the jurisdiction of MoWIE who can delegate its powers to regional states as well as to river basin organizations (such as the Awash Basin Authority). However, in practice the demarcation between its authority and those of the regional governments on issuing permits is not straightforward. According to Hailu, Tolossa, & Alemu (2017, p. 11), "regional states believe that the powers and duties of the basin authority are unconstitutional because they conflict with the powers and duties given to the regional states as stated in the Article 52 of the Constitution".

During the time of field data collection for this study, a restructuring of the governance structures for water allocation was underway. Respondents expressed hopes that these reforms might address problems of unclear division of mandates among regional governments, the Ministry and the Awash Basin Authority. It remains to be seen whether this reform will improve the financial basis of basin authorities, so that they are able to carry out their administrative and regulatory roles (Srigiri et al., 2021).

Environmental regime

Approval of water permits, and overall licensing of hydraulic projects (i.e., dams, reservoirs, irrigation projects), including the allocation of land titles to project developers require the implementation of Environmental Impact Assessments (EIA)[3] and the issuance of Environmental Clearance Certificates by the Federal Commission for the Environment, Forest and Climate Change. Managing the EIA process is shared between the Federal Commission with its EIA Directorate, regional bureaus and the environmental units of sector agencies, while final decisions about projects rest with the respective authorizing agencies, in most cases at the federal level.

Interview respondents as well as previous scholarly studies have rated the Ethiopian EIA system as not being effective (Danyo et al., 2017) as major investment projects (such as highways or railways) commenced without EIAs. The absence and poor quality of EIAs has had negative effects in both ecological and social terms. Sugarcane cultivation areas have been expanded onto the Awash National Park's area or on pastoralists' rangelands despite their negative environmental and social effects (i.e.,). Environmental Impact Studies were not published and, even when they were, they had no influence on decisions taken. Frequently, actors from the federal government intervened in favour of projects, arguing they would promise economic benefits (Damtie and Bayou, 2008). The Ethiopian EIA system is particularly weak concerning the assessment and prevention of adverse social impacts. Mandatory procedures regarding public meetings, consultations with affected communities, information dissemination and disclosure are often not followed. The weak standing of environmental and social concerns in projects targeting economic growth reflects the priorities of GTP II, set at the national level and implemented in a hierarchical mode with little or no inclusion of stakeholders beyond the federal government. The enforcement of the EIA instrument lacks secondary regulations, such as a binding EIA directive, sectoral guidelines, and threshold values, which affects the administrations' capacity to effectively implement this instrument. Capacity constraints are even worse at the lower levels with negative implications for the quality of environmental impact studies, and the EIA process in general. Further, the Commission has a weak status in the government hierarchy. It is a regulatory organ accountable to the Prime Minister's Office, but "most government offices are hierarchically at a higher level than the Environment Protection Authority (EPA, now the Federal Commission), which prevents EPA from regulating the activities of those government offices" (Damtie and Bayou, 2008, p. 41).

Next to regulatory deficits and capacity constraints, the EIA system has been "distorted" (Danyo et al., 2017, p. 10) as the powers of the EPA[4] have been delegated to sector agencies (Janka, 2012, footnote 488), while international practice suggests that the issuance of Environmental Clearance Certificates should be decided by an appointed independent commission in which relevant stakeholders are represented. At the time of our field research (November 2018), the Commissioner had written a letter to the Prime Minister to revoke the power of sector agencies since some obstacles (such as the number of qualified staff) would no longer exist.

Overall, low capacities of state agencies, political will to empower environmental bureaus at different levels, lack of clear operational rules for enforcement of EIAs are resulting in a trade-off between economic growth strategies and the social and environmental dimensions of sustainable development.

Use and management of land and water – Operational choice level

Conflicts over land and grazing resources in the basin are longstanding but intensified with the foundation of large sugar cane and cotton farms (Wonji Shoa Sugar Estate, Nura Era, Metahara) in the 1960s and the establishment of the Awash National Park in 1969 (Ayalew, 2013; Kloos, 1982). Traditional pastoralists have lost dry-season grazing lands, first inside the Awash National Park, then outside the Park where cotton and sugar cane plantations expanded at their expense (Meuer and Moreaux, 2017, p. 42). Recent competition over land is increasing because programmes have been settling mobile pastoralists in irrigation schemes (small-scale and state farms), and because commercial farms intend to acquire additional land from local communities. Pastoralists rely on less productive, marginal land which is also threatened by the expansion of *Prosopis juliflora*, a noxious weed (World Bank and Department of International Development, United Kingdom, 2019). Although the regional Rural Land Use and Administration Proclamations of the regional states of Afar and Oromia, which are both located in the basin, enshrine customary rights of pastoralists and communities to grazing land (communal ownership, communal holdings), policies promote commercial agriculture and provide land titles to individual households (Reda, 2014). Changing land use from rain-fed and rangelands to irrigated agriculture as well as changes in land tenure intervene in the livelihoods of pastoral communities and contribute to the erosion of their way of life, and food security. Here, we observe a trade-off between the strategy of the Ethiopian government to achieve economic growth (SDG 8.1, driven by the expansion of commercial agriculture) and ensuring security of customary tenure rights of pastoralists to grazing land (SDG 1.4.2).

Concerning the management of modern and upgraded traditional irrigation schemes, the responsibilities are devolved to the WUAs since 2014.[5] WUAs are membership-based, non-profit organizations governed by their General Assembly where any member has at least one vote. However, WUAs are not fully self-governed entities but are under the control of MoWIE (Haileslassie et al., 2016) which holds considerable decision-making power. A WUA's mandate embraces water distribution, maintenance of irrigation infrastructure, and the collection of service charges within one's command area but has limited authority and financial means to do so. While the existing regulations do not enable the recovery of recurrent costs of repairs and maintenance entirely from member contributions, they do not specify where financial support has to come from. According to the law, the establishment of WUAs and support for their operations falls under the responsibility of local administrative units, namely the *woredas* and the *kebeles*. However, neither the *woreda* water officers' responsibility nor the *kebele* development agents' mandates at present embrace irrigation advisory services. District and village administrations and the country in general, lack agronomists specialized in irrigation system and on-farm water management. Without the requisite financial means and the human capacities to provide repair and maintenance services of irrigation infrastructure as well as irrigation extension, the actors at the lowest level are unable to fulfil the devolved responsibilities.

Management of the hydraulic headworks down to the service areas of WUA is a complex issue because the irrigation schemes, and the farm units, are technically linked and served by one common water source. There is some evidence that mandates to operate headworks and main conveyance canals are not clearly assigned and that both state agencies and farmer organizations, are overstretched in terms of their capacities. For instance, a canal built by Oromia Irrigation Development Authority was never operational and never delivered water according to the respondents from a WUA in the Ediget Filagot region. The diversion gate of the dam built on Kessem River, a tributary to the Awash River, between 2004 and 2009, has been defective for the last two years. A WUA in the Boset-Fentale irrigation scheme cannot afford to remove sediments from the unlined canals with the consequence that they do not get water. It has yet to be clarified who is responsible for repairs and who bears the associated costs. Further, rules that clearly assign responsibilities of provisioning irrigation services are lacking. This is leading to coordination failures over infrastructure repairs and maintenance.

Interviews conducted and reports reviewed indicate that irrigation schemes – large and small – are not properly managed, and that on-farm water-use efficiency is low. Irrigation schemes lack on- and off-farm drainage systems resulting in the salinization of soils and waterlogged areas, and here and then farmers experience total crop failure and low yields while others abandon their plots (Ministry of Agriculture – Natural Resources Management Directorate, 2011).

Conflicts over water exist between large farms (Wonji Shoa Sugar Estate and Metahara Sugar Estate), and between large and small farms. The Awash Basin Authority's capacities and power to monitor and enforce water allocation plans, including recommendations on on-farm water use, are limited. As a result, large farms withdraw as much water as they can and when they need it. Large farms take advantage of their position and impair water delivery downstream. Further, statements of several interviewees reveal that only large and powerful water consumers (state sugar farms and private agro-industrial fruit plantations) have access to and profit from conflict mediation meetings organized by the Basin Authority whereas smallholders and WUAs are not invited. As one interviewee put it "basically the muscle, the power goes to the main irrigation schemes" (Srigiri et al., 2021). Water is also in short supply in the pastoralist areas because commercial farms hinder access to waterholes and wells (Pastoralist Forum Ethiopia, International Institute of Rural Reconstruction, and The Development Fund, 2010). Under the current institutional framework, which prioritizes economic growth through commercial irrigated agriculture, poor and marginalized groups lose access to land and water resources, which are crucial to ensure their food, income and water securities, thus jeopardizing the social dimension of sustainability.

Indivisibility, Leaving No One Behind, Inclusiveness and Participation – an Assessment

The performance of a governance system is evaluated on the basis of the evaluative criteria applied to both outcomes and processes of the action situations selected, which cover key governance functions of resource abstraction, design of rules, coordination, and knowledge generation in the various action arenas. We adopt three principles of the 2030 Agenda which are applicable to the current study to serve as the evaluative criteria for the outcomes and processes of governance. Below, we provide a brief discussion of the outcomes and processes in light of these three principles: i) indivisibility; ii) leaving no one behind; and iii) inclusiveness.

Indivisibility

The outcomes of land and water governance emerge from the action situations at the operational-choice level, which contribute to the achievement or non-achievement of SDGs and their targets. In its pursuit of agriculture-led economic growth, Ethiopia encourages the expansion of irrigated areas by allocating land obtained through expropriation of pastoralists, and by redistributing state land. While the increase in commercial, large-scale irrigated agriculture leads to economic growth (SDG 8), it alienates pastoralists from rangelands and waterholes, which are crucial resources for pastoral livelihoods although their rights are constitutionally recognized but have not yet been certified. This endangers their food security (SDG 2) and further pushes them into extreme poverty (SDG 1). Furthermore, untreated wastewater from settlements and industries and reverse flows from irrigated commercial agriculture, which are regarded as the engines of economic growth (SDG 8), directly affect water quality and thereby the wetland ecosystems in the basin (SDG 15). A lack of capacities for measuring and monitoring water use, assigning water permits without specified limits, ineffective on-farm irrigation practices, the use of poor-quality irrigation water, and the lack of drainage facilities (particularly at the Tendaho State farm) lead to soil salinization in some areas. Therefore, in the lowlands of the Awash River Basin, we see major trade-offs in outcomes of action situations at operational-choice level which could be attributed to three key interrelated factors: First, the trade-offs result from a lack of or the ineffective coordination and inclusive decision-making in the process of allocating land and water resources. Second, the human and financial capacities of water management units, environmental authorities and *woreda*- and *kebele*-level authorities in effectively managing hydraulic systems, and implementing social and environmental safeguards. Thirdly, rules and practices of expropriating land for public purposes together with inadequate compensation regulations are leading to a trade-off between the economic and social dimensions of sustainable development.

Leaving no one behind

As evident from the key trade-offs between social and economic goals, the marginalized groups in the basin, the pastoral communities and small farmers, are alienated from the pastures and water sources through expropriations (revocation of their rights of use) for expansion of large irrigation schemes. Deprived of their livelihood resources, their food security is endangered and their vulnerability to poverty increased. Although their rights are constitutionally recognized, the key factor driving the marginalization of pastoralists is their lack of legal titles to communal land. Without titles, pastoralists also become ineligible to receive compensation in the event of expropriation, forcing them to intrusions into the National Parks, further leading to trade-offs between social and environmental protection goals. The GTP II and the policies to expand irrigated agriculture could lead to further land expropriations. When the government offers pastoralists, alternative livelihoods such as irrigated agriculture, or also other income opportunities, then these would have to be accompanied by broad support measures.

Inclusiveness and participatory decision-making

Building the integrated visions and strategies needed to support sustainability transformation requires a broad societal consensus that can only be achieved through the engagement and inclusion of all major societal groups. In addition, compliance with the 2030 Agenda's principle of "leaving no one behind" will also require engaging with the full diversity of societal stakeholders, including representatives of marginalized and minority groups (United Nations, 2018). However, as of now, both sub-national levels of government and non-government stakeholders have only been weakly involved in the process of elaborating national plans (GTPs) which are binding for the development of regional development policies and plans and the alignment of GTPs with the SDGs. Currently, neither regional governments nor non-state stakeholders have been able to provide input into the initial phases of the elaboration of the GTP and are only consulted on it at a very late stage. Ethiopian approach to civil society participation in SDG implementation has thus far focused on raising awareness and creating endorsement among stakeholders for the weakly SDG-aligned GTP II, rather than engaging them in genuinely participatory ways.

Conclusion

The institutions and governance mechanisms for water and land were not found to comply with the core principles of the 2030 Agenda, namely, indivisibility, LNOB, and inclusiveness. The research identifies key trade-offs among SDGs in the Lower Awash River Basin of Ethiopia, thereby contributing to the *first research question* of this book. The strategy of the Ethiopian government to achieve sustained economic growth (SDG 8; Target 8.1) through large-scale commercial agriculture is leading to a loss of customary tenure rights of pastoralists to grazing land (Target 1.4.2) exacerbating their poverty and food insecurity (SDG 2; Target 2.1). Further, return flows or drainage water from commercial farms and untreated wastewater from industries and urban settlements into the river system negatively affect the wetland ecosystems (SDG 15; Target 15.1).

Contributing to the *second research question* of the book, the research provides an assessment of existing governance mechanisms for land and water in the basin. With regards to land governance, multiple informal and formal rules co-exist and contradict each other. Poor definition of rules for expropriation and compensation, complexity of recognition and certification process of pastoral lands, and the high demand for fertile land in the basin render the existing regulations ineffective in protecting the customary rights of the pastoralists and gaining compensation payments. In order to counter the power asymmetries, efforts towards recognition and certification of the use-rights of pastoralists to communal lands need to be strengthened. Further, income diversification and livelihoods support for pastoralists and smallholder farmers needs to be designed along with developing appropriate capacities for awarding compensation and resettlement upon expropriation.

With regards to the governance of water, the existing mechanisms are resulting in inefficiencies in distribution and use of water resources, besides being biased towards large and powerful commercial farmers. Further, lack of coordination between different levels of government (national and regional states) and the River Basin Council and River Basin Authority are leading to the ineffectiveness in delivery of key governance functions, namely, the implementation of water allocation plans, and issuing and monitoring water permits. Environmental Impact Assessment and provision of clearance certificates is a key mechanism available to balance the social, environmental, and economic goals, which is also found to be ineffective.

Fostering coordination to avoid trade-offs among different goals requires strengthening institutional, human and financial capacities at different levels. At the operational choice level, capacities of *woredas* and *kebeles* need to be strengthened so that they can provide extension services to farmers to improve on-farm water use efficiency. Further, develop operational guidelines with clear definitions and mandates for water infrastructure operation and maintenance are required for WUAs to fully take over the responsibility.

In response to the *fifth research question* of the book, the study identifies the main political-institutional reasons underlying the ineffectiveness of existing governance mechanisms to minimize the trade-offs mentioned above. The authoritarian regime type in Ethiopia under which the economic and social development policies and strategies are planned and executed, does not allow for effective participation of sub-national and non-governmental actors in the decision-making. Therefore, the existing policies and strategies do not match the interests and capacities of local authorities and communities. Furthermore, the ineffectiveness of governance mechanisms also results from inadequate capacities at different levels of the state for certifying customary land rights, measuring and monitor water permits and use, supporting farm-level water-use efficiencies and formulating supporting laws for environmental impact assessment. Therefore, towards creating enabling political institutional conditions, capacities of river basin authorities need to strengthened, besides strengthening institutions to perform Environmental and Social Impact Assessments. Whereas, at the constitutional-choice level, clear definition of what constitutes a "public purpose" in the legislation for expropriation might be helpful in safeguarding the communal land rights of pastoralists.

Notes

1 Production and provision refer to goods or services. For details, refer McGinnis (2011, p. 57)
2 Key actors interviewed include civil servants from federal ministries, commissions and authorities as well as regional sectoral bureaus; employees of donor agencies and inter-governmental organizations; staff of state sugar farms; members of a water user association and members of the academic community.
3 EIA became a statutory instrument in 2002 (Federal Democratic Republic of Ethiopia, 2002).
4 EPA renamed as Federal Commission for the Environment, Forest and Climate Change. Whether the role of the Federal Commission has been upgraded with the reshuffling of the ministerial landscape in late 2018 remains to be seen.
5 The Proclamation on Irrigation Water Users' Associations No. 841 of 2014, which replaced the Cooperative Societies Proclamation, does not specify whether traditional institutions are recognized.

References

Abdulahi, M. (2007). The Legal Status of the Communal Land Holding System in Ethiopia: The Case of Pastoral Communities. *International Journal on Minority and Group Rights*, 14, 85–125. doi:10.1163/138548707x181566.

Ambaye, D.W. (2015). *Land rights in Ethiopia*, (pp. 27–92). Springer.

Aneme, G.A. (2015). *Ethiopia: Legal and judicial plurality and the incorporation of traditional dispute resolution mechanisms within the state justice system*. In T. Röder, F. Schuppert, and R. Wolfrum (Eds), (pp. 80–99). Palgrave Macmillan.

Anteneh, A. (2007). *The assessment of rural land valuation and compensation practices in Ethiopia: Final Main Report*. Ethiopia Land Tenure and Administration Program.

Ayalew, M.M. (2013). Ethiopia must rethink finance to achieve 2025 vision. Retrieved from: www.iied.org/ethiopia-must-rethink-finance-achieve-2025-vision.

Ayane, A. D. (2014). *Land grab in Ethiopia: Reflection on the regulatory legal framework*. Lambert Academic Publishing.

Breuer, A., Janetschek, H., and Malerba, D. (2019). Translating Sustainable Development Goal (SDG) interdependencies into policy advice. *Sustainability*, 11(7). doi:10.3390/su11072092.

BTI. (2021). *The BTI Transformation Index*. Retrieved from: https://bti-project.org/en/?&d=D&cb=00000.

Damtie, M. and Bayou, M. (2008). *Overview of environmental impact assessment in Ethiopia*. MELCA Mahiber.

Danyo, A. et al. (2017). *Realizing Ethiopia's green transformation: Country environmental analysis, environment and natural resources global practice*. World Bank.

Federal Democratic Republic of Ethiopia. (2002). *Environmental Impact Assessment Proclamation (No. 299 of 2002).* Addis Ababa.

Federal Democratic Republic of Ethiopia. (2017a). *Ethiopia 2017 Voluntary National Review on SDGs: Government commitments, national ownership and performance trends.* Addis Ababa.

Federal Democratic Republic of Ethiopia. (2017b). Pollution Status of Akaki River and its Contamination Effect on Surrounding Environment and Agricultural products: Technical Report. Retrieved from: http://repository.iifphc.org/bitstream/handle/123456789/467/Akaki%20River%20Technical%20Report_2017%20.pdf?sequence=1.

Haddis, Z. (2018). *Pastoral Communities Receive 2.7 million Hectares of Land in Ethiopia.* Retrieved from: https://www.land-links.org/2018/09/pastoral-communities-receive-2-7-million-hectares-of-land-in-ethiopia/.

Haile, T.G. (2015). Comparative Analysis for the SDPRP, PASDEP and GTP of the FDR of Ethiopia. *Global Journal of Business, Economics and Management*, 5, 13–25. doi:10.18844/gjbem.v5i1.61.

Haileslassie, A., Hagos, F., Agide, Z., Tesema, E., Hoekstra, D., and Langan, S.J. (2016). *Institutions for irrigation water management in Ethiopia: Assessing diversity and service delivery.* Retrieved from: https://www.researchgate.net/publication/305868823_Institutions_for_irrigation_water_management_in_Ethiopia_Assessing_diversity_and_service_delivery.

Hailu, R., Tolossa, D., and Alemu, G. (2017). Water institutions in the Awash basin of Ethiopia: the discrepancies between rhetoric and realities. *International journal of river basin management*, 16, 107–121. doi:10.1080/15715124.2017.1387126.

International Council for Science and International Social Science Council. (2015). Review of the sustainable development goals: The science perspective. Paris: International Council for Science.

Janka, D.G. (2012). Environmental Impact Assessment in Ethiopia: Laws and Practices. (PhD thesis). University of Alabama Libraries, Tuscaloosa Alabama. Retrieved from http://acumen.lib.ua.edu/content/u0015/0000001/0001119/u0015_0000001_0001119.pdf.

Kassa, G. (2001). Resource Conflicts Among the Afar of North-East Ethiopia. In. African Pastoralism, ed. Salih, MA Mohamed, 145-171.

Kloos, H. (1982). Development, Drought, and Famine in the Awash Valley of Ethiopia. *African Studies Review, 25(21)*. doi:10.2307/524399.

LANDac. (2018). Food security and land governance fact sheet Ethiopia: Land governance for equitable and sustainable development. Retrieved from Utrecht http://www.landgovernance.org/assets/20160608-Factsheet-Ethiopia.pdf.

Le Blanc, D. (2015). Towards Integration at Last? The Sustainable Development Goals as a Network of Targets. *Sustainable Development*, 23, 176–187. doi:10.1002/sd.1582.

McGinnis, M.D. (2011). Networks of Adjacent Action Situations in Polycentric Governance. *Policy Studies Journal*, 39, 51–78. doi:10.1111/j.1541-0072.2010.00396.x.

Ministry of Agriculture – Natural Resources Management Directorate. (2011). *Small-Scale Irrigation Situation Analysis And Capacity Needs Assessment (A Tripartite Cooperation Between Germany, Israel and Ethiopia).* Addis Ababa.

Napier, A. and Desta, S. (2011). *Review of Pastoral Rangeland Enclosures in Ethiopia: PLI Policy Project Review.* Retrieved from: https://fic.tufts.edu/assets/Tufts-Range-Enclosure-Review-PLI.pdf.

Nega, B., Adenew, B., and Gebre Sellasie, S. (2003). Current land policy issues in Ethiopia. *Land Reform, Land Settlement, and Cooperatives*, 11(3), 103–124.

North, D.C. (1993). *Five propositions about institutional change (Economic History Working Paper 9309001).* Washington University.

Ostrom, E. (1990). *Governing the Commons.* Cambridge University Press.

Ostrom, E. (2005). *Understanding Institutional Diversity*: Princeton University Press.

Páez-Curtidor, N., Keilmann-Gondhalekar, D., and Drewes, J.E. (2021). Application of the Water–Energy–Food Nexus Approach to the Climate-Resilient Water Safety Plan of Leh Town, India. *Sustainability*, 13(19), 10550. Retrieved from: www.mdpi.com/2071-1050/13/19/10550.

Pahl-Wostl, C. (2019). The role of governance modes and meta-governance in the transformation towards sustainable water governance. *Environmental Science & Policy*, 91, 6–16. doi:10.1016/j.envsci.2018.10.008.

Pastoralist Forum Ethiopia, International Institute of Rural Reconstruction, and The Development Fund. (2010). *Pastoralism and Land: Land Tenure, Administration and Use in Pastoral Areas of Ethiopia*. Retrieved from: www.utviklingsfondet.no/files/uf/documents/Rapporter/Pastoralism__Land.pdf.

Rahmato, D. (2011). *Land to Investors: Large-scale Land Transfers in Ethiopia*. Addis Ababa: African Books Collective.

Reda, K.T. (2014). Formal and informal land tenure systems in Afar region, Ethiopia: perceptions, attitudes and implications for land use disputes. *African Journal on Conflict Resolution*, 14, 41–62.

Srigiri, S.R., Breuer, A., and Scheumann, W. (2021). Mechanisms for governing the water-land-food nexus in the lower Awash River Basin, Ethiopia: Ensuring policy coherence in the implementation of the 2030 Agenda: Discussion Paper.

Srigiri, S.R. and Dombrowsky, I. (2022). Analysing the Water-Energy-Food Nexus from a polycentric governance perspective: Conceptual and methodological framework. *Frontiers in Environmental Science. Environmental Economics and Management.* doi:10.3389/fenvs.2022.725116/abstract.

Tamrat, I. (2010, 26–27 April 2010). *Governance of large scale agricultural investments in Africa: The case of Ethiopia*. Paper presented at the World Bank Conference on Land Policy and Administration, April, Washington, DC.

Tura, H.A. (2018). Land rights and land grabbing in Oromia, Ethiopia. *Land Use Policy*, 70, 247–255. doi:10.1016/j.landusepol.2017.10.024.

United Nations. (2018). Working Together: Integration, institutions and the Sustainable Development Goals, World Public Sector Report 2018. Retrieved from: https://publicadministration.un.org/publications/content/PDFs/World%20Public%20Sector%20Report2018.pdf.

UN System Staff College. (2021). The 2030 Agenda for sustainable development. Retrieved from: www.unssc.org/sites/unssc.org/files/2030_agenda_for_sustainable_development_kcsd_primer_en.pdf#:~:text=The%20%202030%20%20Agenda%20%20embodies%20,comprehensive%20%20effort%20%20towards%20%20sustainable%20development.

van Koppen, B. and Schreiner, B. (2018). *A hybrid approach to decolonize formal water law in Africa* (Vol. 173): International Water Management Institute.

Weitz, N., Strambo, C., Kemp-Benedict, E., and Nilsson, M. (2017). Closing the governance gaps in the water-energy-food nexus: Insights from integrative governance. *Global Environmental Change*, 45, 165–173. doi:10.1016/j.gloenvcha.2017.06.006.

Woldegiorgis, S.B. (2018). *Formally Recognizing Pastoral Community Land Rights in Ethiopia*. Retrieved from: https://www.land-links.org/2018/03/formally-recognizing-pastoral-community-land-rights-in-ethiopia/.

World Bank. (2012). Options for Strengthening Land Administration, FDRE, Report No. 61631 – ET. Retrieved from: https://reliefweb.int/sites/reliefweb.int/files/resources/Full_Report_3314.pdf.

World Bank. (2016). *Mexico Public Expenditure Review*.

World Bank and Department of International Development, United Kingdom. (2019). *Poverty and Vulnerability in the Ethiopian Lowlands: Building a More Resilient Future*.

8

IMPLEMENTING THE 2030 AGENDA UNDER RESOURCE SCARCITY

The Case of WEF Nexus Governance in Azraq/Jordan

Ines Dombrowsky

Introduction

The 2030 Agenda with its 17 Sustainable Development Goals (SDGs) sets an ambitious framework for sustainable development. The Agenda's Resolution alludes to five core principles that underpin this ambition: (i) universality; (ii) leave no one behind; (iii) indivisibility and interconnectedness of SDGs; (iv) inclusive governance; and (v) multi-stakeholder partnerships (UN System Staff College, n.d.). An integrated implementation of the 2030 Agenda requires the realization of synergies and the mitigation of trade-offs between economic, social, and ecological dimensions of sustainable development, considering these principles. However, this can be challenging in water scarce areas, where strong interlinkages can be observed between SDG 2 (Zero hunger), SDG 6 (Clean water and sanitation), SDG 7 (Clean energy), SDG 8 (Decent work and economic growth), and SDG 15 (Life on land). These interlinkages have previously been acknowledged by the concept of the water–energy–food (WEF) nexus, which stresses that the achievement of water, energy and food securities strongly relies on interrelated natural resource systems, such as water resources, soils, and biodiversity (Leck, Conway, Bradshaw, and Rees, 2015; Müller, Janetschek, and Weigelt, 2015). However, governing the WEF nexus interlinkages can be demanding and conditions for effective coordination are not yet well understood (Pahl-Wostl, 2017; Weitz, Strambo, Kemp-Benedict, and Nilsson, 2017). As this volume's introduction suggests (Breuer, forthcoming), there is an expectation that governing SDG interlinkages and the WEF nexus can be promoted through dedicated governance mechanisms and policy-mixes that are based on synergetic combinations of hierarchical, market and network-based governance modes (Pahl-Wostl, 2019; Srigiri and Dombrowsky, 2022). Furthermore, as outlined in this volume's introduction, the political-institutional preconditions, such as regime type and state capacity, are likely to influence integrated SDG implementation and WEF nexus governance. With respect

DOI: 10.4324/9781003254683-8

to regime type, it remains an open question whether democracies' relative advantage over autocracies can be affirmed in terms of their ecological sustainability performance (Wurster, 2013). In terms of state capacity, it is assumed that a strong state needs to perform its role "through flexible structures that mobilize non-state actors in innovation networks and connect local and international spaces" (O'Riain, 2004). Against this background, this chapter seeks to contribute to all four overriding research questions of this book, namely: (i) What do we know about the most important interlinkages between the SDGs)? (ii) What governance mechanisms and policy processes are needed to address power and capacity asymmetries between stakeholders form different sectors and levels? (iii) Which policy mixes to increase policy coherence in the implementation of different interlinked SDGs have proven to be effective, socially just and acceptable? (iv) What political-institutional preconditions are conducive to the establishment of effective governance mechanisms to manage SDG interactions?

This chapter pursues these questions through a case study of the Azraq area in Jordan's eastern desert region, where different users compete for shrinking groundwater resources. Jordan, a constitutional monarchy, has been characterized as a moderate autocracy (BTI, 2020) with moderate state capacity (World Bank, 2016). It is also one of the most water-scarce countries in the world (Ministry of Water and Irrigation, 2015). The Azraq groundwater basin currently provides six per cent of national domestic water supply (Ministry of Water and Irrigation, 2018, unpublished)[1]. Considerable amounts of groundwater are furthermore abstracted for farming. The Azraq wetland, a waterfowl habitat protected under the Ramsar Convention, is artificially recharged by groundwater. In consequence, groundwater resources in Azraq are over-abstracted by at least 260 per cent (Ministry of Water and Irrigation, 2018, unpublished). Hence, the case presents a typical WEF nexus situation with competition on water for domestic use, food production as well as ecological needs, while abstraction also relies on energy for pumping. This chapter analyses this WEF nexus situation through the lens of pertinent SDGs, asking for trade-offs and synergies. It then evaluates the government's policy-mix and governance mechanisms and their implementation, asking how governance modes are combined and to what extent groundwater governance in Azraq complies with the 2030 Agenda's core principles. In doing so, it also reflects on the role of the political-institutional context.

The chapter is based on a comprehensive review of primary and secondary literature as well as 67 semi-structured interviews between February and April 2020. Respondent information was anonymized and interviews numbered I1–67 in the order they were conducted. The second section introduces the 2030 Agenda's core principles as evaluative framework for the integrated implementation of the 2030 Agenda. The third section then analyzes SDG interlinkages in the given case. The fourth section presents the policy-mix and governance mechanisms to deal with these interlinkages and evaluates them against the backdrop of the core principles, while considering the political-institutional context. The fifth section concludes the chapter.

The Agenda's Core Principles as Evaluative Framework

Since 2015 an increasing body of literature has emerged to identify inter-linkages among SDGs. Nevertheless, to date, no common definition or operationalization of "integrated implementation" of the 2030 Agenda exists (Breuer, Janetschek, and Malerba, 2019). This chapter uses the 2030 Agenda's core principles of leaving no one behind and inclusiveness, interconnectedness and indivisibility as well as multi-stakeholder partnerships as basis for evaluating the degree to which WEF nexus governance in Jordan contributes towards integration implementation of the SDGs.

The Agenda's core principles run throughout the 2030 Agenda Resolution (Gesellschaft für Internationale Zusammenarbeit, n.d.; UN System Staff College, n.d.). The declaration's introduction mentions the principle of "leaving no one behind" (LNOB) early on:

> "[W]e pledge that no one will be left behind. Recognising that the dignity of the human person is fundamental, we wish to see the Goals and targets met for all nations and peoples and for all segments of society."
>
> *(UN General Assembly, 2015, p. 3)*

LNOB is furthermore grounded in United Nations' normative standards of equality and non-discrimination. These entail: (i) Equality in opportunities and outcomes: both formal equality (procedural rights that protect equality, such as equality before law) and substantive equality (no inequality caused by structural disadvantages or different needs); (ii) Non-discrimination in multiple discriminations, including based on gender, age, ethnicity, disability and indigenous identity and intersecting forms of discrimination; and (iii) the fair treatment of all population groups in society and the fair distribution of costs (UN System Chief Executives Board for Coordination, 2017). Equality in opportunities is also underpinned by SDG 16 which calls for inclusive institutions and SDG 16.7 demanding inclusive decision-making at all levels. The principle of "inclusiveness" refers to equality and non-discrimination through equal access to and participation in decision-making (UN System Staff College, n.d.). We consider this principle covered through the criteria "equality in opportunities" and "non-discrimination" under LNOB.

Several paragraphs in the 2030 Agenda refer to "interconnectedness and indivisibility". The principle stresses the multiple interlinkages – both synergies and trade-offs – between SDGs (Breuer et al., 2019) as well as to their integrated nature: they can only be implemented as a whole. This implies that silo thinking needs to be avoided and various policy areas must be coordinated horizontally (Miola, Borchardt, Neher, and Buscaglia, 2019; Nilsson and Weitz, 2019). In line with this, SDG 17.14 seeks to enhance policy coherence for sustainable development (UN General Assembly, 2015). Furthermore, WEF nexus research has shown that horizontal coordination is not sufficient, but that interrelated policy areas also need to be coordinated vertically across multiple governance levels for sustainable outcomes (Pahl-Wostl, 2017; Weitz et al., 2017). Thus, we assess both horizontal and vertical coordination in policymaking.

The principle of "multi-stakeholder partnerships" features in the preamble's second paragraph: "All countries and all stakeholders, acting in collaborative partnership, will implement this plan" (UN General Assembly, 2015, p. 1). Furthermore, SDG 17.7 encourages "effective public, public-private and civil society partnerships" (UN General Assembly, 2015, p. 27). There is no universally agreed definition of multi-stakeholder partnerships. Some authors refer to cooperative relations between governments, business enterprises and non-profit organizations to fulfil a political purpose (Lindner and Vaillancourt Rosenau, 2000; Treichel, Höh, Biermann, Conze, and HUMBOLDT-VIADRINA Governance Platform gGmbH, 2017). Furthermore, Pattberg and Widerberg (2016) define them as institutionalized cross-border interactions between public and private actors designed to provide collective goods. On this basis, we define multi-stakeholder partnerships as institutionalized interactions between the state, private sector and civil society. Table 8.1 summarizes our criteria for assessing the core principles.

TABLE 8.1 Criteria for operationalizing the 2030 Agenda's core principles

Core principle	*Criteria*
Leaving no one behind including inclusiveness	• Equality in opportunities and outcomes • Non-discrimination of individuals on the ground of gender, age, ethnicity, disability, indigenous identity or other individual characteristics • All population groups receive fair treatment in distributing costs, benefits, and opportunities
Interconnectedness and indivisibility	• Horizontal coordination in policy-making across sectors • Vertical coordination in policy-making across multiple governance levels
Multi-stakeholder partnerships	• Institutionalized interactions between the state, the private sector and civil society

Source: Author's own elaboration

Competition for Groundwater in Azraq in Light of the SDGs

The town of Azraq has about 14,000 inhabitants (Department of Statistics, 2010) and is located in the eastern desert region of Jordan (Figure 8.1). It developed on the fringes of a natural oasis and related wetland at the lowest point of the concave Azraq surface water basin. Government data estimate the safe yield of the Azraq groundwater basin, i.e. the amount that can be used annually without severe negative effects, at 24 million cubic metres per year (MCM/yr) (Ministry of Water and Irrigation, 2018, unpublished). In 2018 governmental wells abstracted 19.7 MCM/yr of water, most of which was conveyed outside the basin for domestic water supply. These wells also provided 0.65 MCM/yr to artificially maintain the remainder of the Azraq wetland as a site for migratory birds. According to government data, private wells abstracted 38.6 MCM/yr for agriculture and 2.9

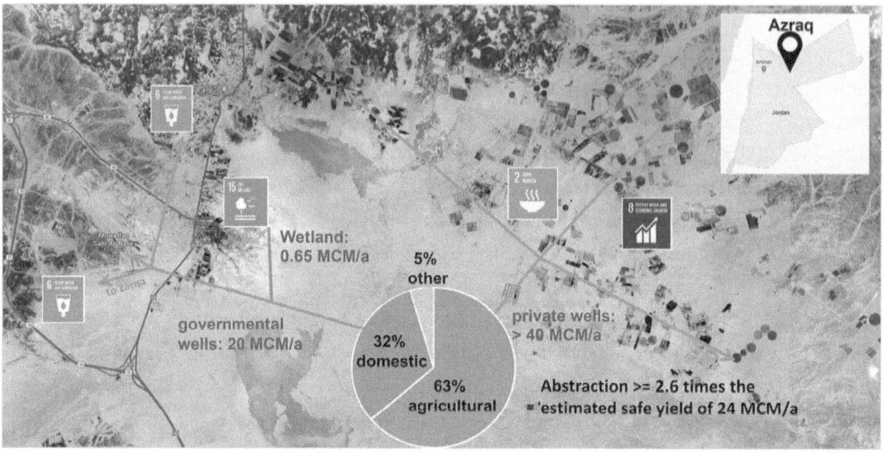

FIGURE 8.1 Competition for groundwater in Azraq
Source: Author based on freevectormaps.com and MWI (2020) data on water abstraction in 2018

MCM/yr for other purposes (Ministry of Water and Irrigation, 2018, unpublished). This implies that the estimated safe yield is over-abstracted by at least 260 per cent. Eighty per cent of the estimated safe yield is used for domestic water supply alone, while agriculture uses at least double the amount provided for domestic purposes. The over-abstraction results in quickly falling water tables and increasing salinization of the groundwater (Ministry of Water and Irrigation and Bundesanstalt für Geowissenschaften und Rohstoffe, 2019), rendering it unfit for various uses. Basically, Azraq is currently experiencing a classical tragedy of the commons (Hardin, 1968) and it is unlikely that the current use pattern can be maintained in the long run (I-05, I-41, I-63) (Al Naber and Molle, 2017b). As one interviewee succinctly mentioned: "Azraq means blue, but the future is brown" (I-41).

In view of the various SDGs and SDG targets related to this groundwater use situation, the following picture emerges:

At first glance there seems to be a direct competition between water use for safe drinking water (SDG 6.1), food security (SDG 2), economic growth (SDG 8) and life on land (SDG 15). However, a more detailed analysis shows a somewhat more differentiated picture.

First, while it could be assume that agricultural water use in Azraq contributes to SDG 2 (Zero hunger) and SDG 8 (Decent work and economic growth), it remains questionable to what extent this is the case. In general, Jordan imports about 80 per cent of its food (Hashemite Kingdom of Jordan, 2015), and agriculture only contributed 3–4 per cent to Jordan's GDP in 2013 (Ministry of Water and Irrigation, 2015). In the Azraq area, the farming community is highly heterogeneous. Many of the original inhabitants of Azraq town, namely, Druze, Chechens and Bedouins, have small or medium-size farms close to town. However, since the 1970s, investment farmers have settled in the desert region east of Azraq (Demilecamps and Sartawi,

2010) and often run large and very large farms. Owners of small- and medium size farms hardly make a living from their farms and might have other occupations besides farming, thus contributing to SDG 2 to a limited extent (I-05, I-50, I-58). Large-scale investment farmers partly export their products, in which case that they might contribute to SDG 8 and, but not necessarily, to SDG 2 in Jordan. Still, some farmers sell their produce at the central market in Amman, which can be seen as a contribution to SDG 2.1 (access to food). Some farmers use innovative technologies, thus contributing to agricultural productivity (SDG 2.3) or fulfilling SDG 8 with respect to economic growth (SDG 8.1) and entrepreneurship (SDG 8.3) as well as increasing efficiency of water use (SDG 6.4), but this remains the exception. Given that most farm employees are migrant workers from Egypt, agriculture in Azraq also does not contribute much to SDG 8.5 on (national) job creation. Nevertheless, agriculture is understood as important factor for cultural identity (I-02, I-57, I-66), and most farmers wish to continue farming in the area.

Second, Jordan's water authorities mainly pursue SDG 6.1 (ensuring access to safe drinking water) by conveying water to urban conglomerates. Jordan's water strategy prioritizes water supply for domestic use followed by uses with the highest economic returns (Ministry of Water and Irrigation, 2015), which is rarely the case for agricultural water use. Hence, SDG 6.1 is a national priority (I-02, I-17) that is pursued at the expense of food security (SDG 2.1) and economic growth (SDG 8.1) through agriculture.

Third, together, domestic and agricultural water use have contributed to the destruction of large parts of the Azraq oasis and wetland. With its rich flora and fauna and variety of aquatic and terrestrial species, the Azraq wetland used to be of high ecological value (Al Naber, 2016; Ramsar Bureau, 1990) and has formerly been described as a "glimpse of heaven" (Mountfort, 1965). The local community traditionally used the wetland for their livelihoods (I-59), engaging in fishing and hunting. In 1977 the wetland and the adjacent mudflat was declared a Ramsar site for migratory birds on the African-Eurasian flyway (Royal Society for the Conservation of Nature, n.d.). As groundwater pumping for domestic and agricultural purposes increased, the four natural springs feeding the Azraq oasis dried out, reaching an alarming rate state in the early 1990s (Ramsar Bureau, 1990) and the decision was made to recharge the remainder of the wetland artificially. Hence, domestic and agricultural water uses have destroyed large parts of the wetland's biodiversity and ecosystem services (SDG 15.1) and conflict even with SDG 6.4 (sustainable withdrawals) and SDG 6.6 (conservation of wetlands) within SDG 6.

SDG 7 (affordable and clean energy) also plays a role. Pumping water is energy intensive. Many Azraqi farmers aim to increase their use of renewable energy (SDG 7.1) because operational costs are lower than diesel or electricity from the national grid. However, this could lead to higher water abstraction than diesel pumps and hence be a trade-off with SDG 6.4 (I-17). In principle, given the vastness of the territory and high radiation levels, solar farming could constitute an alternative to agriculture in Azraq area. However, upfront investment costs for switching from agriculture to solar farming are expected to be high and a pilot project found several regulatory obstacles (Al-Saidi, 2018). Interviewees argued that a switch to solar farming disrupts family traditions (I-18) and lacks the emotional aspect of cultivating trees (I-21, I-29). Still, solar farming holds the potential for a synergetic solution, reducing pressures on water (SDG 6.4), while maintaining livelihoods and income.

WEF Nexus Governance Against the Agenda's Core Principles

Various policies as well as the overarching political-institutional setting affect the actions of the different water abstractors. This section assesses the policy-mix and governance mechanisms regulating access to the resource in relation to the 2030 Agenda principles LNOB, interconnectedness and indivisibility of the SDGs and multi-stakeholder partnerships, taking the political-institutional setting into account, as such addressing research question 2 to 4.

Leaving no one behind

Given water scarcity in Jordan, the Jordanian Water Strategy prioritizes groundwater for domestic needs, followed by uses with the highest economic returns (Ministry of Water and Irrigation, 2015). On this basis, the government abstracts groundwater in Azraq for domestic supply elsewhere. We therefore infer that the Ministry of Water and Irrigation (MWI) sees providing safe drinking water to the entire population including vulnerable groups and refugees as a strategy of complying with the LNOB commitment.

With respect to groundwater use for agriculture, there is in consequence no ambition to provide access for all. Instead, a set of policies seeks to restrict access to groundwater and to incentivize its efficient use. A drilling ban in place since 1992 intends to prevent the drilling of new wells (Al Naber and Molle, 2017b). However, the ban only started to be enforced since 2013/2014 and about 1,300 illegal wells have been closed (I-11). In doing so, the authorities tend to focus on unproductive wells and smallholder farmers (Hussein, 2018). Given that the closing of illegal wells and the seizing of drilling rigs caused significant conflict, enforcement of the ban has slowed down (I-63).

Access to groundwater is furthermore governed through a system of water licenses and permits and related water abstraction fees. Four types of wells can be distinguished in Azraq, depending on the legal status of the land on which the well is located: wells with license, wells with permits, registered illegal wells and unregistered illegal wells (Al Naber and Molle, 2017a, 2017b). Before 1992 landowners who legally owned the land according to government regulations were able to receive a well license. Owners of licensed wells get a certain amount ("block") of water for free and once this quantity is exhausted, the fee is relatively low (Table 8.2). There is no upper abstraction limit associated with the well license, but abstraction depends on the productivity of the well and cost factors (pumping costs and abstraction fees). However, in the Azraq area, many farmers occupy and farm land without legally owning it (Al Naber and Molle, 2016). Some farmers claim land based

TABLE 8.2 Prices for agricultural water abstraction based on well type according to Groundwater By-Law 85–2002 and amendments in 2004 and 2014

Well with license		*Well with permit*		*Illegal, but registered well*	
Quantity [1,000 m^3]	Price (2004) [JOD]	Quantity [1,000 m^3]	Price (2004) [JOD]	Quantity [1,000 m^3]	Price (2014) [JOD]
< 150*	0.000	< 50	0.000	< 10	0.150
150–200	0.005	50–100	0.020	10–30	0.250
> 200	0.060	> 100	0.060	> 30	0.500

* 75,000 m^3 for deepened/substitution wells

Source: Author's own elaboration based on Al Naber and Molle (2017b)

on tribal laws, others just occupy unused land. Water permits were provided for illegal wells drilled before 2005, allowing holders to abstract groundwater based on the permit until their land titles are legalized (Al Naber and Molle, 2017b). Wells with permits only exist in Azraq where the number of illegal wells is particularly high, owing to the shallow nature of the aquifer. Permitted wells have a smaller free block than those with licenses and pay higher fees for subsequent blocks of water. Finally, registered illegal wells are wells drilled after 2005 that are known to the authorities. Owners of registered illegal wells have no free block and pay relatively high fees beginning with the first cubic meter. Finally, owners of unknown illegal wells do not pay anything. Farm size and well type are not correlated, there are both small and large farms with registered illegal wells.

Enforcement of water abstraction fees has long been an issue. In the past, illegal well owners prevented authorities from accessing their land and controlling wells and reading well meters, partly by force (Al Naber and Molle, 2017a). Therefore, in 2014 authorities introduced remote sensing to estimate water demands and bill illegal registered wells on this basis. Our interviews showed that farmers are dismayed about the differentiation of wells and the introduction of remote sensing for billing water abstraction. Farmers regard the unequal treatment of wells and the special prices for permitted wells that only exist in Azraq as well as the prices for illegal wells as unjust and demand equal treatment (I-37, I-38, I-44). They deploy different strategies to reduce their water bills: Some switch to more efficient irrigation techniques to save water and money (I-9, I-21, I-42, I-61), some try to reduce their water costs by harvesting water and recharging the aquifer (I-35), some quit farming altogether (I-42, I-52). However, in particular influential farmers with illegal registered wells on large farms self-organize through social media and organize marches in the front of the MWI to negotiate rebates (I-20. I-30, I-40, I-42, I-47). In particular, powerful actors with personal connections (so called wasta) to authorities have been able to negotiate discounts of up to 70 per cent (Dombrowsky et al., 2022; Oberhauser, Hägele, and Dombrowsky, 2022). Wasta is a social norm of favourably treating member from the same tribe or family, deeply embedded in Jordan's society and political system (Barnett, Yandle, and Naufal, 2013; Brahms and Schmitt, 2017).

Hence, Jordanian water laws and policies primarily aim at increasing the sustainability of groundwater use through abstraction licenses and fees and a drilling ban and thus a combination of regulatory and economic instruments. This policy-mix works to a certain extent, but it also has significant limitations. First, there is no overall cap of allowable abstraction per license. Second, fees are generally rather low when considering shadow prices (I-05). Third, fees and the drilling ban are enforced inconsistently. The question is how this inconsistent enforcement relates to regime type and state capacity respectively. Arguably, the use of remote sensing for billing for water use can be seen as an expression of state capacity. However, informal institutions such as wasta and the lack of trust between national authorities and farmers hinders consistent enforcement. This seems to relate more to the underlying regime type and its associated social contract (Oberhauser et al., 2022).

Our research indicates that the governance access to groundwater for agriculture in Jordan does not comply with the principle of LNOB. First, in terms of inclusiveness, farmers complained that they had no say in the policy-making process and were informed only after a policy reform had been decided (I-21, I-22). Second, there is inequality among farmers in terms of their farm sizes and the status of their wells and thus in the distribution of benefits and costs. For sure, the current policy of differentiated fees according to well status implements a pragmatic first come first serve principle to avoid overuse, but it is debatable whether this can be considered as a fair. Third, farmers with registered illegal wells that have to pay relatively high water prices are treated unequally as often farmers with small farms are forced to pay their bills, while influential farmers with large farms are able to negotiate rebates. A similar observation can be made with respect to the closing of wells when mostly unproductive wells are closed.

Clearly, sustainable use of groundwater also means that there is simply not enough water to meet all the demands. The question is how equality in outcomes, i.e. reducing or avoiding inequality caused by structural disadvantage could be achieved under scarce natural resources. In that sense, it seems that LNOB and sustainable use can be at odds with each other. This raises the question whether the 2030 Agenda misses a core principle of "sustainable natural resource use".

Interconnectedness and indivisibility

Overall, Jordan as a constitutional monarchy is organized as a centralized state (BTI, 2020). Policies are generally made at national level and implemented in a top-down manner through line agencies. Governors in the governorates are appointed by the King and are mainly responsible for security, but not for sector policies (I-49, I-59). Municipalities report to and might express concerns to the Ministry of Local Administrations (I-23), but not to sector ministries. They also have no own development policy (I-23).

In the water sector, the perceived need to administer the allocation of scarce water resources at national level furthermore reinforces top-down governance (I-63). Staff at the government well field in Azraq that pumps water for domestic water supply only receive instructions from above, but do not have a say in the amount to be abstracted (I-41). Local level agencies report data to higher levels, but are not asked for their opinions (I-41). Modest steps towards decentralization in the water sector focus on water supply utilities at the governorate level (Gesellschaft für Internationale Zusammenarbeit, 2018). This implies that vertical coordination is based on a hierarchical, but not a cooperative governance mode. This approach might give the central government a feeling of control over scarce water resources. However, given that the relation between farmers and water authorities is characterized by a high level of mistrust and even hostility, the government is actually not able to control illegal water uses in a comprehensive manner.

In terms of horizontal coordination across policy sectors, interviewees confirmed that overall in Jordan strong silo thinking prevails and pointed to a mentality of postponing interaction with other sectors, of keeping prestigious projects for oneself or of feuding over responsibilities (I-02, I-06, I-60). In line with the WEF nexus perspective, the following paragraphs discuss coordination between the MWI and the Ministry of Agriculture (MoA) and between the MWI and the Ministry of Energy and Mineral Resources (MEMR). We furthermore address the role of the Jordanian national implementation mechanism for the 2030 Agenda.

One would expect close coordination between the MWI, which is responsible for supplying irrigation water to farm gates (in case of surface water supply), and the MoA which is, among other things, responsible for supporting an efficient use of irrigation water at farm level. However, coordination between the two ministries is reportedly weak (I-15, I-17, I-18). The MoA supplies estimates on crop water demand to the MWI on a regular basis (I-11, I-15). However, there is no intersectoral coordination body at working level (I-15, I-18). One consequence can be counterproductive policies in terms of groundwater abstraction by farmers. For instance, the MoA has in the past encouraged the planting olive trees in Azraq (I-42, I-51), which consume relatively large amounts of water, while providing little return (Badran et al., 2018). Hence, from a water perspective this was not a sensible strategy.

Coordination between the MWI and the MEMR shows a mixed picture. For instance, an energy policy that is potentially problematic from a water perspective is that farmers might get subsidies for solar energy (I-55) (Omari, 2014). Solar energy has relatively high investment and but low operational costs. While it helps to combat climate change, it can provide perverse incentives to increase water pumping, owing to low operational costs, once it is installed. More recently, coordination between the MWI and the MEMR has somewhat improved owing to the establishment of the so-called nexus committee (I-18, I-57). Given that the water sector uses 15 per cent of national electricity supply, the committee discusses options to save costs and energy, e.g. through energy storage and load shifting. At the time of the interviews it was still an open question whether the two ministries would come to an agreement (I-57).

As most other signatory states to the 2030 Agenda, Jordan has set up institutional mechanisms for the implementation of the SDGs, which are described in the Voluntary National Review that the country presented to the UN High-Level Political Forum on

Sustainable Development in 2017. This raises the question whether these mechanisms could support coordination across sectors and governance levels. The Ministry of Planning and International Cooperation (MoPIC) serves as a hub for implementing the 2030 Agenda. It is supposed to merge several sectoral Executive Development Programmes elaborated by individual ministries into a National Executive Development Programme (I-13). In addition, a Coordination Committee for the SDGs has been established, but according to our interviewees it has been more a battleground where the various ministries try to push through their own goals as much as possible and as of early 2020 it had not met since 2017 (I-13). Furthermore, some ministries, such as the MWI, have installed SDG focal points to coordinate reporting on particular SDGs to MoPIC (I-13, I-18). This is useful for reporting on SDG indicators, but does not serve as mechanism for cross-sectoral coordination. Each ministry seems to be working on what it views as "its" SDG (I-16). One interviewee expressed the opinion that the government has only adopted the terminology of SDG interlinkages to acquire funds (see also Breuer, Leiniger, Marlerba in this volume), but that it has not improved coordination (I-03). Hence, we did not receive indications that the 2030 Agenda and the national SDG mechanisms enhances inter-ministerial coordination.

Overall, there is much room for further considering the interconnectedness and indivisibility of SDGs related to the WEF nexus in Jordan. While some governance mechanisms for horizontal coordination exist they do not seem effective. This seems to be rooted in overall governance set-up of the monarchy, rather than this relates to a lack of state capacity only.

Multi-stakeholder partnerships

Multi-stakeholder partnerships as institutionalized interactions between the state, the private sector and civil society are not very prevalent in Jordanian politics and we could only identify few fora that bring all three actors groups together. An interviewee reported that the private sector hosts monthly breakfast meetings for the government, private sector, and non-governmental organizations (NGO) to discuss current affairs with respect to energy, water and environment (I-10). With respect to civil society, it is noteworthy that NGOs are often either close to the state or the private sector (I-10). In Azraq, currently, no multi-stakeholder partnership exists with respect to groundwater use and the WEF nexus.

However, in the period 2009 to 2014, the donor-supported Highlands Water Forum (HWF) brought together government official, experts, and farmers (including ten from Azraq) to develop a joint action plan to reduce groundwater use. The first phase of the HWF sought to establish communication channels and build trust between authorities and farmers, which gradually improved over time (I-14, I-17). In the beginning, stumbling blocks included excessive expectations, the aquifer's open-access character and the heterogeneity of the farming community (I-14). The plan foresaw a comprehensive set of activities grouped in four pillars: (i) legal and institutional framework conditions; (ii) on-farm water efficiency; (iii) alternative income opportunities; and (iv) community development (Azraq Groundwater Basin Committee, 2014). After the action plan was developed and several donors had agreed to establish a basket fund for its implementation, several factors impeded implementation (I-14). Against the background of the "Arab Spring" protest movements that overthrew several autocratic governments in the MENA region starting from 2011, the MWI leadership viewed participatory approaches that might encourage people to become politically active as a risk (I-14) (Zawahri, 2012). Furthermore, in view of the influx of 1.3 million Syrian refugees (Department of Statistics, 2016), their water supply became a priority. A new camp for some 36,000 refugees near Azraq that was supplied with local deep groundwater wells undermined the farmers' willingness to conserve water (I-67), even if the water reportedly stems from a deeper aquifer. In 2012 the MWI's new Secretary General was not convinced of "soft" approaches like the HWF (I-14, I-17, I-67). The MWI phased out the HWF without ever implementing the action plan (I-14, I-18, I-57, I-67). Lack of implementation negatively affected famer' relations with government and donors (I-14, I-25). In hindsight, farmers are disenchanted with the HWF: "Let me be frank. Sixty farmers meet with 10 officials in a hotel, the cost of the meeting reaches 20,000 dinars, and we met tens of times with no results" (I-25).

Given that many authors and practitioners view multi-stakeholder partnerships and coordination fora such the HWF as promising for an integrated implementation of the 2030 Agenda it is certainly unfortunate that trust has been lost in the case of the HWF. In judging the HWF's effectiveness, it should also be considered that the change of context conditions was quite significant. However, the example also shows that the trust of the Jordanian government in deliberative methods remains limited which, in turn, indicates that significant progress towards inclusive and participatory decision-making at all levels (SDG 16.7.) is unlikely in the context of autocratic rule.

Conclusion

This chapter explored the implementation of the 2030 Agenda in Jordan, using the case of competition for groundwater in Azraq as an example in which strong interlinkages between WEF nexus related SDGs exist. In view of the research question about synergies and trade-offs, the case exhibits traits of a tragedy of the commons and related to this strong conflicts among several SDGs. In particular agriculture (goals 2 and 8) and domestic water use (SDG 6.1 on access to drinking water) cause strong trade-offs with SDG 6.4 (sustainable water withdrawals), SDG 6.6 (conservation of wetlands) and SDG 15.1 (conservation of ecosystems), while only marginally contributing to goals 2 and 8. Actions related to SDG 7 (clean energy) might affect the situation positively or negatively.

In terms of the research question on the effectiveness of the policy-mix governing access to groundwater, Jordan pursues a stick and carrot approach, combining regulatory measures with economic incentives for water saving. The interviews showed that while farmers do not always care about the regulatory measures, they protest against what they perceive as high water abstraction fees for illegal wells enforced since 2014 through remote sensing. Hence, while recent policy reforms have had some effects on farmers, it would be important to enforce the policy-mix more consistently, reducing the negative influence of wasta. Still, given a high proportion of groundwater's safe yield is dedicated for domestic water supply elsewhere in Jordan, the problem will not be solved based on abstraction fees alone. The case also shows there can be limits towards implementing the Agenda's core principle of LNOB in view of natural resource scarcity. This notwithstanding, the Jordanian policy-making process could be more inclusive and it might be worthwhile to reconsider the fairness implications of the current differentiated system of water abstraction fees, meaning that LNOB could be strengthened.

With respect to the question on governance mechanisms for policy coordination, the case is characterized by a strong silo approach underpinned by centralized decision-making and hierarchical vertical coordination across governance levels. Related to this, cross sectoral policy coordination is also weak, in particular between the water and agriculture sector. The water-energy nexus committee is still relatively new and hence it is too early to assess its effectiveness. Some agricultural and energy policies provide perverse incentives in terms of water abstraction. A past attempt of fostering participatory groundwater management through the Highlands Water Forum was undermined by the Syrian refugee crisis and changes in ministry leadership, further reducing farmer's trust in the government. Hence, the 2030 Agenda core principles of LNOB including inclusiveness, interconnectedness, and indivisibility and multi-stakeholder platforms are hardly reflected by groundwater governance in Jordan. While there is no expectation that Jordan would adjust its groundwater policies, owing to the 2030 Agenda, the case study still reflects a divergence between the Agenda's aspirations and realities on the ground.

By presenting a case of a moderate autocracy with moderate state capacity, the case also provides insights into the research question how regime type and state capacity might affect the 2030 Agenda's implementation. The case study shows that Jordan's formal rules tend to be dominated by a hierarchical governance mode, also in an attempt to control the allocation of scarce resources. However, formal rules' effectiveness is partly undermined by the social norm of wasta, which is deeply embedded in the Jordanian society and system of ruling. In consequence, the Jordanian autocracy does not seem to be particularly conducive towards ecological sustainability. Related to this and a lack of trust, we also saw considerable limits in its capacity to mobilize non-state action for sustainable water use. While politically demanding, it still seems that a possible way forward would require initiating a broader societal discourse on fair and sustainable water use (Dombrowsky et al., 2022).

Acknowledgments

Research for this paper was carried out in the context of the 55th Postgraduate Training Programme of the German Institute of Development and Sustainabilityand contributes to the Bundesministerium für wirtschaftliche Zusammenarbeit und Entwicklung-funded research project "Growth, Environment, Inequality, Governance: Implementation of the 2030 Agenda". The author would not have been able to write this chapter without the excellent and manifold research contributions by the research team consisting of Lukas Behrenbeck, Thomas Bollwein, Mirjana

Koeder, Daniel Oberhauser, and Ronja Schamberger, her research team co-lead Ramona Hägele and her Jordanian research partners Majd Al Naber (West Asia-North Africa Institute), Marwan Al Raggad (Inter-Islamic Network on Water Resources Development and Management) and Elias Salameh (University of Jordan). These contributions highly appreciated! Further thanks goes to all interview partners and the local field guides, translators and transcribers. Anita Breuer and Srinivasa Srigiri provided helpful comments on the manuscript.

Note

1 M. Al Dwairi (personal communication, March 31, 2020). On March 31, 2020, Mohammed Al Dwairi, Assistant State Secretary at MMI, sent us an email with a table showing the unpublished official groundwater statistics of Jordan for the year 2018 including for the Azraq basin.

References

Al-Saidi, M. (2018). *Water-Energy-Food Nexus Assessment of Solar Energy Farming Interactions. The Azraq Case in Jordan with Insights from India.* Bonn and Eschborn: Nexus Regional Dialogue Programme, Gesellschaft für Internationale Zusammenarbeit.

Al Naber, M. (2016) Jordan-Azraq Basin Case Study. In *International Water Management Institute (IWMI) (Series Ed.)*: Vol. 12. *IWMI Project Report: Groundwater Governance in the Arab World* (pp. 1–107). Colombo: International Water Management Institute, USAID.

Al Naber, M. and Molle, F. (2016). The politics of accessing desert land in Jordan. *Land Use Policy*, 59, 492–503. doi:10.1016/j.landusepol.2016.09.026.

Al Naber, M. and Molle, F. (2017a). Controlling groundwater over abstraction: state policies vs. local practices in Jordan highlands. *Water Policy*, 19(4), 692–708. doi:10.2166/wp.2017.127.

Al Naber, M. and Molle, F. (2017b). Water and sand: Is groundwater-based farming in Jordan's desert sustainable? *Groundwater for Sustainable Development*, 5, 28–37. doi:10.1016/j.gsd.2017.03.005.

Azraq Groundwater Basin Committee. (2014). *Participation in Decision Making. Highland Water Forum - Azraq Groundwater Management Action Plan.*

Badran, I., Taimeh, A., Takruri, H., Abdullah, F., Bader, H., Khasawneh, M., and Zreigat, G. (2018). *Achieving SDG 2 in Jordan: A National Strategic Review.* Amman: The Higher Council for Science and Technology (The National Center for Research and Development).

Barnett, A., Yandle, B., and Naufal, G. S. (2013) Regulation, trust, and cronyism in Middle Eastern societies: The simple economics of "wasta". In *Institute for the Study of Labor (IZA) (Series Ed.)*: Vol. 7201. IZA Discussion Papers. Bonn: Institute for the Study of Labor.

Brahms, K.S. and Schmitt, M. (2017). "It's All About Something We Call Wasta": A Motivated Moralization Approach to Favoritism in the Jordanian Workplace. *Social Justice Research*, 30(2), 145–170. doi:10.1007/s11211-017-0285-2.

Breuer, A. (forthcoming). Introduction/conceptual framework. In A.Breuer, D.Malerba, P. Balasubramanian, and S.R. Srigiri (Eds), *Governing the Interlinkages between the SDGs: Approaches, Opportunities and Challenges* .Taylor & Francis.

Breuer, A., Janetschek, H., and Malerba, D. (2019). Translating Sustainable Development Goal (SDG) Interdependencies into Policy Advice. *Sustainability*, 11(7). doi:10.3390/su11072092.

BTI. (2020). *Country Report Jordan.* Retrieved from: www.bti-project.org/content/en/downloads/reports/country_report_2020_JOR.pdf.

Department of Statistics. (2010). *Report of Poverty Status in Jordan. Based on the Household Expenditures & Income Survey data 2008.* Retrieved from: www.dos.gov.jo/dos_hom e_a/main/Analasis_Reports/poverty_rep/Jordan_%20Poverty_%20Jordan_eng_2008.

Department of Statistics. (2016). *Results of the General Population and Housing Census (2015).* Retrieved from http://dosweb.dos.gov.jo/censuses/population_housing/census2015.

Dombrowsky, I. et al. (2022). *Natural Resource Governance in Light of the 2030 Agenda. The Case of Competition for Groundwater in Azraq, Jordan.* IDOS Studies 106, German Institute of Development and Sustainability (IDOS), Bonn, doi:10.23661/is106.2022.

Gesellschaft für Internationale Zusammenarbeit. (2018). *40 Years of German-Jordanian Cooperation in the Water Sec*tor. Eschborn.

Gesellschaft für Internationale Zusammenarbeit. (n.d.). The 2030*Agenda: Five Principles for Implementation*. Retrieved from: www.giz.de/en/aboutgiz/40669.html.

Hardin, G. (1968). The tragedy of the commons. *Science*, 162, 1243–1248.

Hussein, H. (2018). Tomatoes, tribes, bananas, and businessmen: An analysis of the shadow state and of the politics of water in Jordan. *Environmental Science and Policy*, 84, 170–176. doi:10.1016/j.envsci.2018.03.018.

Leck, H., Conway, D., Bradshaw, M., and Rees, J. (2015). Tracing the Water–Energy–Food Nexus: Description, Theory and Practice. *Geography Compass*, 9, 445–460. doi:10.1111/gec3.12222.

Lindner, S.D. and Vaillancourt Rosenau, P. (2000). Mapping the Terrain of the Public-Private Policy Partnership. In P. Vaillancourt Rosenau (Ed.), *Public-Private Policy Partnerships* (pp. 1–18). Cambridge, MA: MIT Press.

Miola, A., Borchardt, S., Neher, F., and Buscaglia, D. (2019). *Interlinkages and policy coherence for the Sustainable Development Goals implementation: An operational method to identify trade-offs and co-benefits in a systemic way*. Luxembourg: European Union.

Mountfort, G. (1965). *Portrait of a desert: the story of an expedition to Jordan*. Boston, MA: Houghton Mifflin Company.

Müller, A., Janetschek, H., and Weigelt, J. (2015). Towards a governance heuristic for sustainable development. *Current opinion in environmental sustainability*, 15, 49–56. doi:10.1016/j.cosust.2015.08.007.

Ministry of Water and Irrigation. (2015). *National Water Strategy: 2016–2025.*

Ministry of Water and Irrigation. (2018, unpublished). *Groundwater Statistics 2018.* M. Al Dwairi, Assistant State Secretary at MMI (personal communication, March 31, 2020).

Ministry of Water and Irrigation and Bundesanstalt für Geowissenschaftern und Rohstoffe. (2019). *Groundwater Resource Assessment of Jordan (2017).*

Nilsson, M. and Weitz, N. (2019). Governing Trade-Offs and Building Coherence in Policy-Making for the 2030 Agenda. *Politics and Governance*, 7(4), 254. doi:10.17645/pag.v7i4.2229.

O'Riain, S. (2004). *The Politics of High Tech Growth. Developmental Network States in the Global Economy*. Cambridge: Cambridge University Press.

Oberhauser, D., Hägele, R., and Dombrowsky, I. (2022). Unravelling hidden factors explaining competition for and overuse of groundwater in Azraq, Jordan: Digging deeper into a network of action situations Sustainability Science. Published 7 April 2022. doi:10.1007/s11625-022-01135-w.

Omari, R. (2014, 28-1-2014). Project offers loans to help farmers go solar. *Jordan Times*. Retrieved from: www.jordantimes.com/news/local/project-offers-loans-help-farmers-go-solar.

Pahl-Wostl, C. (2017). Governance of the water-energy-food security nexus: A multi-level coordination challenge. *Environmental Science and Policy*, 92, 1–12. doi:10.1016/j.envsci.2017.07.017.

Pahl-Wostl, C. (2019). The role of governance modes and meta-governance in the transformation towards sustainable water governance. *Environmental Science and Policy*, 91, 6–16. doi:10.1016/j.envsci.2018.10.008.

Pattberg, P. and Widerberg, O. (2016). Transnational multistakeholder partnerships for sustainable development: Conditions for success. *Ambio*, 45, 42–51. doi:10.1007/s13280-015-0684-2.

Ramsar Bureau. (1990). *Ramsar Advisory Missions: Azraq Oasis, Jordan (17)*. Retrieved from: www.ramsar.org/sites/default/files/documents/library/ram17e_azraq_jordan_from_old_website_0.pdf.

Royal Society for the Conservation of Nature. (n.d.). Azraq Wetland Reserve. Retrieved from: www.rscn.org.jo/content/azraq-wetland-reserve-0.

Srigiri, S.R. and Dombrowsky, I. (2022). Analysing the Water-Energy-Food Nexus from a polycentric governance perspective: Conceptual and methodological framework. *Frontiers in Environmental Science. Environmental Economics and Management*. doi:10.3389/fenvs.2022.725116/abstract.

Treichel, K., Höh, A., Biermann, S., Conze, P., and HUMBOLDT-VIADRINA Governance Platform GmbH. (2017). *Multi-stakeholder partnerships in the context of Agenda 2030. Practiced-based analysis of potential benefits, challenges and success factors*. Bonn and Eschborn: Partnerships 2030.

UN General Assembly. (2015). *Transforming our world: the 2030 Agenda for Sustainable Development (A/RES/70/1)*.

UN System Chief Executives Board for Coordination. (2017). Leaving No One Behind: Equality and Non-discrimination at the Heart of Sustainable Development. Retrieved from https://unsceb.org/sites/default/files/CEB%20equality%20framework-A4-web-rev3.pdf.

UN System Staff College. (n.d.). *The 2030 Agenda for Sustainable Development*. Retrieved from: www.unssc.org/sites/unssc.org/files/2030_agenda_for_sustainable_development_kcsd_primer_en.pdf.

Weitz, N., Strambo, C., Kemp-Benedict, E., and Nilsson, M. (2017). Closing the governance gaps in the water-energy-food nexus: Insights from integrative environmental governance. *Global Environmental Change*, 45, 165–173. doi:10.1016/j.gloenvcha.2017.06.006.

World Bank. (2016). Worldwide Governance Indicators. Government Effectiveness. Retrieved from: https://datacatalog.worldbank.org/dataset/worldwide-governance-indicators.

Wurster, S. (2013). Comparing ecological sustainability in autocracies and democracies. *Contemporary Politics*, 19(1), 76–93. doi:10.1080/13569775.2013.773204.

Zawahri, N. (2012). Popular protests and the governance of scarce fresh water in Jordan. *The Arab World Geogrpaher*, 15(4), 267–301.

9

TO GROW OR NOT TO GROW?

Revisiting Economic Growth as a Sustainable Development Goal in Light of the Degrowth Debate

Daniele Malerba and Yannick Oswald

Introduction

The Sustainable Development Goals (SDGs) merge the three dimensions of sustainable development: environmental, social and economic (Breuer, Janetschek, and Malerba, 2019). This constitutes a significant departure from the Millennium Development Goals (MDGs), where the environmental dimension was of secondary importance. Another novelty of the SDGs is the inclusion of an inequality goal (SDG 10), recognizing inequality as potentially the greatest social challenge of the 21st century. In addition, the SDGs apply to all countries and not just to the so-called "developing" nations as the MDGs did; all countries are called to reach socio-economic objectives respecting environmental boundaries.

The three pillars of sustainable development have also given birth to frameworks that relate them and assess progress in relation to the SDGs. For instance, there are now SDG indices (Diaz-Sarachaga, Jato-Espino, and Castro-Fresno, 2018) where all goals have the same importance. Conversely, there are approaches that explicitly distinguish between means and ends. For example, the "wedding cake approach" or doughnut economics (Raworth, 2017) state how resource factors are means to achieve societal goals, which in turn have to be achieved within environmental boundaries

Considering this literature, the chapter focuses on the specific goal of economic growth (SDG 8) and the controversy surrounding it. Is economic growth a desirable goal in itself or only a means to an end (social outcomes and even green transformations)? Is it even feasible to include growth as a goal while also staying within planetary boundaries? Some previous literature reached the conclusion that it is not. On the other hand, most climate modelling scenarios assume continuous economic growth. Here we test various growth and redistribution scenarios, and subsequently feed the results to the Kaya decomposition.

DOI: 10.4324/9781003254683-9

Furthermore, we implicitly discuss whether degrowth should be integrated into the SDG agenda or if it even turns out be necessary. Our approach is to remain open about the outcome and assess either goal "pure growth" or "pure degrowth" critically and keeping in mind the agrowth perspective as a third option (van den Bergh, 2017). It is important to clarify that degrowth is not simply about reducing GDP. Degrowth is a planned reduction of energy and resource throughput designed to bring the economy back into balance with the living world in a way that reduces inequality and improves human well-being (Kallis et al., 2018). Degrowth proponents also have argued that absolute decoupling of GDP from emissions can be achieved starting by replacing fossil fuels with renewable energy (Lamb et al., 2021) but that it cannot be done quickly enough to respect carbon budgets for 1.5°C and 2°C if the economy continues to grow at usual rates (Haberl et al., 2020). Finally, we also want to make clear that we focus on CO_2 emissions, but that there are other environmental pressures that need to be considered.

In summary, this chapter explicitly addresses the first research question that this book deals with, namely the interlinkages between the SDGs. We mainly look at the relationship between SDG8 (economic growth) and environmental goals.

SDG indicators: Growth and Inequality

We start by exploring the SDGs of interest: economic growth, inequality, environmental footprints and poverty. The SDGs comprise the following target on economic growth (target 8.1): "*Sustain per capita economic growth in accordance with national circumstances and, in particular, at least 7 per cent gross domestic product growth per annum in the least developed countries*". In comparison, the MDGs had no explicit goal for economic growth. They had one target (1.b) on "*achieving full and productive employment and decent work for all*" (which is still part of the SDGs).

The SDGs have a goal dedicated to inequality (SDG 10). This is a departure from the MDGs, where we can just find the goal of halving extreme poverty, which is a considerably more minimalistic compared to reducing inequality. If the focus is on extreme absolute poverty (the current US$1.90 a day), poverty reduction could be achieved through economic growth with conventional trickle-down economics as it would be necessary that the lowest income households minimally benefit from the increasing economic pie. Reducing inequality as it is, on the other hand, questions the entire current distribution and economic order, with potentially large implications for international and national political and economic dynamics. In addition, MDG 1 focused on the poorest, leaving out a large share of the world's population in moderate (or vulnerable to) poverty, including all poor in high-income countries not considered by the MDGs.

Despite being a highly relevant addition, this inequality target needs to be interpreted carefully (Fukuda-Parr and McNeill, 2019). SDG target 10.1 states that "*By 2030, progressively achieve and sustain income growth of the bottom 40 per cent of the population at a rate higher than the national average*". Two main considerations can be made. First, the focus is on the bottom of the distribution rather than the top; its focus then is very different from the one pursued by some of the popular indices of inequality, such as the Gini Index. In addition, target 10.1 does not comply with some of the most important axioms such as the transfer axiom, which states that inequality should increase if there is a transfer from a poorer to a richer household. In fact, if a household in the 60th percentile gives income to a household in the 90th percentile (richest 10%), that would have no consequence to the inequality measure of the SDGs (as average income would grow at the same rate and the growth of the bottom 40% would also be unaffected). It also contradicts using the Gini in indicator 10.4.2 to measure the redistributive impact of fiscal policy. The second problem with SDG 10.1 (and specifically indicator 10.1.1) chosen for inequality is that it is clearly related to a growth agenda. In fact, it frames the decrease in inequality in terms of different growth rates rather than explicit redistribution of income and wealth. This can be problematic especially for high-income countries, where GDP levels are high already and significantly impact environmental outcomes. One proposed solution is to use absolute rather than relative inequality measures. Relative inequality indicators measure the differences in income in proportional terms; for example, if an individual A has income of US $100 and individual B income of US$150, individual B would be defined as having 50% more income than individual A. Relative inequality remains unchanged if all income grows at the same rate. On the other hand, absolute inequality, considers the absolute differences in income. The previous example would be framed in terms of individual A having US$50 (rather than 50%) more than individual B, and absolute inequality would increase if all incomes grow at the same rate. If all incomes grow by 10%, the absolute difference between A and B would be of 55 instead of 50, while the relative difference would stay the same. (Relative) inequality goals can also be found elsewhere. For example, poverty target SDG 1.2 is basically an inequality goal in the context of high-income countries,[1] where national poverty lines are relative lines, measured as a share (such as 50%) of the median income: therefore, the decrease of national poverty by 50% stated in SDG target 1.2 is an inequality reduction target.[2]

Relative inequality indicators have been by far the most widely used in empirical economic analysis, but, based on economic theory and empirical evidence, it is far from clear that we should favour relative over absolute notions of inequality.[3] The evidence suggests that many people do perceive absolute differences in incomes as being an important aspect of inequality (Amiel, Creedy, and Hurn, 1999). These insufficiencies in the specifications of SDG 10 and 8 have been pointed out before. For example, it has been critiqued that the goals are only about relative inequality but not about the absolute gap between the richest and poorest (Hickel, 2018) and just emphasize the necessity of higher growth rates for LDCs. Critically, the goals do not state any (permanent or temporary) income caps for rich nations. This is an obvious blind spot in the SDG framework as income caps for already rich regions naturally would limit international inequality and, according to established models, also resource use. There are a number of important points made about ensuring fair international legislation but it remains unclear which institutions and laws in particular require revision.

The only quantified SDG measure to reduce inequalities therefore remains a characterization of growth in poor segments (regions) happening faster than in richer ones. Why are national and international redistribution not proposed on the agenda more explicitly? One reason might be that in public discourse "taking away from the wealthy" is predominantly seen as negative and not righteous because a common neoliberal narrative is that wealthy people are wealthy because they earned it. Evidence suggests reality is more complex including structural path dependencies and privileges. For instance, we know that intra-generational income group mobility is low and that "where" you are born (i.e. in what country and even what neighbourhood) has immense influence on later success (Plewis and Bartley, 2014).

In sum, the inequality targets seem to be aligned with the economic growth ones (SDG 8) but diametrically opposed to any degrowth notion, as summarized in Table 9.1. There is no reference to caps on consumption or GDP levels and no reference to the degrowth aligned idea of "sufficiency". This is also reflected on other environmental SDGs. For example, SDG 12 on sustainable consumption, has no stated targets but rather recommendations and a focus on relative indicators (intensity/efficiency rather than volume). Similarly, indicators in SDG 8 on material consumption and material footprints state relative goals but no absolute value: they focus on intensity rather than caps. Therefore, the SDGs as a whole portray a vision of continuous economic growth and consumption with no indicator and target limiting consumption indefinitely. As a consequence, it seems that the wedding cake and doughnut economics approaches are not pursued by the SDGs. In addition, this is contrary to the Paris goals, which states limits to emissions that are not reflected here. One reason could be the significant reliance on negative emission technology in most climate scenarios.

Data and Methods

Redistribution for reducing inequalities and growth dependence

Since economic growth so far comes with large ecological costs (Wiedmann, Lenzen, Keyßer, and Steinberger, 2020), and redistribution lessens the need for growth, redistribution has the potential to mitigate trade-offs between ecological

TABLE 9.1 SDG approach vs. degrowth approach

Example issues	*SDG/Mainstream approach*	*Degrowth approach*
Growth	Explicit standalone goal	Non desirable in itself
Inequality reduction	Goal, but considering relative inequality	Goal, sufficientarian (absolute inequality)
Poverty	Growth	Redistribution

Source: Author's own elaboration.

stability and social welfare. Redistribution as a lever for development has been partly investigated before in (Oswald, Steinberger, Ivanova, and Millward-Hopkins, 2021), although the focus was on the relationship between inequality and energy demand structure. It concluded that redistribution from the top 1% income earners (measured in GDP per capita) can alleviate extreme poverty. It however did not approach the question of how much more economic growth is required in order to ensure high living standards and high well-being around the world. Such question has been addressed by Roser (2021), who concluded that substantial global growth of at least five times is necessary, even in a "minimum" scenario which is specified as all countries reaching the daily consumption expenditure level of Denmark (~$ 55/day/capita). It remains open why spending levels in Denmark constitute the "minimum" scenario. Reasons might include the outstanding performance of Denmark in social and health indicators such as life expectancy and trust among citizens. In principle other countries can serve as development models. For instance, Costa Rica or Czech Republic also score high on social and health metrics but exhibit considerably less consumption expenditure per capita. Therefore, here we test the sensitivity of the need for economic growth based on choosing specific countries as a benchmark.

We implement a simple static scenario model based on disposable income data from the World Bank and its Poverty calculation (Povcal) unit as well as population forecasts from the United Nations. This is similar to what Roser has done but we conduct a sensitivity analysis dependent on the income target and go beyond his results by considering redistributive implications, degrowth possibilities for high-income nations and environmental impacts. Our model relies on household income data for the year 2017 and assumes this to be representative of current conditions and uses medium population projections for the year 2030. We choose population in 2030 to underline the connection to the Sustainable Development Goal agenda whose time horizon ends then. This analysis is easy to reproduce and already illustrates the main point that the sensitivity to income targets is high. In reality, population growth rates interact with income growth rates and thus realistic dynamic trajectories of growth necessity are far more complex to model than what we attempt here. We just set a specific global income towards which all countries converge. For instance, if we specify $50/day/capita as necessary to live well, then all countries with a daily household income below scale up to this level and all countries above that threshold scale down to it.

Two alternative candidates for desirable income levels other than Denmark (mean daily income $56) are the Czech Republic ($31) and Costa Rica ($24). Costa Rica is occasionally brought forward in the literature as a role model of non-western alternatives in economic development with very low environmental footprint per capita yet high life expectancy (Hickel, 2019b). Czech Republic is brought forward as European alternative where GDP per capita is also relatively high but household disposable income much lower.

The left part of Figure 9.1 illustrates the linear relationship between the total global growth necessary (expressed as factor of the current size of the economy (2017)) and the aspired global daily income in 2030 (x-axis). This relationship takes redistribution already into account. A value of 1 means that the economy would

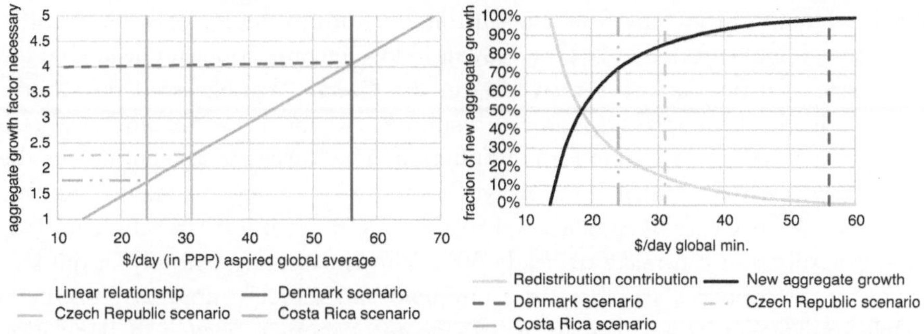

FIGURE 9.1 Linear relationship between aspired global average income and growth factor necessary (left) and non-linear relationship between aspired global average income and potential redistribution contribution (right)

Source: Author's own elaboration

Note: A full colour version of this figure is available open access under: https://www.routledge.com/Governing-the-Interlinkages-between-the-SDGs-Approaches-Opportunities/Breuer-Malerba-Srigiri-Balasubramanian/p/book/9781032184654

only need to be exactly as large as it is now. A value below 1 would mean that the economy can shrink, above 1 that it has to grow. If we want all countries converge to the household income level of Denmark the world requires an economy four times as big as today or ~300% growth (corroborating Roser's conclusion but not with the same numeric result, owing to different assumptions on population size: we calculate in terms of population in the year 2030, Roser in terms of projected population in the year 2100). Given environmental constraints and that ecological impact is tightly coupled to aggregate growth this might prove disastrous (which we elaborate on further down in the Kaya analysis). If we lower the aspiration to Czech Republic's income level, the world economy still requires around more than 100% growth, or in other words more than doubling the size of the economy. In the Costa Rica scenario less than doubling the economy is required. In sum, we affirm that if the world is to catch up to Northern or Central European income levels on average, then large economic growth is necessary. The Costa Rica case demonstrates that there are scenarios thinkable in which the need for economic growth is less than doubling while still reducing poverty.

Another perspective is thinking about how much of the new growth could be covered by moving income from rich nations to poor nations – or in other words international redistribution of income. In reality, that would most likely not mean physically moving income (or rather income generating wealth/capital stocks) but downscaling it in one place and upscaling it in another. While the relationship between the overall growth factor and the daily income level specified is linear, the trade-off between redistribution and "new" growth is highly non-linear.

In the right side of Figure 9.1 the trade-off is illustrated. The x-axis displays the income target; the y-axis displays the fraction of the total additional growth required that is either covered by "truly new" growth or in contrast by "redistribution" of income from high-income nations to poorer ones. Redistribution only delivers around 2% of the growth in low-income countries needed in the Denmark scenario but more than 15% in the Czech Republic scenario. This is because in the Denmark scenario few countries are above the income target, so there is little income available for redistribution. The lower the chosen income the more sensitive the redistribution-growth trade-off. Relaxing the desired income constraint from Czech Republic level to Costa Rica level increases the contribution redistribution can make to nearly 30%. This is because, obviously in this scenario, more countries are above the threshold and there is more to redistribute while simultaneously less growth in low-income nations is needed. If population remains constant at 2017 levels, redistribution could contribute even much more at lower income targets. For instance, at 2017 population levels, redistribution can make up nearly 60% of all new growth required in the Costa Rica scenario. This is important to understand because it illustrates that a substantial fraction of the necessary expansion of the economy comes from population growth rather than from increasing the affluence of people. However demographic projections are relatively robust so constant population would be too unrealistic to consider.

So far we have exclusively studied the necessity of global growth but the growth requirements per country also vary with the scenario. The per country growth requirements are highly diverse and this is one central point often made by degrowth scholars. Degrowth in high-income nations could provide leeway for growth in low-income ones. Many countries in the Global South still require a lot of growth independent of the scenario chosen. For instance, South Sudan nearly requires an economy 40 times as big as today to catch up with Denmark and still 15 times as big as today in order to catch up with Costa Rica. If we consider income reductions in high income nations as degrowth (for simplicity) then we can compare the global potential for degrowth with the global need for growth. There are many more countries in the world that likely require substantial growth than there are countries that have potential for degrowth. This does not imply however that growth is more important than degrowth. On the contrary, the conclusion is that a few countries can give vast room for additional growth just by degrowing their economies. A critical limitation to this conclusion is that degrowth is usually not understood as income reduction in the first place by degrowth scholars but as a likely reduction in economic output accompanying deliberate reductions in resource use (Keyßer and Lenzen, 2021).

Figure 9.2 illustrates the growth (or degrowth) factor needed in the Denmark and the Costa Rica scenario. It highlights the factor requirements for South Sudan, Rwanda, Luxembourg, and the United States in the Denmark scenario but also that the number of the countries that could implement degrowth as a goal (the degrowth zone) substantially increases in the Costa Rica scenario.

Yet another caveat is that analysing income alone is almost meaningless. Countries differ drastically across other dimensions, even countries at the same income level often exhibit high variation across indicators. Therefore, aspiring to the high-income level of Denmark is not per se a guarantee for quality of life and aspiring to the much lower income level of Czech Republic or Costa Rica not necessarily bad. In Table 9.2 we

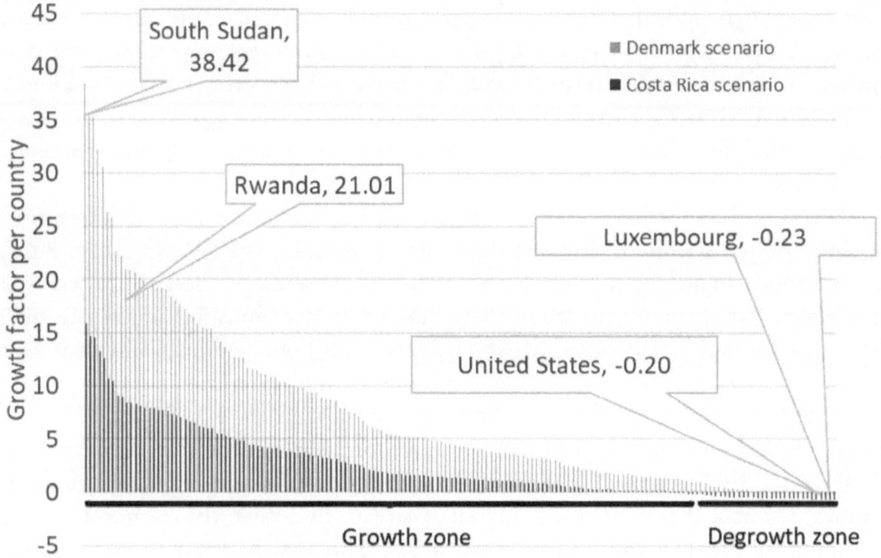

FIGURE 9.2 Per country (de)growth factors
Source: Author's own elaboration

TABLE 9.2 Key indicators across countries

	Denmark	*Czech Republic*	*Costa Rica*
GDP per capita, PPP, 2019	58,000	41,000	20,300
Gini index	0.28	0.25	0.48
Life expectancy	81	79	80
Carbon emissions tonpc (territorial)	6	10	2
World Happiness report (self-reported, 0 to 10)	7.6	6.9	7.1
Homicide rate (per 100,000)	1.0	0.6	11.3
Covid fatality rate per 1million (22/12/2021)	533	3,305	1,422
Disposable household income $/day/capita	56	31	24

Source: Author's own elaboration.

depict key characteristics of the countries chosen for the scenarios above. This data provides an intuition for what life looks like in those countries beyond income and enables more holistic reasoning. We do not propose that only Denmark, Costa Rica or Czech Republic are worth emulating but that any role-model for economic development should be evaluated across a range of indicators.

For example, it is noticeable that all of the here used countries represent a substantial improvement in life expectancy compared with the current global average, despite their drastic income gap. Even Costa Rica with a disposable income less than half of Denmark has nearly the same life expectancy and subjective life evaluation. It is difficult to judge how much such marginal improvements should be worth to society but considering that huge economic growth might cause planetary breakdown which in turn could reverse much progress made, any growth proposal needs careful consideration and should not be propagated without highlighting the link between output (measured by GDP) and environmental degradation.

Denmark does perform well on average across all dimensions. Its territorial greenhouse gas emissions dropped drastically in recent years (World Bank, 2021) yet are much higher than Costa Rica's or the world average. Costa Rica seems at first glance like a country to aspire to because of its low emissions and outstanding health performance. However, it exhibits very high homicide rates and very high inequality, characteristics shared with many other Latin American countries (World Bank, 2021). The question is if there are countries that exhibit similarly low emissions and good health without the drawbacks? Or can the social indicators in Costa Rica be improved without the environmental trade-off? Unfortunately studies evaluating this question in detail so far find that this is not the case (Fanning, O'Neill, Hickel, and Roux, 2021). The Czech Republic has remarkably an even lower homicide rate than Denmark but did perform very poorly during the Covid-19 crisis with the highest death rates in the entire world. This could represent momentary political mismanagement or hint at deeper problems in governance.

All in all, this analysis demonstrates that economic growth needed to satisfy basic needs and achieve many of the SDGs are not necessarily larger than a factor of 2.5, but also that it is very unlikely that there is no growth at all needed. The need for growth is highly dependent on what world and what level of affluence we aspire to and varies for countries individually. Therefore, the growth vs. degrowth debate might welcome the "agrowth" perspective (van den Bergh, 2017) because it does not view growth necessarily as good or bad and is more flexible in adopting to the development needs of individual countries.

Is the suggested economic growth compatible with emission targets? Using the Kaya Identity

This section links the projected economic growth for distinct aspired income levels to emission reduction targets, therefore exploring the implication between economic and environmental SDGs. We do this by exploring the Kaya identity (González-Torres, Pérez-Lombard, Coronel, and Maestre, 2021). The Kaya identity, represented in equation (1), decomposes CO2 emissions as the product of population (P), GDP per capita, and the CO_2 intensity of GDP.[4]

$$CO_2 = P * \left(\frac{GDP}{P}\right) * \left(\frac{CO_2}{GDP}\right) = P * \left(\frac{GDP}{P}\right) * \left(\frac{E}{GDP}\right) * \left(\frac{CO_2}{E}\right) \quad (1)$$

As outlined in equation (1), the carbon intensity of GDP can be further decomposed in the product of the energy (E) intensity of GDP and the emission intensity of energy $\left(\left(\frac{CO_2}{GDP}\right) = \left(\frac{E}{GDP}\right) * \left(\frac{CO_2}{E}\right)\right)$. As we are interested in dynamics, Equation (1) can be re-written in terms of changes: proportional changes (%Δ) in CO_2 emissions can be decomposed into population growth, GDP per capita growth, changes in the energy intensity of GDP and changes in the emission intensity of energy, such as:

$$\%\Delta CO_2 = \%\Delta P + \%\Delta\left(\frac{GDP}{P}\right) + \%\Delta\left(\frac{CO_2}{GDP}\right) + interaction\ tems \quad (2)$$

Where $\%\Delta\left(\frac{CO_2}{GDP}\right) = \%\Delta\left(\frac{E}{GDP}\right) + \%\Delta\left(\frac{CO_2}{E}\right)$. By using the Kaya equation, we aim to provide an intuition of whether the estimated economic growth target can be realistically achieved while decreasing emissions, and what are the requirements for the decreases in the energy intensity of GDP and the CO_2 intensity of energy.

Global Kaya identity

Figure 9.3 below shows the trends in global energy, GDP, population and CO_2 (panel (A)) and their changes (panel (B)) between 1990 and 2019, using data from the International Energy Agency (IEA) on energy, emissions, GDP and population. There has been an increase in emissions despite an improvement in the energy/GDP ratio. This means

(A)

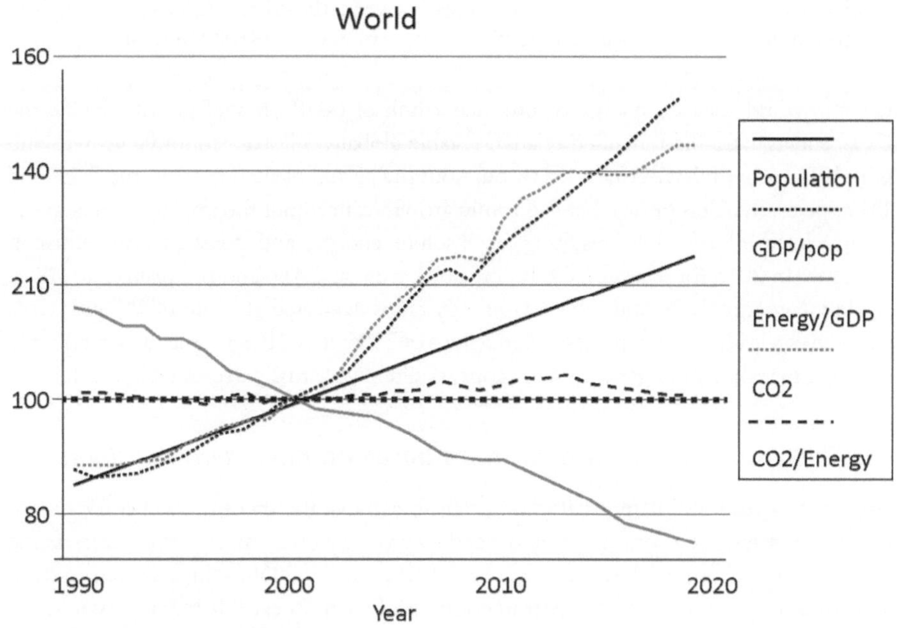

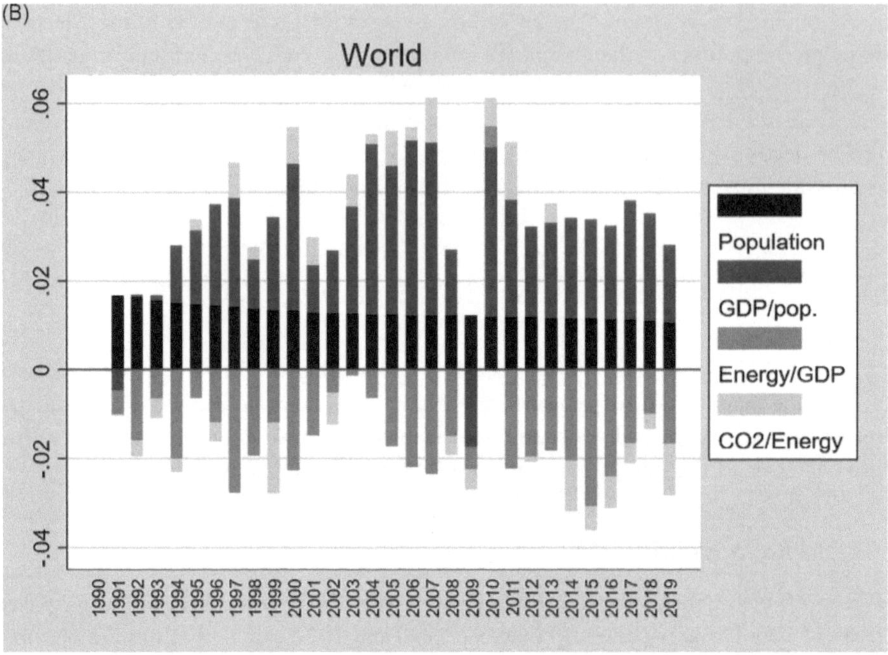

FIGURE 9.3 Trends in Global Kaya identity components. Figure 9.3A represents levels (2000 = 100); Figure 9.3B shows annual proportional changes

Source: Author's own elaboration

Note: A full colour version of this figure is available open access under: https://www.routledge.com/Governing-the-Interlinkages-between-the-SDGs-Approaches-Opportunities/Breuer-Malerba-Srigiri-Balasubramanian/p/book/9781032184654

that the world uses less energy to produce a unit of GDP on average. In parallel, the CO_2/energy ratio has remained relatively stable globally. Global economic growth and population growth have continued throughout the period of analysis at around 2.2% and 1% respectively. This means that economic growth cancels out the improvements in the energy intensity of GDP, resulting in absolute energy, and consequently, emission increases. In 2018 for example GDP per capita grew at 2.4% (and population at 1.1%), while the energy/GDP and the CO_2/energy ratios decreased at a rate of 1% and 0.3%, respectively, leading to an increase of emissions of 2.2%. If GDP per capita was steady (still allowing for population growth), emissions would have actually decreased by 0.2%.

Feasibility of estimated growth rates based on Kaya decomposition

An earlier section estimated that if we want to achieve the income level of Denmark for all countries, the economy would need to grow by 300% and be four times as big as today. Let us assume that the aim is to achieve this by 2050, which corresponds to an economic growth of 4% per year in compound terms (see Table 9.3; this would rise to 10% if the deadline would be 2030, the final year of the SDGs); this is also the

TABLE 9.3 Annual (compound) changes in the Kaya factors, by scenario

	Variable	*Scenarios*		
		Costa Rica	*Czech Republic*	*Denmark*
a	Emission reduction target	-7.6%	-7.6%	-7.6%
b	Population growth	c.a. 1%	c.a. 1%	c.a. 1%
c	Compound annual growth rate (2030 target)*	4.3%	6.4%	11.3%
d	Compound annual growth rate (2050 target)*	1.7%	2.4%	4.3%
e	Initial total world GDP (trillions US$)	121.80	121.80	121.80
f	Final total world GDP (trillions US$) (2030 target)	212.00	273.84	494.67
g	Growth factor necessary (= f/e)	1.74	2.25	4.06
h	Necessary reduction CO2/GDP (= a-b) (2030 target)	-11.9%	-14%	-18.9%
i	Necessary reduction CO2/GDP (= a-b) (2050 target)	-9.3%	-10%	-11.9%

*it merges the first two terms of the Kaya equation, population growth and GDP per capita.
Source: Author's own elaboration

upper bound of growth rates assumed in the Shared Socio-economic Pathways and climate model scenarios (Dellink, Chateau, Lanzi, and Magné, 2017). In addition, population growth of around 1% per year (following population projections) is already included in the total GDP growth estimate (that then merges the first two terms of the Kaya equation, namely population growth and GDP per capita growth), and, most importantly, that the needed compound emission reductions rate is 8% per year (UN Environment Programme, 2019). Following equation (2), this means that the overall CO_2/GDP ratio has to decrease by 12% per year (19% with a target for 2030); this can be decomposed in the sum of changes in the Energy/GDP and CO_2/energy ratios.

How many times has this occurred globally? In short, never. Years in which global emissions have decreased are rare. For example, this happened in 2009, owing to the economic crisis, where GDP per capita decreased by 1.7% while CO_2 emissions decreased by 1.5%. In 2020 the reduction as a result of the pandemic was larger, of around 5.2% (International Energy Agency, 2022), but there was already a rebound in 2021 (6% increase). There were also smaller reductions in earlier years, such as in 1992. Therefore, apart from 2020, no year witnessed emission reductions close to the ones needed to achieve the Paris goals.

Since 1990 changes in the carbon intensity of GDP have gone in the right direction: the ratio has decreased in all years apart from two instances (where it increased just by 1%). Nonetheless, the improvements where far from the required 12%. The maximum value was a decrease of 3.6% in 2015. Changes were (slightly) higher than 3% also in 2014 and 2016. In terms of the single dimensions, improvements in the energy/GDP ratio have also been steady, with a maximum decrease of 3.1% in 2015. However, a decrease in the CO_2/energy ratio has been slower, with a maximum of 1.6% in 1999;

in 2014 and 2019 the ratio also decreased by more than 1%. But the CO_2/energy ratio increased in half of the years considered, with a maximum increase of 1.3% in 2011.

In summary, judging on past trends, it is clearly unrealistic that the CO_2 intensity of GDP can decrease by more than 12%. Similarly, it is more feasible to approach one of the intermediate options that is Czech Republic or Costa Rica levels. In these cases, the CO_2/GDP ratio would need to decrease by 10% and 9% per year respectively (which cumulatively over many years is a much easier goal already).

Heterogeneity in the Kaya identity

Even if globally we have seen that the required emission reductions and the consequent reductions in carbon and energy intensities have never been met before, what about at the group or country level? Are there (groups of) countries that we can use as examples? In a recent paper, Lamb, Grubb, Diluiso, and Minx (2021) show that since 1970, apart from Jamaica, only countries in the Global North have managed to reduce emissions, with just a few countries averaging more than 3% per year (e.g. Greece, Italy, Finland, Denmark, Ukraine). In parallel, Lamb et al. (2021) show that the only region witnessing a decline in emissions since 2010 has been Europe (−0.8% per year), yet far from the reductions needed. When decomposing these reductions into the Kaya factors, they find that the energy intensity of GDP fell almost in all regions (but insufficiently), while the carbon intensity of energy fell just in some regions. Therefore, also at the regional level there is no region to be taken as a blueprint to achieve emission reductions needed while pursuing economic growth rates.

We complement their analysis by first looking at different income groups (panel (A) of Figure 9.4). We use recent IEA data for 131 countries and aggregate them by income group (31 high-income, 38 low income, 35 lower-middle, and 27 upper-middle). For high-income countries we see a decrease in emissions in most years recently, especially after the 2010 rebound from the financial crisis, but far from the rates needed. This was driven by improvements of energy/GDP (around 2% per year) and also CO_2/energy (slightly less than 1% per year). These improvements are able to counterbalance economic growth (less than 2% per year) and low population growth (0.5%). Nonetheless, improvements in energy/GDP and CO_2/energy have always been far off from the ones outlined in the previous sub-section.

For middle-income countries we have similar improvement in energy/GDP (around 2% per year). On a negative note, the CO_2/energy ratio actually increases in some years for upper-middle income countries, and for most years in lower-middle income ones. In addition, compared with high-income countries, economic growth is much stronger and has a dominant effect (more than 5% per year), as well as population growth being significant especially in lower-middle income countries (2%). Low-income countries have increased emissions, while also increasing their CO_2/energy ratio.

In summary, the energy/GDP ratio is falling in all groups since 1990, with the richest countries showing the lowest decrease. But the average decrease is low compared to the needed one; lower income countries where the ratio is falling the most never saw an average annual decrease of 5% or more. The CO_2 intensity of

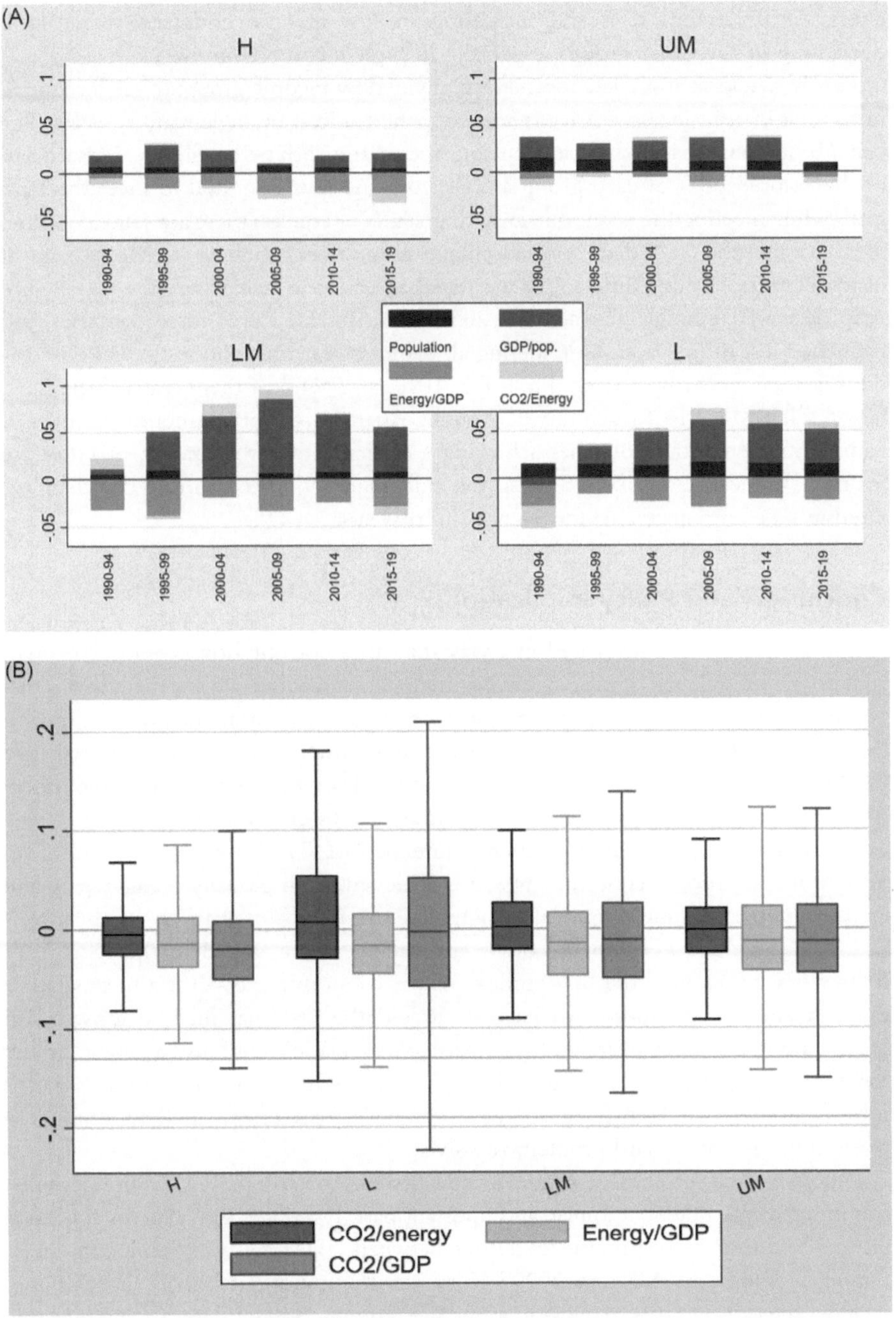

FIGURE 9.4 Changes in Kaya identity components (Figure 9.4A) and distribution of annual changes (Figure 9.4B) by income group

Source: Author's own elaboration

Note: A full colour version of this figure is available open access under: https://www.routledge.com/Governing-the-Interlinkages-between-the-SDGs-Approaches-Opportunities/Breuer-Malerba-Srigiri-Balasubramanian/p/book/9781032184654

energy is diverging. It is still increasing in low-income countries, while it is decreasing in the other groups, especially in high-income countries.

Finally, we look at single countries, still divided by income group, to see if required reductions in emissions and Kaya factors happened at least in single country-years. The panel B of Figure 9.4 shows that even countries in the 25th percentile (the bottom part of the colored box) of each group are far from the level required to meet the Paris goals while meeting the economic growth goals. Nonetheless there are some countries that have met the GDP decarbonization level required in certain years. More than half of the countries in the sample (73) witnessed at least one year where the CO_2/GDP ratio decreased enough to reach the 2050 goal. But for 22 of these countries, this happened just in one year. Considering all country-year observations (3,930 observations in total, given by 131 countries for 30 years), 5.24% of them reached the 2050 target, while just 2.16% reach the 2050 target. Apart from North America and Europe and Central Asia, the countries achieving a sufficient decarbonization of GDP are mainly low-income ones. On the negative side, Figure 9.4 also confirms that the CO_2 intensity of energy increased for many country-years.

Conclusion and Policy Implications

While the SDGs are formulated in a way that does not question economic growth and does not put any limits to environmental degradation, we have shown that continuous economic growth is very problematic for environmental dimensions. In fact, reaching the Paris Goals is nearly impossible under high economic growth and past trends of decoupling rates. We have shown this by developing a simple model of economic growth and Kaya decomposition. Rates of decarbonization of energy and past improvements in the energy intensity of GDP are far from the ones needed if continuous economic growth is achieved, both globally on aggregate and across countries. These empirical findings question the sustainability of the SDGs and emphasize the need to use other approaches, more linked to the Doughnut framework rather than the SDG index, where economic growth is a means rather than an end. Furthermore, we have illustrated the fact that income growth still grants enough degrees of freedom for particular social dimensions to evolve in any direction, but yet the relationship between income and social outcomes overall is robust enough that "satisfying all (development) goals under all (environmental) constraints" is a very hard problem to solve.

Here we have shown that whatever new systems are to emerge, economic redistribution has the potential to play an important part. It reduces the need for aggregate growth and thus helps mitigate trade-offs between collective well-being and environmental constraints (Malerba, 2020). However, it is important to note that this only includes the direct effects of redistribution. We did not consider systemic and dynamic implications of greater equality. For instance, let us assume the world is completely equitable at ~$PPP 20,000 GDP per capita per year: We do not know what policies it requires to ensure a steady state at that level of income (one can think through the same argument in terms of material throughput). It is possible that widespread high living standards encourage further entrepreneurial activity for instance and thus might trigger feedback towards renewed and more economic growth.

What are the policy implications? Despite the limitations discussed, a first implication for policy makers is to focus on international and national redistribution. Inequality must be lowered through economic transfers and progressive taxation but potentially also through more heterodox approaches such as income and wealth caps. Internationally, redistribution should move beyond development aid alone and focus on setting up fair trade systems between countries of the Global North and the Global South in the first place. For instance, this could include giving Global South countries threatened by climate change higher voting power (possibly proportional to population or the specific issue at hand) in international financial institutions like the International Monetary Fund (IMF, 2022). In the long term, a more equitable international governance will then likely contribute to an international levelling of income, although we cannot make any precise statements about such futures.

Second, it is important to note that in any case, if we want to raise prosperity to much higher standards (let us say Costa Rica household expenditure levels as in our minimal scenario) for everyone on the world within the next decades, technological improvements and societal adoption of these technologies must proceed very fast. After all, the necessary GDP-carbon decoupling rates we have found are very high even in the minimal high prosperity scenario (Costa Rica levels). Therefore, it is very important to further push the diffusion of renewable energy and energy efficiency by carbon pricing and subsidies for green and renewable energy wherever possible.

A third policy implied through international redistribution of income and wealth is that a better consideration of degrowth needs to happen, including improvements in its scientific understanding through scenario-modelling. For example, just a few scenarios and Integrated Assessment Models (IAMs), including the ones that the Intergovernmental Panel on Climate Change uses, include decreasing energy use. This is mainly a normative issue; in fact, scenarios for the future all envisage economic growth, such as the Shared Socioeconomic Pathways. Reduced energy use in those scenarios is a consequence of very large increases in energy efficiency. But there are limits to energy demand reduction especially in economically growing regions focusing on industrialization. While fully decarbonizing the supply of energy is possible, it has been estimated that a 10% increase in GDP per capita corresponds to an 8.9% increase in primary energy, with the relationship flattening at very high level of economic growth (Semieniuk, Taylor, Rezai, and Foley, 2021).

Recent research has shown that degrowth would reduce many risks in achieving emission targets (Keyßer and Lenzen, 2021), especially linked to the issue of using negative emission technologies that is present in many models, however, it remains critical to understand the political economy barriers that make degrowth unpopular. One solution focuses on wording (Drews and Antal, 2016). It has been found that focusing on the promotive rather than preventive nature of degrowth increases its support; one example is framing degrowth as a future of "radical abundance" in contrast to the popular perception of "scarcity" implied (Hickel, 2019a). In addition, neo-classical economics that are the basis for IAM models advocate for technological and top-down solutions (which are easier to model and less normative), while degrowth requires more bottom-up socio-cultural change (Semieniuk et al., 2021). It would be naïve to simply replace the very high and unprecedent technological change rates in current scenarios by very high rates of socio-cultural change as degrowth proponents suggest. Both theories of change on their own are possibly overly idealistic. However, there are now large international social and climate movements pushing for change (e.g. Fridays for Future). Such movements constitute one critical aspect of system themselves, the social force that brings about change. Scenarios can do justice to the technological and social progress, and in fact to the whole of the Sustainable Development Goals, if they further investigate the conditions which bring about rapid technological change and social change, rather than just assuming it.

Notes

1 With the exception of the USA, for example.
2 Other targets directly or indirectly apply to policies to reduce poverty and inequality. In terms of social protection, SDG 1 and SDG 10, include (relatively vague) statements on social protection systems and more progressive fiscal policies. In parallel, a couple of goals (SDG1 and 17) mention official development assistance flow and other international flows.
3 In practical terms, the absolute Gini index can be thought as " the expected payment necessary to equalize incomes among two randomly chosen individuals" (Bowles and Carlin, 2020).
4 There is also an "extended" Kaya identity that we do not consider here. https://unfccc.int/sites/default/files/2.4_cicero_peters.pdf.

References

Amiel, Y., Creedy, J., and Hurn, S. (1999). Measuring attitudes towards inequality. *The Scandinavian Journal of Economics*, 101(1), 83–96.
Bowles, S. and Carlin, W. (2020). Inequality as experienced difference: A reformulation of the Gini coefficient. *Economics Letters*, 186, 108789. doi:10.1016/j.econlet.2019.108789.
Breuer, A., Janetschek, H., and Malerba, D. (2019). Translating Sustainable Development Goal (SDG) interdependencies into policy advice. *Sustainability*, 11(7). doi:10.3390/su11072092.
Dellink, R., Chateau, J., Lanzi, E., and Magné, B. (2017). Long-term economic growth projections in the Shared Socioeconomic Pathways. *Global Environmental Change*, 42, 200–214. doi:10.1016/j.gloenvcha.2015.06.004.
Diaz-Sarachaga, J.M., Jato-Espino, D., and Castro-Fresno, D. (2018). Is the Sustainable Development Goals (SDG) index an adequate framework to measure the progress of the 2030 Agenda? *Sustainable Development*, 26(6), 663–671.
Drews, S. and Antal, M. (2016). Degrowth: A "missile word" that backfires? *Ecological Economics*, 126, 182–187. doi:10.1016/j.ecolecon.2016.04.001.
Fanning, A.L., O'Neill, D.W., Hickel, J., and Roux, N. (2021). The social shortfall and ecological overshoot of nations. *Nature Sustainability*. doi:10.1038/s41893-021-00799-z.
Fukuda-Parr, S. and McNeill, D. (2019). Knowledge and Politics in Setting and Measuring the SDGs: Introduction to Special Issue. *Global Policy*, 10, 5–15. doi:10.1111/1758-5899.12604.
González-Torres, M., Pérez-Lombard, L., Coronel, J.F., and Maestre, I.R. (2021). Revisiting Kaya Identity to define an Emissions Indicators Pyramid. *Journal of Cleaner Production*, 317, 128328. doi:10.1016/j.jclepro.2021.128328.

Haberl, H. et al. (2020). A systematic review of the evidence on decoupling of GDP, resource use and GHG emissions, part II: synthesizing the insights. *Environmental Research Letters*, 15(6), 065003. doi:10.1088/1748-9326/ab842a.

Hickel, J. (2018). *The divide: global inequality from conquest to free markets*: W.W. Norton & Company.

Hickel, J. (2019a). Degrowth: a theory of radical abundance. *Real-World Economics Review*, 87(19), 54–68.

Hickel, J. (2019b). Is it possible to achieve a good life for all within planetary boundaries? *Third World Quarterly*, 40(1), 18–35.

IMF. (2022). IMF Members' Quotas and Voting Power, and IMF Board of Governors. Retrieved from: www.imf.org/en/About/executive-board/members-quotas.

International Energy Agency. (2022). What is energy security?

Kallis, G., Kostakis, V., Lange, S., Muraca, B., Paulson, S., and Schmelzer, M. (2018). Research on degrowth. *Annual Review of Environment and Resources*, 43, 291–316.

Keyßer, L.T. and Lenzen, M. (2021). 1.5 C degrowth scenarios suggest the need for new mitigation pathways. *Nature communications*, 12(1), 1–16.

Lamb, W.F., Grubb, M., Diluiso, F., and Minx, J.C. (2021). Countries with sustained greenhouse gas emissions reductions: an analysis of trends and progress by sector. *Climate Policy*, 1–17. doi:10.1080/14693062.2021.1990831.

Lamb, W. F. et al. (2021). A review of trends and drivers of greenhouse gas emissions by sector from 1990 to 2018. *Environmental Research Letters*, 16(7), 073005. doi:10.1088/1748-9326/abee4e.

Malerba, D. (2020). The Trade-off Between Poverty Reduction and Carbon Emissions, and the Role of Economic Growth and Inequality: An Empirical Cross-Country Analysis Using a Novel Indicator. *Social Indicators Research*, 150(2), 587–615. doi:10.1007/s11205-020-02332-9.

Oswald, Y., Steinberger, J., Ivanova, D., and Millward-Hopkins, J. (2021). Global redistribution of income and household energy footprints: a computational thought experiment. *Global Sustainability*, 4.

Plewis, I. and Bartley, M. (2014). Intra-generational social mobility and educational qualifications. *Research in Social Stratification and Mobility*, 36, 1–11. doi:10.1016/j.rssm.2013.10.001.

Raworth, K. (2017). A Doughnut for the Anthropocene: humanity's compass in the 21st century. *The lancet planetary health*, 1(2), e48–e49.

Roser, M. (2021). How much economic growth is necessary to reduce global poverty substantially? Retrieved from: https://ourworldindata.org/poverty-minimum-growth-needed#:~:text=A%20five%2Dfold%20increase%20is,to%20reduce%20global%20poverty%20substantially.

Semieniuk, G., Taylor, L., Rezai, A., and Foley, D.K. (2021). Plausible energy demand patterns in a growing global economy with climate policy. *Nature Climate Change*, 11(4), 313–318. doi:10.1038/s41558-020-00975-7.

UN Environment Programme. (2019). *Emissions Gap Report 2019*. Nairobi.

van den Bergh, J.C.J.M. (2017). A third option for climate policy within potential limits to growth. *Nature Climate Change*, 7(2), 107–112. doi:10.1038/nclimate3113.

Wiedmann, T., Lenzen, M., Keyßer, L.T., and Steinberger, J.K. (2020). Scientists' warning on affluence. *Nature Communications*, 11(1),3107. doi:10.1038/s41467-020-16941-y.

World Bank. (2021). *World Development Indicators*. Retrieved from: https://databank.worldbank.org/source/world-development-indicators.

10

GENDER AND EDUCATION

Sarah Khan

The objective of this chapter is to explore gender inequality in education, its causes, and its implications for economic outcomes in the low- and middle-income countries. The chapter will specifically analyse the interlinkages between SDG 4 (Quality Education), SDG 5 (Gender Equality), and SDG 8 (Decent Work and Economic Growth).

Even though education equality is a pivotal social policy for achieving women's empowerment, it alone is not sufficient in reaching overall gender equality. A key challenge in implementing SDG 5 is how gender interacts with other dimensions of inequalities beyond education, such as inequalities in the labour market, political participation, welfare and legal systems, and regressive social norms, which can mute the effects of progresses achieved in other areas. In recent decades, there has been a promising upward trend in girls' enrolments in low and middle-income countries (LMIC)/low-income countries (LIC). Yet, there remains a large gap by gender, where boys are more likely to complete all forms of education. These differences by gender are then translated into unequal distribution of quality work. Many countries which have seen improvements in female education do not see a similar trend for women's inclusion in the labour market. For the women that do make it to the labour market, their employment is mostly concentrated in highly segregated industries with a high pay gap. Moreover, these trends in education also do not translate into improvements in women's health. Skewed sex-ratios, gender-based violence, and discriminatory practices are still prevalent in many of the LMIC/LICs that have shown improvements in education.

Effective implementation and integration of SDG 4 with SDG 5 will require gender-inclusive policies and institutional processes which prioritize gender equality in all sectors. This includes effective institutional measures for implementing and monitoring, including systems for gender mainstreaming and robust gender statistics to track progress (Un Women, 2018). There is scope for institutionalizing gender

DOI: 10.4324/9781003254683-10

equality in SDG-specific structures. This can include inter-ministerial coordination offices and commissions for the effective assessment and implementation of SDG policies across government departments. Finally, feminist organizations have been crucial in advocating and implementing for gender equality policies.

Achieving the SDG for gender equality also need a boost in financing and investment for these plans to materialize. UN-Women shows that, on average, only 1 percent of the national budgets were allocated gender-based programs and policies (UN Women, 2018). There needs to be international standards to define statistical concepts and methodologies, along with more detailed coverage of intersectionality, such as location, income level, race, age or gender identities/sexualities.

This chapter addresses research question 1 'What do we know about the most important linkages between the SDGs?' by particularly focusing on the trends and causes behind the gaps in education through a gendered lens and subsequently the long-term impacts of gender inequality in education for women's access to decent work and achieving overall economic growth. The main question in this section addresses how economic growth and education interact with one another, and how this relationship between gender equality and economic growth is asymmetric, where causality is stronger in one direction than the other.

Trends in Education in LMICs/LICs

Over the last few decades, there has been a remarkable progress in school enrolment of boys and girls, and the gender gap in education participation has diminished. By 2015 most regions have reached close to 100 percent enrolment in primary education, with 85 percent in Latin America and the Caribbean, Central Asia, and East Asia. South Asia and Middle East and North Africa (MENA) are also catching up to the rest of the world with around 70 percent. This increase in enrolment rates is higher for girls than boys in secondary schooling, reducing gender gaps in all regions except Sub-Saharan Africa (SSA). At the tertiary level, the female-male ratio of enrolment is above 1 (Klasen, 2017, 2019).

However, the improvement in enrolment rates has not translated into improvement in learning at schools. Learning poverty[1] (the share of children who are not able to read proficiently at age 10) is about 4 percentage points lower for girls than boys, with the average being 55 percent for girls and 59 percent for boys in LMICs. In low-income countries, this gap becomes smaller but the overall average of learning poverty reaches over 90 percent for both genders[2].

Also, counter to expectations, the progress in education has not led to a commensurate improvement in employment outcomes of women. Trends in female labour force participation rate have been heterogeneous across different regions, with some countries experiencing a decline. Except for MENA where female labour force participation (FLFP) is rising very slowly, most of Asia has experienced declining FLFP in recent years, with South Asia having the lowest FLFP rates at 24 percent. Similarly, SSA is showing a slow growth in FLFP. Only Latin America experienced a positive trend, but only until 2005. As shown in Figure 10.1, the

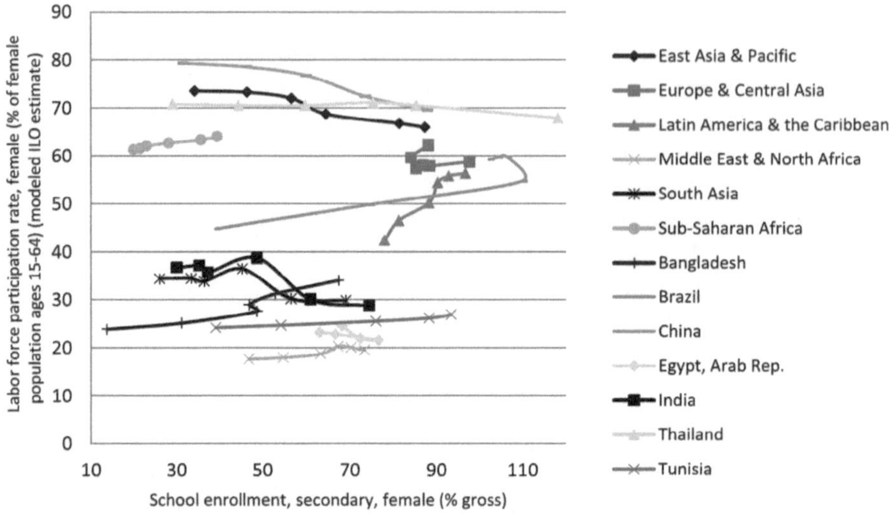

FIGURE 10.1 Female secondary enrolment rate and female labour force participation rate (ages 15–64) by region and selected countries, 1990–2015
Source: Klasen, S., 2019, World Development Indicators

association between increases in female education and FLFP is also heterogenous. Only Latin America shows a strong positive correlation between female education and participation. Except for SSA, Bangladesh, Tunisia, and Brazil, the rest of the regions show a negative or no correlation between FLFP and education. In South Asia, there is a decrease in LFPF in relation to education (70 percent female secondary enrolment) with very low FLFP rates (30 percent), and the same is true for East Asia, at a substantially higher level of FLFP (Klasen, 2019). Besides, women continue to be employed predominantly in certain sectors or occupations that have lower wage rates, perpetuating the gender wage gap in the labour market and having an overall impact on the quality of their lives (Seguino and Braunstein, 2019).

A closer look into the education sector reveals that the progress in participation masks large gender disparities in access to good quality education that is often manifested through the choice of schools or study streams. For instance, underrepresentation of women in science, technology, engineering, and mathematics related fields plays a significant role in subsequent labour market outcomes.

Causes Behind Gender Gap in Education

It is rather difficult to establish a causal relationship between gender inequality in education and its drivers, due to various endogenous factors such as economic growth, fertility, and social norms (Bertocchi and Bozzano, 2020; Duflo, 2012). Below, I discuss some of the widely accepted determinants of the inequality of education by gender:

Cultural factors

Cultural practices have a significant influence on parental preferences regarding investments in children's education, particularly girls' schooling. Some of the practices include Son bias, polygamy, patrilocality, parental and marital transfer (Bau, 2021; Chiappori, Salanié, and Weiss, 2017).

Son-Bias (SDG 5 Gender equality)

Son-bias is a practice where parents value their sons over their daughters over certain dimensions, and thus make different choices for investments in children (e.g. in health and education) on the basis of these preferences. Parents view boys as the more lucrative investment and consider girls as net-economic burden, and hence invest more in their sons. These practices have long-term impacts on girls' education outcomes and women's health outcomes (Milazzo, 2018).

In India, gender bias in the enrolment decision has greatly fallen but bias in the conditional expenditure decision has significantly risen over time, i.e. households have only changed the way they practice gender bias. From 1995, the probability of lower enrolment for girls has diminished overtime, but now the gender bias appears as higher expenditure on boys schooling compared to girls. This is due mainly to the decision to send sons to private schools while the daughters are enrolled in free public schools, leading to a gendered choice for schooling (Datta and Kingdon, 2021).

However, these recent trends are not shared across all of South Asia: Bangladesh has experienced a decline in fertility together with an improvement in child sex-ratios (Kabeer, Huq, and Mahmud, 2014). Recent estimates show a decrease in declared son-preference among mothers. India and Nepal have shown similar trends, however to a lesser extent (Asadullah, Mansoor, Randazzo, and Wahhaj, 2021).

There seems to be an inter-generational effect of empowerment of mothers, such as in Pakistan and Ethiopia, where more empowered mothers spend more on their daughters' (Alvi and Dendir, 2015; Khan, 2021). Similarly, in Bangladesh, son-bias is reduced for mothers with more education, pointing to an intergenerational effect to reducing education gaps (Asadullah et al., 2021).

Marital customs (early marriages and returns to education in the marriage market)

Marital customs are an important determinant of parental education decisions that are not always considered in policies to address gender inequality. Besides credit-constraints, cultural and social norms can also influence parents' decisions to withdraw their daughters from school entirely, and marry off their daughters at an early age, ending their academic careers.

The tradition of making marriage payments, such as dowry and bride price, can also be an economic driver for early marriages. In South Asia, where the practice of dowry is more prevalent, delaying girls' marriages for more years of schooling comes at the cost of paying a higher dowry. Hence, owing to financial and social pressure, parents marry off their daughters as young as possible, and women thus attain less schooling (Field and Ambrus, 2008).

However, the story of the custom of bride price is contrary, for instance, in Indonesia and Zambia, groups that practice the marital custom of bride price, the value of bride price payments that the parents receive tend to increase with their daughter's education. As a consequence, the probability of a girl being educated is higher among ethnic groups that have the custom of bride price. Ashraf, Bau, Nunn, and Voena (2020) find that families from bride price groups in Indonesia were the most responsive to policies aimed at increasing female education. The INPRES school construction programme in Indonesia, as well as a similar programme in Zambia, show large effects only in communities that partake in the tradition of bride price.

Of course, this does not rule out the interaction of credit constraints and economic shocks that could drive households to use early marriages of girls as a coping strategy. Low-income households are twice as likely to practice child marriage for their daughters, compared to girls from higher-income households. Owing to patrilocality, the married daughter exits her parents' household, thus reducing the number of economically inactive dependants (Hoogeveen, Klaauw, and Lomwel, 2011). Moreover, in regions where there is an absence of credit markets, early marriages of daughters can act as a consumption-smoothing insurance, as a younger bride is expected to receive a larger marital asset (Corno and Voena, 2016).

These findings emphasize the importance of the marriage market as one of the drivers of educational investment, and provide evidence for the importance of cultural context in the effectiveness of development policy.

Gendered choice for schooling

In LMICs, especially in South Asia, low-cost private schools are associated with better quality of education, with lower teacher absenteeism, better test scores, higher expected earnings, and acquisition of higher levels of cognitive skills (Sahoo, 2017). These schools, however, are not free like public schools, and come with a small tuition fee. Public schools, on the other hand, are perceived as low quality, with higher teacher absenteeism, lack of resources, higher class sizes. Households with credit constraints, then make a choice on which child to send to the costlier private schools. Boys get a preference for the "better" school, as parents expect a higher return in the labour market for their sons rather than their daughters. This translates into lower human capital accumulation, and subsequently lower labour market prospects for girls in public schools compared to their brothers in private schools, even within the same households (Maitra, Pal, and Sharma, 2013).

Existing empirical research provides strong evidence on the gender dimension of private school enrolment in South Asia. In one large "Young Lives" program in Andhra Pradesh, 3000 children were tracked for over 15 years. The program increased uptake of private schools by 20 to 40 percent for the period 2002–2009, girls and children from rural areas, and lower socioeconomic backgrounds continued to be under-represented. Besides the reduced likelihood to be enrolled in a "better" private school than their brothers, girls also have lower within-household educational expenditures (Aslam, 2009; Khan, 2021). Reducing credit constraints does seem to improve enrolment rates of girls in private schools. A school-fee waiver that reduced the fees of private schools from $13 to $0 increased private school enrolment by 7.5 percentage points for girls and 4.2 percentage points for boys. i.e. expenditure on girls' schooling is an obstacle in school choice (Carneiro, Das, and Reis, 2016).

In SSA, as indicated by the MICS survey data, this discrepancy in private-school enrolment by gender is not observed. MICS6 survey for 2017–2019 shows that in seven countries, there is no gender disparity in private school enrolments. In fact, female enrolment in private secondary schooling is higher in five of the seven countries. This is likely due to a limited range of school choices in most sub-Saharan African countries, where public schools are not considered worse than low-cost private schools. Private schools are mostly there to "soak up" excess demand for schools and provide more of a supportive role. Moreover, data on schooling is not regularly collected, which makes it difficult to draw robust generalizations on the quality of private and public schools. Finally, disparity by location in access to private schools also plays an important role giving less access to poorest households in rural areas (Bennell, 2022).

Choice of field of study

Another area of a visible gender discriminatory gap is in the choice of field of study at secondary and especially tertiary education. At tertiary level of education, men are over-represented in STEM, creating a gender gap in employability, occupational choices and earning profiles.

A recent study by UNESCO showed that during the period 2014–2016, globally, male students over-represented in enrolment in engineering and infotech by 72 percent, while girls in arts and humanities, health and welfare, and other social sciences, by 60 to 70 percent (Hill, Corbett, and St. Rose, 2010). Not only are there fewer women in STEM fields, there is also a higher rate of attrition of women from these fields. Once girls do enter STEM fields, there is also worse progression and lower learning achievement (UNESCO, 2017).

A potential explanation given for the under-representation of female students choosing STEM is that boys have a comparative advantage in these subjects. Empirical studies have found that this is not the case, and inherent gender difference in cognitive ability is almost non-existent; rather, other societal, psychosocial and preference related factors play a larger role in explaining the under-representation of women in math-intensive STEM subjects (Buser, Niederle, and Oosterbeek, 2014; Friedman, 2016; Zafar, 2013). In Israel, gendered choices remain unperturbed even after conditioning on differences in perceived mathematical ability captured by prior achievements. In India, even though there is a 20-percentage-point gender disparity in girls studying technical streams, (STEM) and commerce, this is not driven by gender specific differences in mathematical ability (Sahoo and Klasen, 2021). However, this evidence in limited in the context of LMICs and LICs, owing to data constraints.

Long-term conflict effects

Between 2015 and 2019 women and girls were directly targeted or more exposed to attacks because of their gender in at least 21 of the 37 countries profiled: Afghanistan, Burundi, Colombia, Democratic Republic of Congo (DRC), Egypt, India, Iran, Iraq, Libya, Myanmar, Nicaragua, Nigeria, Pakistan, the Philippines, Somalia, South Sudan, Sudan, Syria, Turkey, Venezuela, and Yemen. These attacks generally took the form of sexual violence or violent repression of women and girls' education (Global Coalition to Protect Education from Attack, 2019).

Regarding gender differentials in educational outcomes in conflict-hit regions, there is substantial evidence which shows that girls' schooling suffers disproportionately, owing to conflict.

Households in affected regions reallocate labour; using child labour to compensate for income losses during war time. As a result, children are removed from schools, reducing the stock of human capital, and resulting in loss of long-term welfare for the family (Justino, 2011). A reduction in educational attainment and enrolment is observed in countries undergoing civil war, as well as higher rates of child labour (Merrouche, 2011; Shemyakina, 2014). This can also lead to redistribution of education investments by gender. Owing to the destruction of industries and infrastructure during war periods, job opportunities for skilled labour generally become scarce. Households might redistribute their resources away from investments with lower returns and invest more in the education of boys rather than girls, as boys might have a higher probability of finding better paid jobs (Shemyakina, 2014).

Moreover, fear of physical attacks and sexual violence can act as an obstacle for school enrolment for girls in conflict-affected regions. Conflict-ridden countries often experience direct attacks on teachers and students or the destruction of school buildings. A multi-country study on the impact of attacks on education of women and girls found that female students and teachers were targeted because of their gender, and often suffered long-term consequences, such as loss of education, early pregnancy, child and forced marriage, and stigma associated with sexual violence, and psychological trauma (Global Coalition to Protect Education from Attack, 2019).

In Afghanistan, for example, girl school children and female staff have been directly targeted by the Taliban. In Nangahar province in 2018, there were 13 incidents of attacks on girls' schools, with 80 more schools closing, owing to fear of attacks (Global Coalition to Protect Education from Attack, 2019). In Northern Pakistan, girls' education was banned for the year that the Taliban were in control of Swat districts. There was a short-term reduction in enrolment rates in districts where the Taliban were not directly in power (Khan and Seltzer, 2016). In Nigeria and the DRC, girls were less likely than boys to return to school following the conflict, as credit-constrained families prioritize boys' schooling. Families also reported higher fears sexual violence and general insecurity for their children currently attending school (Global Coalition to Protect Education from Attack, 2019).

Impact of the COVID-19 on schooling

The COVID-19 pandemic has brought new challenges for global efforts to reduce gender inequalities, especially in education. There is a lingering risk that a lot of the progress in reaching SDG 4 and 5 goals might be undone by the challenges brought forward by the pandemic.

In the early months of the pandemic, schools were closed to about 1.6 billion children in 188 countries., with over one billion living in LMICs and LICs. When the schools reopened, many did not offer face-to-face classes several months into 2021, with many institutions adopting online or hybrid models of instructions Findings from Bangladesh, Côte d'Ivoire, Kenya, Mali, and Pakistan show that remote learning further exacerbated the gender gap in education (UNESCO, 2021b).

The effects of COVID-19 on girls' schooling are again showing a pattern of son-preference, where boys get precedent in times of need. As in case of any monetary shock, the COVID-19 shock is also making parents prioritize boys learning, while girls now have fewer opportunities of remote learning. Findings from the listed countries showed that the closure of schools led to an increase in girls' time spent in unpaid domestic responsibilities. Boys, on the other hand, only participate in paid work outside of the house. Overall, the pandemic has led to an increase in child labour, owing to school closures, in some places by 30 percent (UNESCO, 2021b).

The pandemic has also exacerbated the gender digital divide in developing countries. For countries with data, there was a large discrepancy is access to internet and mobile phones by gender in the first year of the pandemic. In Pakistan, only half of girls reported owning their own mobile phones, whereas boys almost always had a phone. As for returning to school during low-infection months, there are also gender gaps (Tyers and Binder, 2021). In Kenya, about 16 percent of girls did not re-enrol once the school opened in 2021, while only 8 percent boys did not re-enrol. Girls also seem to be disproportionately suffered from mental-health issues related to the pandemic, reporting more stress, anxiety and depression than boys in 15 countries (UNESCO, 2021a).

Long-term Implications of Gender Inequality in Education for Women's Access to Decent Work and Achieving Overall Economic Growth: How Growth and Education Interact

Kabeer and Natali (2013) point out, the relationship between gender equality and economic growth is asymmetric, where causality is stronger in one direction than the other. Reduction in gender gaps in employment and education leads to higher economic growth, not the other way around. Most of the positive impact of gender equality on growth comes from its impact on women's empowerment at the individual and household level.

A recent study estimates the effect of female education on inclusive economic growth by linking to human development index, poverty rates, infant mortality rates, among other indicators, for 144 countries from 2000 to 2010. The authors find that reducing gender gaps in education and increasing female education have a significantly positive effect on inclusive growth[3] (lower poverty and infant mortality rates, and improved health and environmental effects), but there is no significant effect on GDP growth (Hong, Park, and Sim, 2019).

Furthermore, the relationship between gender equality in education on growth can evolve over time as the economy grows. Gender inequality in education is more persistent at primary levels of schooling in both LMICs/LICs and high-income countries (HIC). Owing to its immediate impact of on household health and fertility outcomes, gender equality in primary education and literacy have a larger impact on growth than secondary or tertiary education at low levels of development (Hong et al., 2019).

The link between female labour force participation and growth is not as clear as the impact of education on growth and depends on the type of employment opportunities for educated women and the level of job segregation by gender. For both LMICs and HICs, higher share of female workers in formal non-agricultural occupations leads to higher economic growth (Klasen and Lamanna, 2009).

Even though increasing education rates across the world have improved access to jobs, equitable access to jobs and quality of jobs has still not been achieved. As pointed out by Seguino and Braunstein (2019), there is a scarcity of "good" jobs, which can lead to unequitable distribution of opportunities by gender. The blue solid line in Figure 10.2 shows women's employment-to-population rate relative to men's employment. The red dotted line shows women's relative concentration in industrial employment, i.e. women's relative access to "good jobs". Women's relative concentration in industry has a lower mean value and is more widely dispersed than women's overall employment. As shown in T101.1, during the period 1991–2013, women's employment in industrial jobs, relative to men, fell from an average of 70.2 percent to 47.2 percent. This is mostly observed in the African region and East Asia, even in regions with more jobs in the industrial sector.

Owing to a preference for male workers, firms could exclude women from better jobs, or employ them only in low-quality or unpaid jobs. The authors find that this trend is especially pronounced in low income countries, which are exhibiting increased job-segregation by gender, where more women than men are crowded into low-quality jobs.

Role of Policy in Reaching Sustainable Development Goals for Gender Equality in Education

As discussed in this chapter, gender equality in education and employment, and reducing gender gaps in human capital are interlinked with economic development.

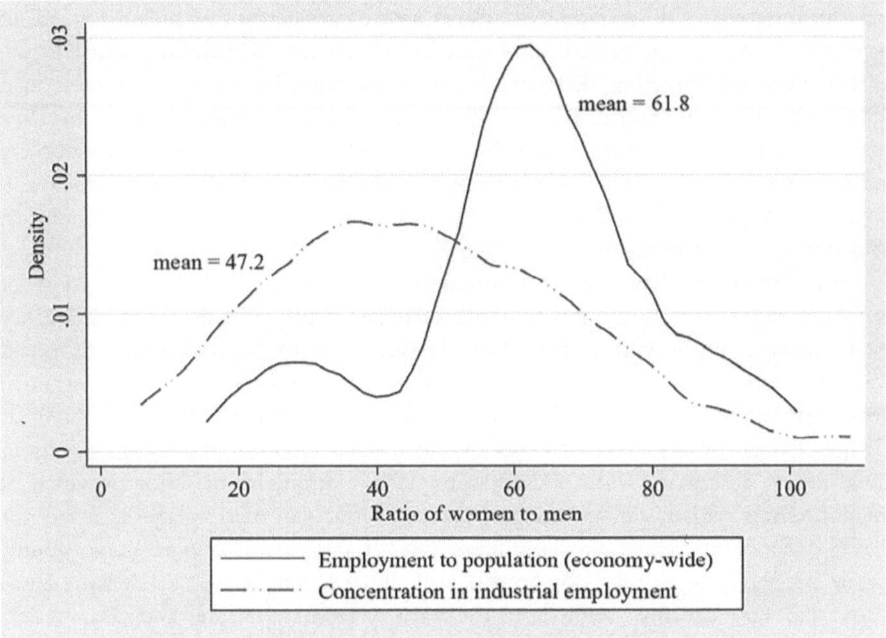

FIGURE 10.2 Distribution of developing countries by women's to men's economy-wide employment rates and shares of industrial sector jobs, 2013
Source: Seguino and Braunstein (2019), based on ILO data, extracted from the World Bank WDI database

SDGs 4 and 5 aim to "ensure that girls and boys, women and men not only gain access to and complete education cycles, but are empowered equally in and through education" (UNESCO, 2016). The existing policies, however, have not been fully effective in achieving gender equality goals, as a result of ignoring the implications of cultural and societal constraints on women.

Programs targeting education in developing countries have been mainly motivated by SDG 4, which aims at increasing school enrolment and daily attendance among students, and improvements in skills and human capital.

Existing programs that target girls' attendance include demand-side interventions (Information-based interventions, conditional cash transfers [CCT], scholarships etc.), and school inputs (access to schools, teaching materials, provision of food, and medical services). CCT programs provide monetary payments to parents if their children are enrolled in school and have a high rate of attendance. Programs specifically targeting girls have had some successes in developing countries. CCT programs in Malawi and Pakistan have had a moderate effect in increasing girls' attendance. In Malawi, CCT programs have shown to increase girls' daily attendance by 8 percentage points for over two years, while enrolment also improved by almost one year (Baird, McIntosh, and Ozler, 2013). In Pakistan, CCT programs showed that a $3 per month cash transfer improved girls' monthly attendance rate by 80 percent or higher for girls in grades 6–8, and also increased enrolment by 8.7 percent (Chaudhury and Parajuli, 2010). Programs targeting skills and learning outcomes have had less success than improving enrolment and attendance. In rural Kenya, girls' scholarship programs have shown to improve participation, but the same programs did not improve the schooling outcomes. Girls who received $6.40 at the beginning of the school year in grade 6 had higher participation by 3.2 percentage points, but no effect on grade completion (Friedman, 2016; Kremer and Thornton, 2009).

Other programs have tried to improve educational inputs in the form of increased access to schools and teaching material, improvement in teacher quality and quantity, and provision of food and medical services. Such programs have had variable degree of success in developing countries. In Pakistan, the construction of low-cost private schools for girls of primary-school age showed an improvement of 25 percentage points in enrolments in urban areas, and 15 percentage points in rural areas (Alderman, Kim, and Orazem, 2003). Similarly, in Ghor province of neighbouring Afghanistan, opening of primary schools in rural villages that did not have a school, increased the enrolment rate of girls by 51.5 percentage points, while 35 percentage points for boys (Burde and Linden, 2013). In Burkina Faso, providing "girl-friendly" schools, in the form of separate latrines and clean water, increased the enrolment rate of all children by 18.5 percentage points (Kazianga, Levy, Linden, and Sloan, 2013).

Besides opening new schools, there are also programs that aim at providing school resources such as teaching material, flipcharts, or textbooks. In rural Kenya, providing flip charts in rural Kenyan schools did not improve student outcomes (Glewwe, Kremer, Moulin, and Zitzewitz, 2004). Similarly, programs that provided textbook also find limited increases in learning outcomes in rural Kenya, as English used in the textbooks was not appropriate for rural children who have limited exposure to the language (Glewwe, Kremer, and Moulin, 2009).

Programs have also aimed to improve teaching quality by reducing class sizes or hiring more teachers.

These programs have shown mixed results with high costs and have little external validity. In Chile, reducing class size have shown to improve student test scores in math and language courses (Urquiola and Verhoogen, 2009). In Kenya, lowering class size by adding more centrally hired civil service teachers did not improve student learning outcomes. Instead, existing teachers reduced their effort in response to the new hires and helped to get their relatives hired into a significant portion of these new teaching slots (Duflo, Dupas, and Kremer, 2015). In India, programs that provided extra teachers and additional educational materials to very small primary schools increased student's primary school completion rates, but it was not possible to determine how much of this effect is due to the extra teacher and how much is due to the additional educational materials (Chin, 2005).

Finally, there are programs which have specifically targeted girls' health. In Nepal, providing menstrual cups for school-age girls found no effect on reducing absenteeism (Oster and Thornton, 2011). A similar program in Kenya found that menstrual cups were not effective in reducing absenteeism (Benshaul et al., 2019). However, providing menstrual pads as well as menstrual education has found to have some positive effects on girls' absenteeism. In rural Uganda and Western Kenya, such programs have had modest but positive effect on girls' schooling (Benshaul et al., 2019; Montgomery et al., 2016). A similar state-level program in India found that attendance gains from providing sanitary pads disappear in response to a temporary withdrawal of the program. This suggests that the impact might be driven by access to a previously unaffordable input rather than a change in social norms (Agarwal, 2021). Besides promoting menstrual hygiene, there are several programs for water, sanitation, and hygiene (WASH). A recent systematic review found that the evidence on the programs seems positive in improving enrolment and attendance outcomes for girls. However, it is difficult to isolate the effect of WASH components, as these programs were rolled out with several other educational components (Psaki, Haberland, Mensch, Woyczynski, and Chuang, 2022).

Conclusion

This chapter explores how gender disparities in education vary across different socio-cultural, economic, and institutional settings, and reflect on their long-term implications for women's access to decent work and achieving overall economic growth.

Considerable progress has been done in identifying and understanding determinants and implications of gaps in gender equality. However, there are still obstacles that slow down the momentum of the closing these gaps. First, even though there is a reduction in gaps in education at a global level, many LMICs and LICs still have disparities by gender in access to schooling and in basic literacy. Second, girls are still opting out of more lucrative careers in the field of STEM, from as early as secondary school. The choice of field of study further translates into worse labour outcomes for women, and women are mostly concentrated in low-paying, often unskilled occupations.

Besides labour market factors, there are social processes that are hindering progresses in gender equality. The preference for male children leading to a rationing of resources, especially low-income households. Parents might invest in the more lucrative asset, by sending their boys to costlier "better" schools, further widening the human capital gap by gender. Deemed as net economic burden, girls from impoverished households are at a higher risk of school dropout and early marriages.

Also, in war and conflict settings, the impact of violent conflict is especially pronounced for girls, as they face different risks than boys, and would require different strategies for their return to schooling and rehabilitation.

Lastly, the ongoing COVID-19 pandemic has brought forward new challenges for girls in schooling. If policies do not incorporate gender-related barriers to studying during school closures, the pandemic could potentially undo some of the progress made in the most-vulnerable communities, where girls are disproportionally affected by school closures.

From the perspective of SDG 8 to promote sustained, inclusive and sustainable economic growth, there is a scope for the promotion of gender equality in education. It has been established that expanding girls and women's educational opportunities promotes growth and development. These policies targeting gender equality in education however need to be custom-fitted to the country-specific context. For LMICs and LICs, policies that promote primary education and literacy, along with skill development would lead to more sustainable growth. Existing policies that promote women's work only loosely focus on raising participation and earning capacity, but do not address unpaid/reproductive work and structural obstacles. Besides enforcement of equal pay and equal opportunity laws, other family-oriented reforms should also be integrated into macro-economic policies. Public programs for childcare and parental leave can reduce the triple burden of work on women, and subsequently lead to better schooling and employment outcomes.

Notes

1 The share of children who are not able to read proficiently at age ten.
2 www.worldbank.org/en/topic/girlseducation#1.
3 Inclusive growth, unlike the traditional growth measure, in embracing equity of health and nutrition, inequality and poverty, social protection, environmental quality, and food security, as well as traditional GDP growth, facilitates investigation of social returns to education not captured by private returns.

References

Agarwal, G. (2021). *Improving school attendance for adolescent girls-Evidence from a policy experiment in Delhi's public schools*. Delhi: Indian Statistical Institute.

Alderman, H., Kim, J., and Orazem, P.F. (2003). Design, evaluation, and sustainability of private schools for the poor: the Pakistan urban and rural fellowship school experiment. *Economics of Education Review*, 22, 265–274.

Alvi, E. and Dendir, S. (2015). Parental Education and Children's School and Work Status in Urban Ethiopia: A Note on Gender Bias. *South African Journal of Economics*, 83(1), 101–116.

Asadullah, M.N., Mansoor, N., Randazzo, T., and Wahhaj, Z. (2021). Is son preference disappearing from Bangladesh?. *World Development*, 140, 105353.

Ashraf, N., Bau, N., Nunn, N., and Voena, A. (2020). Bride price and female education. *Journal of Political Economy*.

Aslam, M. (2009). The relative effectiveness of government and private schools in Pakistan: are girls worse off? *Education Economics*, 17(3), 329–354.

Baird, S.E.G., McIntosh, C., and Ozler, B. (2013). The heterogeneous effects of HIV testing. *Journal of Health Economics*.

Bau, N. (2021). Can Policy Change Culture? Government Pension Plans and Traditional Kinship Practices. *American Economic Review*, 111(6), 1880–1917.

Bennell, P. (2022). Private schooling in sub-Saharan Africa: An egalitarian alternative? *International Journal of Educational Development*, 88, 102533.

Benshaul, T.A. et al. (2019). Pupil Absenteeism, Measurement, and Menstruation: Evidence from Western Kenya. Retrieved from: https://cdep.sipa.columbia.edu/sites/default/files/cdep/WP74-2Tolonen.pdf.

Bertocchi, G. and Bozzano, M. (2020). *Gender Gaps in Education*. Springer International Publishing.

Burde, D. and Linden, L. (2013). Bringing Education to Afghan Girls: A Randomized Controlled Trial of Village-Based Schools. *American Economic Journal: Applied Economics*, 5(3), 27–40.

Buser, T., Niederle, M., and Oosterbeek, H. (2014). Gender, competitiveness, and career choices. *Quarterly Journal of Economics*, 129, 1409–1447.

Carneiro, P.M., Das, J., and Reis, H. (2016). The value of private schools: Evidence from Pakistan. *SSRN*, 2786044.

Chaudhury, N. and Parajuli, D. (2010). Conditional cash transfers and female schooling: the impact of the female school stipend programme on public school enrolments in Punjab, Pakistan. *Applied Economics*, 42(28), 3565–3583.

Chiappori, P.A., Salanié, B., and Weiss, Y. (2017). Partner Choice, Investment in Children, and the Marital College Premium. *American Economic Review*, 107(8), 2109–2167.

Chin, A. (2005). Can redistributing teachers across schools raise educational attainment? Evidence from Operation Blackboard in India. *Journal of Development Economics*, 78(2), 384–405. Retrieved from: https://EconPapers.repec.org/RePEc:eee:deveco:v:78:y:2005:i:2:p:384-405.

Corno, L. and Voena, A. (2016). Selling Daughters: Age of Marriage, Income Shocks and the Bride Price.

Datta, S. and Kingdon, G.G. (2021). *Gender Bias in Intra-Household Allocation of Education in India: Has it fallen over time?*. Quantitative Social Science, Social Research Institute, University College London.

Duflo, E. (2012). Women empowerment and economic development. *J Econ Lit*, 50, 1051–1079.

Duflo, E., Dupas, P., and Kremer, M. (2015). School governance, teacher incentives, and pupil–teacher ratios: Experimental evidence from Kenyan primary schools. *Journal of Public Economics*, 123, 92–110.

Field, E. and Ambrus, A. (2008). Early marriage, age of menarche, and female schooling attainment in. *Journal of Political Economy*, 116(5), 881–930.

Friedman, S. (2016). Gender streaming and prior achievement in high school science and mathematics. *Economics of Education Review*, 53, 230–253.

Friedman, W.K. (2016). Education as liberation? *Economica*, 83(329), 1–30.

Glewwe, P., Kremer, M., and Moulin, S. (2009). Many children left behind? Textbooks and test scores in Kenya. *American Economic Journal: Applied Economics*, 1(1), 112-35.

Glewwe, P., Kremer, M., Moulin, S., and Zitzewitz, E. (2004). Retrospective vs. Prospective Analyses of School Inputs: The Case of Flip Charts in Kenya. *Journal of Development Economics*, 74(1), 251–168.

Global Coalition to Protect Education from Attack. (2019). *Impact of Attacks on Education on Women and Girls*.

Hill, C., Corbett, C., and St. Rose, A. (2010). *Why so few? Women in science, technology, engineering, and Mathematics*.

Hong, G., Kim, S., Park, G., and Sim, S.-G. (2019). Female education ext
ernality and inclusive growth. *Sustainability*, 11(12), 3344.

Hoogeveen, J., Klaauw, B.V.D., and Lomwel, G.V. (2011). On the timing of marriage, cattle, and shocks. *Economic Development and Cultural Change*, 60(1), 121–154.

Justino, P. (2011). Violent conflict and human capital accumulation. *IDS Working Papers*, 2011(379), 1–17.

Kabeer, N. and Natali, L. (2013). Gender equality and economic growth: Is there a win-win? *IDS Working Papers*, 2013(417), 1–58.

Kabeer, N., Huq, L., and Mahmud, S. (2014). Diverging Stories of "Missing Women. *South Asia: Is Son Preference Weakening in Bangladesh? Feminist Economics*, 20(4), 138–163.

Kazianga, H., Levy, D., Linden, L., and Sloan, M. (2013). The effects of girlfriendly schools: evidence from the BRIGHT school construction program in Burkina Faso. *American Economic Journal: Applied Economics*, 5(3), 41–62.

Khan, S. (2021). *Female education and marriage in Pakistan: The role of financial shocks and marital customs*. WIDER Working Paper.

Khan, S. and Seltzer, A. (2016). *The Impact of Fundamentalist Terrorism on School Enrolment: Evidence from North-Western Pakistan, 2004–2016.*

Klasen, S. and Lamanna, F. (2009). The impact of gender inequality in education and employment on economic growth: new evidence for a panel of countries. *Feminist Economics*, 15(3), 91–132.

Klasen, S. (2017). Gender, institutions, and economic development. In *The Handbook of Economic Development and Institutions.*

Klasen, S. (2019). What explains uneven female labor force participation levels and trends in developing countries? *The World Bank Research Observer*, 34(2), 161–197.

Kremer, M. and Thornton. (2009). Incentives to learn. *The Review of Economics and Statistics.*

Maitra, P., Pal, S., and Sharma, A. (2013). *Returns to schooling, English skills and gender gap in private school enrolment: Evidence from India.* SSRN Working Paper.

Merrouche, O. (2011). The long term educational cost of war: evidence from landmine contamination in Cambodia. *The Journal of Development Studies*, 47(3), 399–416.

Milazzo, A. (2018). *Why are adult women missing? Son preference and maternal survival in India.* World Development.

Montgomery, P., Hennegan, J., Dolan, C., Wu, M., Steinfield, L., and Scott, L. (2016). Menstruation and the cycle of poverty: a cluster quasi-randomised control trial of sanitary pad and puberty education provision in Uganda. *Plos one*, 11(12), 0166122.

Oster, E. and Thornton, R. (2011). Menstruation, sanitary products, and school attendance: evidence from a randomized evaluation. *American Economic Journal: Applied Economics*, 3(1), 91–100.

Psaki, S., Haberland, N., Mensch, B., Woyczynski, L., and Chuang, E. (2022). Policies and interventions to remove gender-related barriers to girls' school participation and learning in low-and middle-income countries: A systematic review of the evidence. *Campbell Systematic Reviews*, 18(1), 1207.

Sahoo and Klasen. (2021). Gender Segregation in Education: Evidence From Higher Secondary Stream Choice in India. *Demography*, 58(3), 987–1010.

Sahoo, S. (2017). Intra-household gender disparity in school choice: Evidence from private schooling in India. *The Journal of Development Studies*, 53(10), 1714–1730.

Seguino and Braunstein. (2019). The costs of exclusion: Gender job segregation, structural change and the labour share of income. *Development and Change*, 50(4), 976–1008.

Shemyakina, O. (2014). *Labor market outcomes in post-conflict Tajikistan.* Georgia Institute of Technology.

Tyers, C. and Binder. (2021). *What we know about the gender digital divide for girls: A literature review.* UNICEF Gender and Innovation Evidence briefs.

UNESCO. (2016). *Education 2030: Incheon declaration and framework. ED-2016/WS/28.* Paris: UNESCO.

UNESCO. (2017). *Cracking the code: Girls' and women's education in science, technology, engineering and mathematics (STEM).* Paris: UNESCO.

UNESCO. (2021a). *The state of the global education crisis: A path to recovery*: World Bank Group.

UNESCO. (2021b). *When schools shut: gendered impacts of COVID-19 school closures.* Paris: UNESCO.

UN Women. (2018). Gender Equality and the Sustainable Development Goals in Asia and the Pacific - Baseline and pathways for transformative change by 2030.
Urquiola, M. and Verhoogen, E. (2009). Class-Size Caps, Sorting, and the Regression Discontinuity Design. *American Economic Review*, 99(1), 179–215.
Zafar, B. (2013). College major choice and the gender gap. *Journal of Human Resources*, 48, 545–595.

11

MODELLING THE INTERACTION BETWEEN CLIMATE MITIGATION AND INCOME INEQUALITY

The use of Integrated Assessment Models and the case of India

Johannes Emmerling and Daniele Malerba

Introduction

This chapter explores the critical relationship between inequality (SDG 10) and climate change mitigation (SDG 13) for the 2030 Agenda, focusing on the use of long-term climate models. There are, in fact, several reasons why inequality and climate mitigation are strictly linked.

On the political economy side, potential negative distributional effects of climate mitigation measures hinder social acceptability, and in turn implementation of needed emission reduction policies. Climate mitigation policies are in fact assumed to hurt proportionally more lower income households as they spend a higher share of their budget on energy. For this reason, ongoing research is focusing the distributional implication of climate policies and how to best compensate the people most impacted by them and the most vulnerable. From an inequality perspective, carbon pricing is actually found to be neutral to slightly progressive in most low-income countries, contrary to what has been found for industrialized ones; but much depends on the context and on households' energy use patterns, which vary across countries (Steckel et al., 2021).

On the climate policy modelling side, within-country inequality is still not sufficiently accounted for. This is critical as climate models, such as Integrated Assessment Models (IAM), are used to derive optimal climate policy (such as carbon pricing through the estimation of the social cost of carbon, SCC) recommended to policy makers. Given the level of aggregation of these models, which optimize intertemporal welfare, national distribution and compensation measures are usually not modelled. This omission might alter the derived optimal climate policies; for example, given the type of social welfare functions used in the models and that climate mitigation policies and climate change effects hurt the poorest the most, optimal climate policies might result in significantly more

DOI: 10.4324/9781003254683-11

stringent outcomes, if inequality is considered. This can therefore alter policy strategies on how to reach the overall SDG agenda. In addition, IAMs could also model potential effects of climate change (mitigation) on inequality in the long term. Some findings from studies of revenue recycling have started to also be incorporated also at the global level. However, the bottom-up modelling of inequality and poverty and different carbon revenue is still only nascent with few exceptions such as Soergel et al. (2021).

Given this background, the chapter has three aims. First, it summarizes and discusses the main challenges to be solved in terms of integrating national inequality and redistribution in climate modelling; it also considers the research on revenue recycling mechanisms from micro-macro level studies on climate policies (Malerba, Gaentzsch, and Ward, 2021). Second, the chapter then provides an empirical estimation using a case study. We explore the case of India, based on the availability of data, the prevalence of inequality today and the importance of mitigation and in particular of energy system decarbonization. Moreover, climate impacts are expected to be potentially large making a comprehensive distributional analysis of climate change an important task. In addition, we consider several recycling scheme of carbon tax revenues. Alongside a neutral case (whereby the distribution is not affected) and an equal per capita "climate dividend", we model a "targeted" redistribution of the tax revenues which uses existing targeting schemes of social protection. For the latter we use the holders of the Below the Poverty Line (BPL) card; we redistribute the carbon revenues on top of existing schemes exclusively to these households. This provides an even more dedicated pro-poor climate policy option.

The chapter than concludes on why is critical to account for the aforementioned relationship in terms of reaching the overall 2030 Agenda (the third aim). The chapter stresses how the interaction between climate policies and inequality needs to be sufficiently considered if a policy maker has the objective to mitigate global climate change and inequality simultaneously as in the SDGs.

How to Account for Inequality in IAMs? Current Evidence and Gaps

This section describes how IAMs are currently integrating inequality; it also highlights the main research gaps, some of which we will try to address in the next section. In this paper we consider monetary (income/consumption) inequality, as it is the main inequality dimension that has been so far explored in IAMs. But we do recognize that other inequality dimensions need to be taken into account, including gender, education health or race (Emmerling and Tavoni, 2021). While being important, these dimensions have been less explored in IAMs.

First of all, IAMs can be divided into "Detailed process-based" models, which focus on how energy, land use, and economic systems need to transform in order to achieve climate stabilization; and "Benefit cost" models, which do not model those detailed processes but estimate the total cost of climate change from an economic perspective, such as the SCC. Benefit–cost IAMs include climate impacts, but lack mitigation strategy details and focus solely on monetary impacts. In addition, most models focus on income dimensions, with other critical social dimensions only emerging (Emmerling and Tavoni, 2021).

IAMs are used to develop long-term pathways in two main ways. One way is to use a cost-benefit analysis and estimate the level of CO_2 emissions that maximize the intertemporal welfare function. This is used for example in the estimations of the SCC, and it relies significantly on the values of intertemporal (and intra-temporal) discount rates. A second way is to estimate the carbon budget/emission targets and then build a pathway that minimizes mitigation costs.

One main point related to the aim of the chapter is that the majority of both methods rely on the idea of an average citizen or "representative agent" of the population, behaving as a social planner; there are many reasons for this, one being that there is a need to solve the complex equations of IAMs within the limits of available computational power. Therefore, using representative agents facilitates this goal. But this of course does not favor the analysis of the effects on inequality and poverty as the distribution of the population is not modelled directly.

One first step to study inequality was to look at between-country inequality. This has been possible as, while original models like the DICE were global models (considering the World as a whole), recent models such as the Regional Integrated model of Climate and the Economy (RICE), AD-RICE, PAGE, FUND, CWS, MICA, C3IAM, and STACO, disaggregate the global economy in many countries and macro-regions (Weyant, 2020). This resolution allows only to partially capture the variation in the costs and benefits of climate action. Moreover, it obfuscates the difference between within- and between-country inequality. While inequality between countries has dominated until around 1980, latest results suggest that the share of within-country inequality (as measured by the decomposable General Entropy index) is increasing and according to some estimates it now makes up about 70% of global inequality and could make up the larger share of inequality in future decades (Chancel, Piketty, Saez, and Zucman, 2021a).

Disaggregating the global economy in regions/countries allows to consider transfers between countries. While this was avoided in the past through the use of Negishi weights, that were introduced to restrict redistribution and keep the focus on climate policy, the recent focus on inequality calls for change. In fact, in recent years, there has been increasing consideration of inequality in IAMs for two main reasons. First, the importance of inequality for climate policy. On one side, the poorest might be the ones suffering more from the effects of climate change (despite being the ones contributing the least); on the other, they could be the ones proportionally paying more in mitigating emissions. This is crucial as it might hinder public acceptability of climate policy. A second trend has been the development of more disaggregated climate models and availability of household data. Therefore, the following questions have been started to be explored in more detail:

- Does the inclusion of within country inequality affect the optimal climate policy?
- Does optimal climate policy increase inequality?
- What is the role of redistributive policies?

Current evidence

There are two ways in which IAMs have explored these issues of inequality. Some models use exogenous Gini projections (such as the ones associated with Shared Socioeconomic Pathways, SSPs)[1], and further investigate how climate change (mitigation) affects inequality. Taconet, Méjean, and Guivarch (2020) estimate inequality between countries, by looking at results from different IAMs and adding some regression results of damage functions for different SSPs and RCPs. They find evidence of how climate change affects inequality between countries under a large variety of scenarios where inequality between countries is calculated as the Gini coefficient of the average income across countries. Anthoff and Emmerling (2019) also showed that between-country inequality raises the SCC in most cases by a factor of two to three. Similarly, Gazzotti et al. (2021) show that even optimal climate policy increases (between-country) inequality, especially owing to the effects of climate change.

Soergel, Kriegler, Bodirsky, et al. (2021) consider, on the other hand, within-country inequality. They use Gini simulations from Rao, Sauer, Gidden, and Riahi (2019) and assume log-normal income distributions; they assess the effects of mitigation by looking at the loss of GDP and the increase in food and energy costs plus revenue recycling.

Another set of papers, on the other hand, has started to explore different ways in which distributional implications within countries can be explicitly modelled in IAMs and how this changes optimal climate policy. Dennig, Budolfson, Fleurbaey, Siebert, and Socolow (2015) were the first to model within-country inequality in a large-scale IAM. They use the RICE model and add quintiles for each region, thus constructing the Nested Inequalities Climate Economy (NICE) model; by using values for the quintiles for each region they estimate Gini coefficients. On the limitation side, their model included no redistribution (also across regions) and mitigation costs proportional to income. Their main finding is that considering inequality within countries significantly increases the stringency of optimal climate policy. Adler et al. (2017) also employed the RICE model and a Prioritarian welfare function, and showed substantial differences in the estimates of the SCC compared to the use of the usual Utilitarian social welfare function. They consider inequality between countries, as the RICE model does not consider within-county distribution.

But, while assessing inequality, this latter set of papers do not take into account redistribution from a carbon tax (which is the main policy considered in IAMs). Kornek, Klenert, Edenhofer, and Fleurbaey (2021) also use the NICE model, employing a global social welfare function when equality is preferred at the global level; the national level allocates consumption between households and compensates them for climate change damages and abatement costs. They explicitly exclude transfers between countries to focus on the more plausible case of different SCC across countries, and hence country-specific carbon taxes (traditional optimal taxation theory models only the national level). One limitation of the framework is that governments rely on first-best lump-sum transfers for redistributing the carbon tax revenues. They did this abstraction to highlight the importance of accounting for household heterogeneity when calculating the SCC in the simplest possible way. They show that climate and distributional policy can generally not be separated and that without compensation for poor households, the SCC tends to increase globally. In a stylized global setup, Budolfson et al. (2021) show that an equal per capita refund of carbon tax revenues implies that achieving a 2 °C target can pay large and immediate dividends for improving well-being, reducing inequality and alleviating poverty

In summary, the consideration of inequality within countries is still at an early stage, in particular considering important drivers of inequality including skill premia, differentiated saving rates, wealth accumulation, and consumption patterns. Yet, both climate policies and climate impacts are increasingly being found to affect households differently. Therefore, the role of additional policies notably the redistribution of carbon revenues is becoming an important topic in policy and research likewise.

Revenue Recycling in LMICs

In addition to a better inclusion of inequality in IAMs and climate models more in general, a better understanding of recycling schemes is therefore needed. In fact, most models, owing to computational limitations and simplicity, use idealized recycling schemes (such as per capita equal transfers). While this is surely a possibility and there are many discussions around the implementation of a Universal Basic Income also in low- and middle-income countries (LMIC) (Banerjee, Niehaus, and Suri, 2019), no such country is currently implementing universal redistributive policies. The significant increase in the number of cash transfers schemes for poverty reduction (social assistance) is represented by targeted programs; this means that targeted beneficiaries are identified using different methods, which all present significant targeting errors as not all intended beneficiaries are reached. While the heterogeneity depends more on how a targeting mechanism is designed and implemented rather than which mechanisms is used (Coady, Grosh, and Hoddinott, 2004; Devereux et al., 2017), some general lessons can be outlined. Means testing, which entails using direct estimates of household income, usually performs well, but it is expensive and used rarely (such as in Brazil), owing to informational constraints. Many countries in Latin America use proxy means testing (PMT), meaning that the income of a household is estimated through other information/variables of the household. This is done as income can be under-reported by households to get enrolled in cash transfer programs. The targeting performance of PMT is heterogeneous and depends on the variables used; in some cases, it witnesses sizeable exclusion (and inclusion) errors (Brown, Ravallion, and Van de Walle, 2018). Where poverty is widespread such as in sub-Saharan Africa (SSA), the use of household targeting mechanism is not advised (Ellis, 2012), and geographical targeting can be an effective solution. Evidence from the SSA region also shows that other targeting mechanisms, such as community targeting, could also be used in effectively reaching the poorest despite issues with elite capture. Programs using self-targeting are also employed to reach the intended beneficiaries and avoid inclusion errors, but show various degree of targeting performance and have not delivered as promised. For example, employment guarantee schemes such as the Productive Safety Net Programme in Ethiopia, have not reached many individuals in poverty, especially in poor areas, owing to the lack of jobs and corruption (Ravallion, 2019).

Given the urgency and relevance of climate policies and carbon pricing, it is important to understand if existing transfer schemes can be employed and adapted in the short term to complement climate policies. In fact, while carbon pricing mechanisms are considered easy to administer, capacity issues (delivery chain and payment methods, institutions and governance, information systems platform, citizen interface, performance monitoring, evaluation, learning, and adaptation) are a main obstacle in implementing transfer schemes (Yemtsov and Moubarak, 2018). The use of unified and other national registries, as seen in the current COVID-19 response (Gentilini, Almenfi, Orton, and Dale, 2020) is particularly crucial. Single registries for multiple programs can also decrease targeting costs, even if they are expensive to set up and update. From existing data (Barrientos, 2018), it can be seen that around 32% programs used unified registries in 2015. Illustrative positive examples come from Malawi, Cambodia, Brazil, Colombia, or Chile. But there was significant regional heterogeneity. While around half of programs in Latin America and East Asia report using them, just around 20% of programs take advantage of national registries in Middle-East, North Africa, and SSA.

Therefore, to considering existing schemes in modelling is more realistic as it will take some time for many LMICs to develop universal social protection programs, especially owing to financial and aforementioned technical issues (Barca, 2018). In particular, we are interested in assessing if the current systems are adaptable to increasing needs in the short term; this would mean, for example, just the need to adapt the benefit levels. Or, conversely, if new schemes and architecture need to be built to increase coverage. In the similar case of energy and fossil subsidy reforms, Yemtsov and Moubarak (2018) show that for the case of major reforms, the majority of countries introduced a new program, or significantly altered existing ones to especially increase coverage. In few cases an increase of current benefits was done. This showed that current programs were not easily adaptive.

The level of inadequacy/non-universality of current transfer programs in LMICs can be also inferred through cross-country data. For example, one interesting question to answer is whether countries where lower-income individuals would be hit the hardest by a carbon tax (based on their carbon footprint), are the ones with already adequate social assistance systems in place in terms of (i) coverage and (ii) benefit levels. The former is more critical as benefit levels can be easier scaled up with tax revenues for example. Using data from the World Bank (2022) and Bruckner, Hubacek, Shan, Zhong, and Feng (2022), top part of Figure 11.1 shows that the coverage of social assistance (left) and overall social protection (right), and per capita footprints of individuals (horizontal axis) for the lowest quintile are

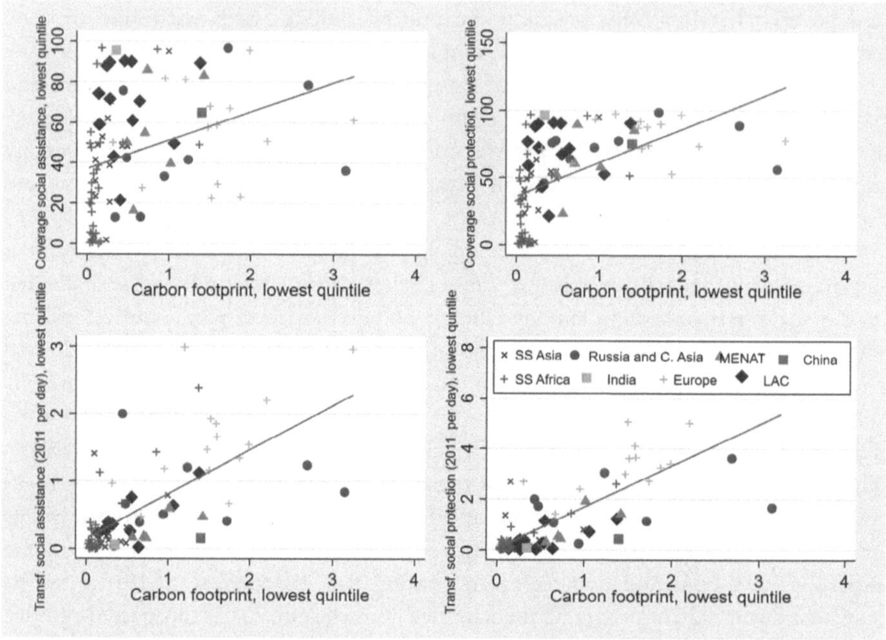

FIGURE 11.1 Relationship between social assistance (social protection) coverage (upper part) and transfer size (lower part) and carbon footprints, for the lowest quintile

Source: Author's own elaboration

weakly positively correlated. This indicates that countries where lower income groups would be hit harder in absolute terms by carbon taxes (owing to level, and carbon intensity, of consumption) are also the ones that are performing better in term of targeting social assistance to the poorest. The figure also shows that many countries in Africa have coverage of less than 5% of the lowest quintile, owing to the poor targeting performance of many programs, and their limited geographical coverage.

This positive narrative is reinforced from the bottom side of Figure 11.1, showing the correlation between per capita benefit levels and the carbon footprints of lowest quintiles. The two variables are also weakly positively correlated. The bottom part of Figure 11.1 also reinforces the view that many programs in SSA need to improve their size and targeting mechanisms or new programs need to be implemented. This could also help using more efficiently (in terms of poverty reduction) the available resources. Therefore, programs in the region do not seem adaptive in the short term.

In summary this section has shown how universal and stylized recycling mechanisms modelled in IAMs are in reality not feasible in the short term. Taking into account the current architecture of social policies is therefore important.

The case of India

As previously mentioned, the targeting efficacy in many countries is low, and a large part of the poorest do not benefit from the programs (exclusion errors), while many richer households actually receive the transfers from the programs even if they are in principle not eligible (inclusion errors). This problem is also relevant for the case of India. The country has gradually expanded its system and reach of social protection since the early 2000s. The system is composed of many small programs, but a few main ones can be identified. One of the main programs is the Indira Gandhi National Old Age Pension Scheme (IGNOAPS) (Unnikrishnan and Imai, 2020). IGNOAPS is an unconditional cash transfer program targeted to the elderly. The eligibility criterion requires an individual to belong to a BPL household; and to be above 60 years (before the age cutoff was 65). Another main program is the Mahatma Gandhi National Rural Employment Guarantee Act (MGNREGA). MGNREGA is arguably the largest public workfare program in the world. Under this act, any adult from a household living in rural areas is entitled to be employed for at least 100 days a year on public works, performing unskilled manual labour at statutory minimum wage. One critical feature of the programme is its decentralized nature, meaning that the administration and allocation of works are executed by the elected local authorities of the respective villages (Das, 2015). Finally, there is the Public Distribution System (PDS) that was replaced by the Government with a targeted version (TPDS) in 1997. BPL households were initially entitled to 20 kg of rice or wheat per month at Rs.5.65 and Rs.4.15 per kg, respectively. Later, in December 2000, 10 million of the poorest households were identified as beneficiaries of the Antyodaya Anna Yojana. They would get 35 kg of wheat or rice per month at Rs.2 or Rs.3 per kg. On the other hand, the above poverty line households have to pay higher prices The Indian government introduced TPDS to control the growing food subsidy and to improve its targeting. However, even after this change, improvements were not significant (Kishore & Chakrabarti, 2015). From November 2016, the TPDS was gradually replaced by the world's largest food aid and social assistance program, the so-called National Food Security Act (NFSA). The NFSA aimed to scale up the existing distribution system by extending its inclusion criteria; 67% of the nation (covering on average 50% of the urban and 75% of the rural population) is now entitled to highly subsidized wheat, rice and coarse grains delivered through the TPDS (Kozicka, Weber, and Kalkuhl, 2019). In addition, there are several minor/smaller programs (Kapur and Nangia, 2015; World Bank, 2019).

Nonetheless there are continuous discussions of reforming social assistance/ protection systems and the introduction of a universal basic income (UBI) (Shah, 2017). One of the main reasons of discontent is that a large share of current subsidies ends up to the non-poor while the eligible poor are denied their benefits, due mainly to leakage and corruption.[2] The results depict the striking extent of misallocated welfare funds: districts where poverty is most prevalent tend to be given the fewest financial resources. The districts where 40 percent of India's poor reside get only 29 percent of this total funding. As a response, *The Economic*

Survey 2016–17 proposed a UBI, as well as a near universal program to avoid giving transfers to the richest.[3] *The Economic Survey* argues that a UBI could be a relatively light administrative burden, eliminating rent-seeking and lowering exclusion errors; and would inject administrative efficiency and transparency. It was also underlined that a UBI is a potential poverty-fighting tool that can perform significantly better than the country's 950 centrally sponsored schemes. Rather, the survey finds that a budget-neutral quasi-UBI could not materialize without rolling back India's existing subsidies and social welfare schemes.

Given this background, comparing currently targeted (through the BPL card) and per capita recycling can enrich also the debate on potential revenue recycling in India.

Empirical Result on Climate Policy and Climate Change Incidence

Climate policy incidence

This section presents the empirical analysis, exploring how climate change and its mitigation affect inequality, also adding redistributive policies. Cross-country empirical estimates of the incidence of climate policies and climate change include Steckel et al. (2021) and Vogt-Schilb et al. (2019). This literature shows that in LMICs the effect of an economy-wide carbon tax is neutral or slightly progressive, especially for lower income countries. In addition, the redistribution of the carbon tax revenues or permit auctions could decrease poverty compared with the status quo, while also further decreasing inequality. Therefore, taking into account revenue recycling is critical.

We do this by linking a global IAM with bottom-up energy system model and a coupled land-use model (WITCH, see Emmerling et al., 2016), with a household-based optimization model based on deciles including differentiated consumption, income, wage, and wealth distribution. Therefore, this allows us to study household behavior and the impact of price increases in particular for energy and food, wage and skill premia, savings and wealth dynamics. We run a business-as-usual and carbon tax scenario (of US\$30 per ton of CO_2 increasing at 5% a year) and use population and GDP projections of the "Middle of the Road" Shared Socio-economic Pathway SSP2 (Riahi et al., 2017). The level of global warming by the end of the century reaches about 3.2°C in the business as usual (BAU) scenario and 1.8°C in the carbon tax scenario (see Figure 11.2).

We use the case of India, which is significant for both climate and social goals. India is in fact the country with the highest number of people in poverty, despite experiencing high economic growth, with a four-fold increase in average incomes since 1990 that has lowered the share of the population living in absolute poverty from 45% to 20%. Inequalities are also large (Balasubramanian, Kumar, and Loungani, 2020). In addition, while its per capita emission levels are low, in national terms they are big and will increase in the future. It is critical then to understand if India can decrease its emission levels while also addressing poverty and inequality.

In our model, consumption, income, wealth shares and saving rates are calibrated to the *2011/12 National Sample Survey* (NSS) 2011–2012 (68th round) – a

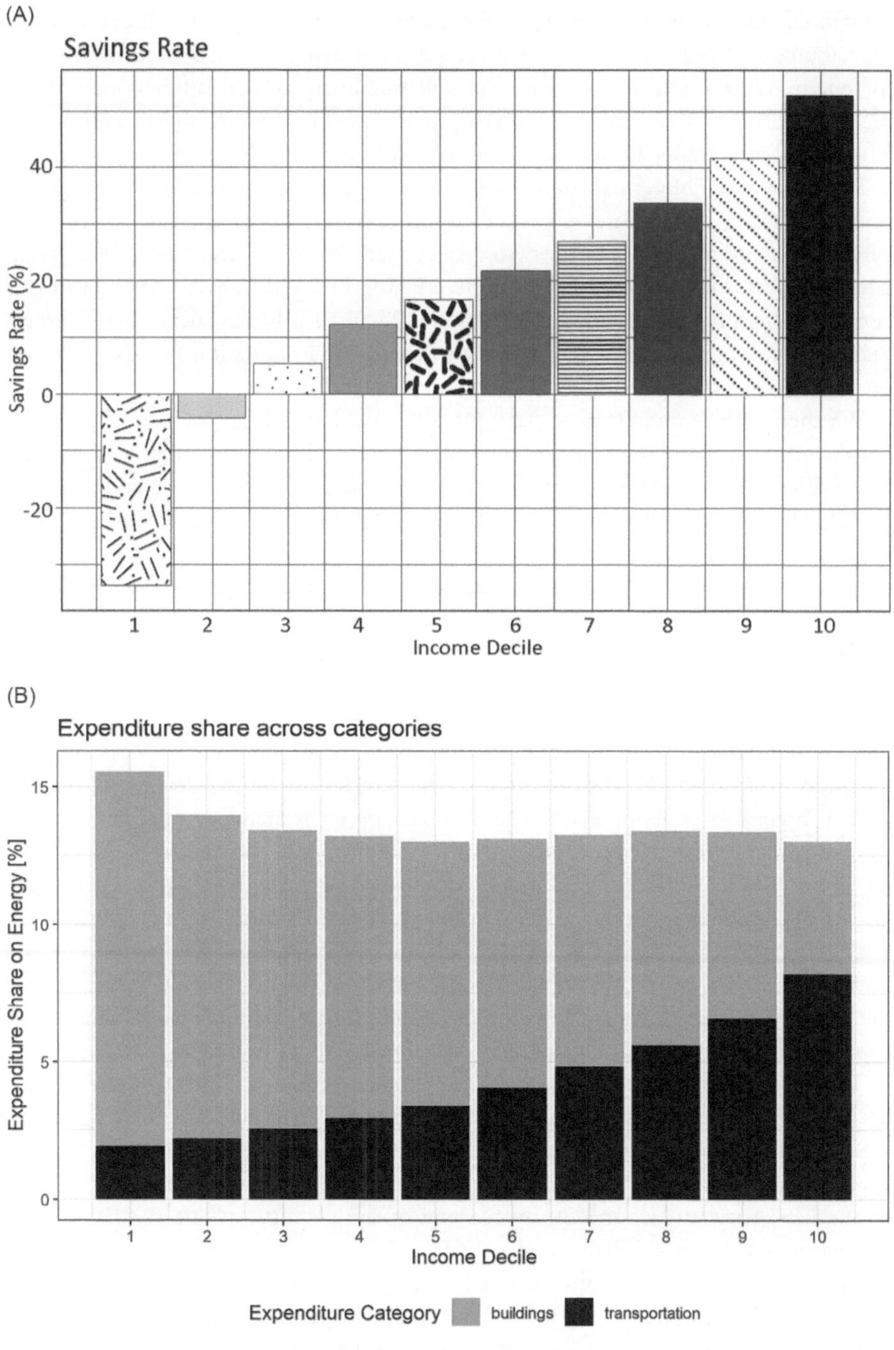

FIGURE 11.2 Saving rate (Figure 11.2A) and expenditure shares (Figure 11.2B) across deciles in the survey year 2011

Source: Author's own elaboration

Note: A full colour version of this figure is available open access under: https://www.routledge.com/Governing-the-Interlinkages-between-the-SDGs-Approaches-Opportunities/Breuer-Malerba-Srigiri-Balasubramanian/p/book/9781032184654

household survey with Consumer Expenditure data - and the India Human Development Survey. Moreover, we calibrate the demand function of households for energy consumption (transportation and buildings) and all other goods, while income is driven by overall wage development based on empirical skill premia and educational attainment projections from IIASA (Samir and Lutz, 2017).

Households at different income levels are very differently affected by changes in prices and taxes, including energy and carbon taxation. To begin with, saving rates vary substantially with the poorest two deciles even exhibiting negative or zero saving rates per month, as we also find for India (see Figure 11.2.) Moreover, expenditures on energy are high for the lowest decile, and in particular for home energy expenditures highly regressive, i.e., higher for poorer households. Transportation fuel expenditures on the other hand are higher for richer households in India.

We then analyze the effects of different pathways and policies on inequality, looking at carbon taxes and revenue recycling. In particular, we use two different scenarios. One with a per capita targeting: this means that all/part of the revenue is recycled on a per-capita basis. The other scenario reflects a targeted approach currently implemented by the government (as in the other LMICs).

Based on the carbon tax, we find a higher tax incidence on poorer households: in 2030, the overall consumption loss is about 2.5% for the poorest decile and decreases to around 1.5% for the richest decile. That is, climate policies without compensation have a higher burden on poorer households. This picture however completely changes when the redistribution of the carbon tax revenues is not distributional-neutral (see panel (A) of Figure 11.3): for equal per capita transfers, consumption increased by around 19% for the poorest decile while the richest decile will suffer a slightly larger loss of about 2%. These numbers are even higher with targeted transfers, in particular for the bottom 20% of the population. This indicates the utmost importance of redistribution schemes which can by far outweigh the regressive impact of climate policies per se. The relative gains are also that large given the high level of poverty and inequality at the bottom of the income distribution: At the lowest decile, an equal per capita based "carbon dividend" would result in net transfers of 56$ per equivalent household member leading to a substantial increase of over 25% in consumption as pictured in Figure 11.3 under the targeting scheme.

The results can also be seen on the Gini index, where the effect of the carbon tax without redistribution on the Gini index is very small, and far outweighed by the redistribution schemes. First, based on wealth dynamics and low educational attainment convergence, we find that the Gini index is slightly increasing in India of about 63 today to around 70 by the end of the century, in line with the SSP2 projection for India of Rao et al. (2019). Yet, we find that the redistribution scheme can have an important impact on inequality based on a carbon tax of US $30/ton of CO_2 increasing at 5% per year. While without redistribution, inequality is slowly increasing, it can be reduced by around 0.5 percentage points for an equal per capita scheme and even more for targeted transfers almost doubling this effect. Note that for 2025, the estimated transfers amount to about US$82 billion. While transfers initially rise with a rising carbon price, as emissions are reduced tax

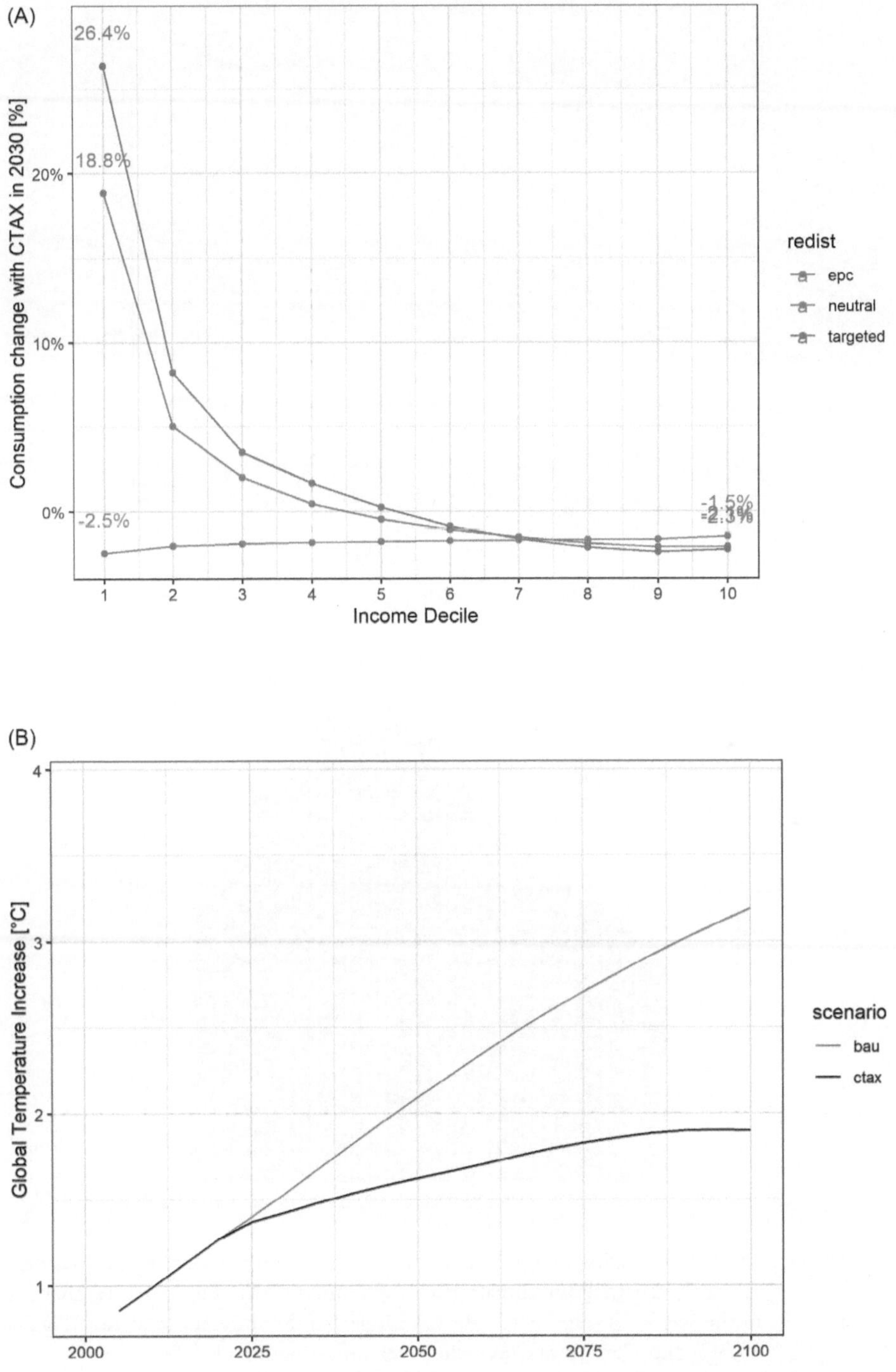
(A)
26.4%
18.8%
-2.5%
-1.5%
-2.3%
Consumption change with CTAX in 2030 [%]
20%
10%
0%
1
2
3
4
5
6
7
8
9
10
Income Decile
redist
epc
neutral
targeted
(B)
Global Temperature Increase [°C]
4
3
2
1
2000
2025
2050
2075
2100
scenario
bau
ctax

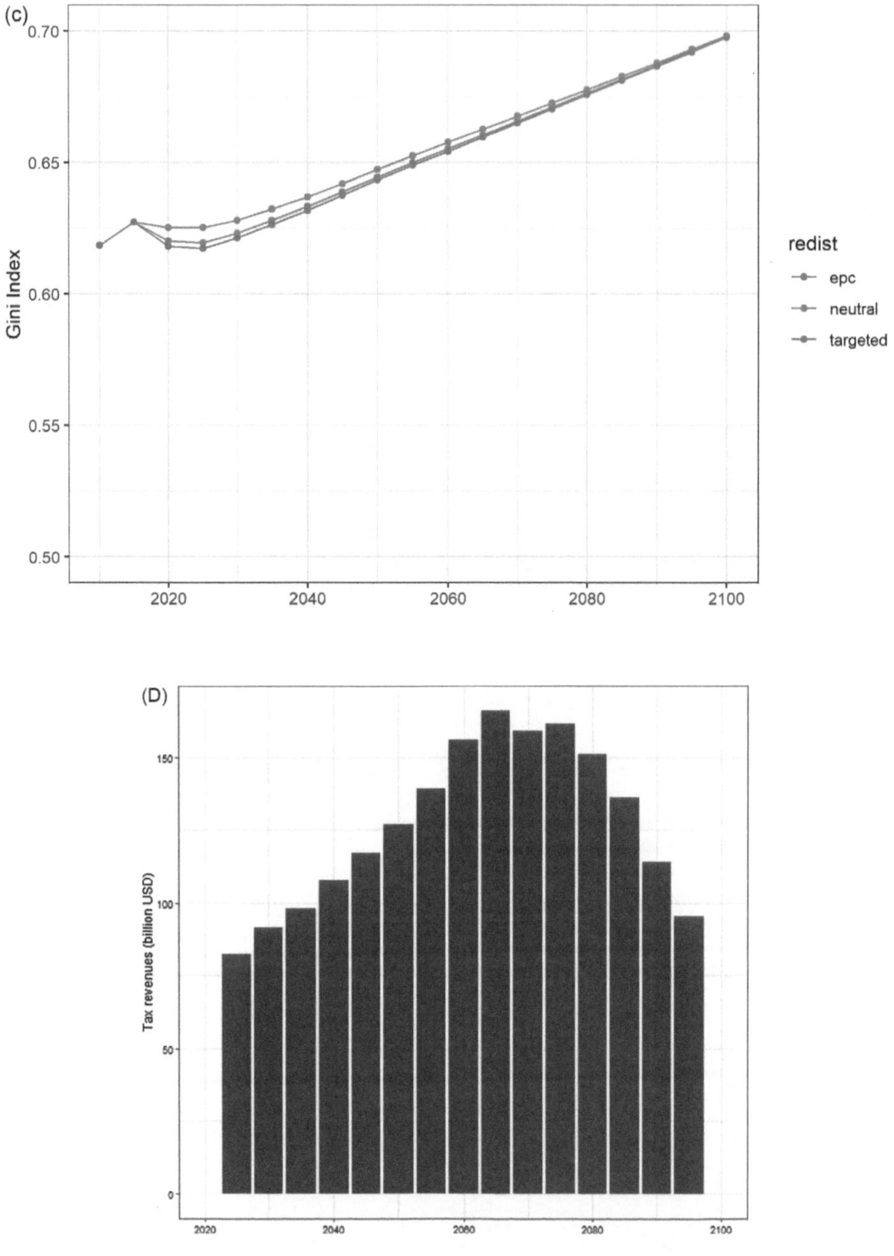

FIGURE 11.3 Modelling results for India. Consumption loss with respect to the business as usual case without climate policy (Figure 11.3A), Temperature changes in the scenarios (Figure 11.3B), Gini index in the different scenarios (Figure 11.3C) and Carbon tax revenues over time (Figure 11.3D)

Source: Author's own elaboration

Note: A full colour version of this figure is available open access under: https://www.routledge.com/Governing-the-Interlinkages-between-the-SDGs-Approaches-Opportunities/Breuer-Malerba-Srigiri-Balasubramanian/p/book/9781032184654

revenues get lower eventually after 2080 starting to decline. This points also to the fact of initial and immediate welfare gains from redistribution, as argued also for in Budolfson et al. (2021). Moreover, it confirms also the finding of Soergel, Kriegler, Bodirsky, et al. (2021) and Soergel, Kriegler, Weindl, et al. (2021) that mitigation increases inequality while redistribution can even overcompensate this effect, and notably leads to a reduction also in poverty.

The Gini coefficient decreases as the total carbon tax revenue is re-distributed across a smaller number of households, most of which are poorer (see Table 11.1 for the share of each decile with BPL card, which get revenue from the carbon tax). Therefore, income for those households increases substantially. But there are also substantial targeting errors, meaning that some of the poorest do not receive compensation, while also even households in the highest deciles receive the transfer.

Implications for the SDGs and Conclusion

The SDGs comprise the tree dimensions of sustainable development: economic, social, and environmental. In addition, compared to the MDGs, one critical addition was to have a specific goal for inequality (SDG 10). Some research has started to show links between inequality and environmental pressures; for example, recent analysis has estimated that the top 10% represent 50% of global emissions (Chancel, Piketty, Saez, and Zucman, 2021b). This chapter has added to this debate by showing the importance of considering inequality when analyzing climate policies in long-term models. In fact, considering inequality in terms of climate policies, is not just a matter of public acceptability and fairness (Klenert et al., 2018), but it also affects the choice of optimal climate policies. More specifically, the chapter has underlined that inequality is being increasingly included in models and especially in IAMs, but major gaps remain. Two of them are (i) to better model within country income distribution and inequality; and (ii) to consider recycling mechanisms of

TABLE 11.1 Share of people with cards for social assistance programs, by decile

Decile	*BPL_card*	*Antyoda card*	*Other card*
1	20.18%	4.42%	7.73%
2	20.56%	3.14%	12.46%
3	19.22%	3.02%	17.37%
4	19.90%	2.58%	18.56%
5	19.06%	2.27%	22.63%
6	16.83%	1.89%	24.72%
7	14.59%	1.23%	27.08%
8	12.26%	0.97%	28.80%
9	8.89%	0.78%	32.03%
10	4.19%	0.41%	32.15%

Source: Author's own elaboration.

carbon tax revenues that go beyond stylized ones, such as per capita transfers. This is crucial especially when we consider LMICs as the current architecture of social protection is far from being universal and reaching everyone; therefore, universal transfers seem unrealistic at least in the short term.

The empirical analysis, which merged a household survey and a state-of-the-art IAM (WITCH) through an inequality module, explored these issues for the case of India. It showed that, redistributing revenues through targeted schemes decreases inequality more compared to per capita recycling.[4] The analysis showed how a more disaggregated analysis can shed light on the distributional implications of climate policies in the long-term; and how important it is to consider limitations of the current architecture of social protection schemes in LMICs. A way forward could be to systematically link inequality to also the selection of optimal climate policy through feedback in the Integrated Assessment Model directly (e.g., RICE50+, NICE, or ReMIND).

In addition, this chapter adds to the literature that uses IAMs to address SDGs and their links. By linking environmental and socio-economic systems, IAMs are in a good position to assess the long-term prospects of the 2030 Agenda. For example, van Soest et al. (2019) look at how IAMs cover multiple SDGs. They conclude that IAMs cover the SDGs related to climate because of their design, and especially environmental ones. They also underline how socio-political and equality goals, and others related to human development and governance, are still not well represented. Most issues are related to modelling limitation and difficulties, including the need to facilitate a better representation of heterogeneity (greater geographical and sectoral detail) by using different types of models (e.g. national and global) and linking different disciplines (especially social sciences) together. In this sense recent research has made important contributions and progress. In a recent paper focusing on links between specific SDGs (climate protection, poverty reduction and reduced inequalities), Soergel, Kriegler, Bodirsky, et al. (2021) estimate that without redistributing revenues progressively, climate policies in the form of a carbon tax would push an estimated 50 million people into poverty; but an equal per capita climate recycling of the revenues dividend would actually slightly decrease global poverty by about 6 million. They also find that international redistribution would make poverty reduction even more substantial, especially in Sub-Saharan Africa.

Other papers go beyond specific SDGs, to look at the 2030 Agenda more broadly. Soergel, Kriegler, Weindl, et al. (2021) look at a broad list of SDGs, covering 56 indicators or proxies across all 17 SDGs. They estimate the effects of ambitious climate policies, alongside general interventions including economic development, education, technological progress and less resource-intensive lifestyles. They find that these interventions are insufficient to reach the targets. In a second step they model more targeted interventions (a sustainable development package), including international climate finance, progressive redistribution of carbon pricing revenues, sufficient and healthy nutrition and improved access to modern energy, all of which enable a more comprehensive sustainable development pathway. They show that these interventions significantly improve the long-term prospects, despite still leaving important gaps; for example, some 180 million people will remain in extreme poverty in 2030.

In summary, IAMs can be a critical tool to understand how to implement sustainable development and the SDG Agenda in an integrated way. Nonetheless, improvements are needed especially in terms of better modelling social outcomes. We have shown in this chapter, for example, the importance of better representing inequality and recycling mechanisms. It is also important to try to include feedback loops from social dimensions (inequality and poverty) on the selection of optimal climate policies, that is, to consider the feedback inequality and poverty dynamics have on total mitigation costs for instance. For example, using Gini or income decile projections, can be informative but does not capture the full essence of the links between social and environmental dimensions of the SDGs. For example, it has been shown how social unrest can block the implementation of climate policies.

From the point of view of the implementation of the SDGs and the overall research questions of the current book, this chapter has focused on policy instruments and policy mixes to steer the implementation of climate mitigation and inequality. It has been shown that also in the short term, climate policies accompanied by social protection measures can make emission reductions more socially acceptable alongside reaching other SDGs. Given the potential sizable amount of carbon revenues and/or transfers, this provides notably a large potential for significant improvements in terms of poverty eradication (SDG 1) and inequality reduction (SDG 10).

Notes

1 Shared Socioeconomic Pathways (SSPs) are scenarios of projected socioeconomic global changes up to 2100. They are used to derive greenhouse gas emissions scenarios with different climate policies. The scenarios are SSP1: Sustainability (Taking the Green Road); SSP2: Middle of the Road; SSP3: Regional Rivalry (A Rocky Road); SSP4: Inequality (A Road divided); and SSP5: Fossil-fuelled Development (Taking the Highway).
2 For example, in 2011–12, 40 percent of those in the bottom 40 percent of India's income distribution were denied their PDS benefits and 65 percent of this population were denied their MGNREGA benefits.
3 The well-off could be excluded using predefined, verifiable exclusion criteria like automobile ownership or a certain bank account balance.
4 As the analysis uses deciles, it could not be analyzed more in detail the number of people losing out in each decile, linked to the leave no one behind principle of the SDGs and the 2030 Agenda.

References

Adler, M., Anthoff, D., Bosetti, V., Garner, G., Keller, K., and Treich, N. (2017). Priority for the worse-off and the social cost of carbon. *Nature Climate Change*, 7, 443–449. doi:10.1038/nclimate3298.

Anthoff, D. and Emmerling, J. (2019). Inequality and the Social Cost of Carbon. *Journal of the Association of Environmental and Resource Economists*, 6, 243–273.

Balasubramanian, S., Kumar, R., and Loungani, P. (2020). Inequality and locational determinants of the distribution of living standards in India. *Center for Open Science*. Retrieved from: https://ideas.repec.org/p/osf/socarx/rmcej.html.

Banerjee, A., Niehaus, P., and Suri, T. (2019). Universal basic income in the developing world. *Annual Review of Economics*, 11, 959–983.

Barca, V. (2018). *Integrating data and information management for social protection: social registries and integrated beneficiary registries*. Retrieved from: https://ideas.repec.org/p/ipc/opager/390.html.

Barrientos, A. (2018). *Social Assistance in Developing Countries Database*. Retrieved from: https://socialprotection.org/discover/publications/social-assistance-developing-countries-database-0.

Brown, C., Ravallion, M., and Van de Walle, D. (2018). A poor means test? Econometric targeting in Africa. *Journal of Development Economics*, 134, 109–124.

Bruckner, B., Hubacek, K., Shan, Y., Zhong, H., and Feng, K. (2022). Impacts of poverty alleviation on national and global carbon emissions. *Nature Sustainability*. doi:10.1038/s41893-021-00842-z.

Budolfson, M. et al. (2021). Climate action with revenue recycling has benefits for poverty, inequality and well-being. *Nature Climate Change*, 11, 1111–1116. doi:10.1038/s41558-021-01217-0.

Chancel, L., Piketty, T., Saez, E., and Zucman, G. (2021a). *World Inequality Report 2022*. World Inequality Lab.

Coady, D., Grosh, M., and Hoddinott, J. (2004). Targeting Outcomes Redux. *The World Bank Research Observer*, 19, 61–85. doi:10.1093/wbro/lkh016.

Das, U. (2015). Can the Rural Employment Guarantee Scheme Reduce Rural Out-migration: Evidence from West Bengal, India. *The Journal of Development Studies*, 51, 621–641. doi:10.1080/00220388.2014.989997.

Dennig, F., Budolfson, M.B., Fleurbaey, M., Siebert, A., and Socolow, R.H. (2015). Inequality, climate impacts on the future poor, and carbon prices. *Proceedings of the National Academy of Sciences*, 112, 15827–15832. doi:10.1073/pnas.1513967112.

Devereux, S., Masset, E., Sabates-Wheeler, R., Samson, M., Rivas, A.-M., and te Lintelo, D. (2017). The targeting effectiveness of social transfers. *Journal of Development Effectiveness*, 9, 162–211. doi:10.1080/19439342.2017.1305981.

Ellis, F. (2012). 'We Are All Poor Here': Economic Difference, Social Divisiveness and Targeting Cash Transfers in Sub-Saharan Africa. *Journal of Development Studies*, 48, 201–214. doi:10.1080/00220388.2011.625408.

Emmerling, J. et al. (2016). *The WITCH 2016 model-documentation and implementation of the shared socioeconomic pathways*.

Emmerling, J. and Tavoni, M. (2021). Representing inequalities in integrated assessment modeling of climate change. *One Earth*, 4, 177–180. doi:10.1016/j.oneear.2021.01.013.

Gazzotti, P. et al. (2021). Persistent inequality in economically optimal climate policies. *Nature Communications*, 12. doi:10.1038/s41467-021-23613-y.

Gentilini, U., Almenfi, M., Orton, I., and Dale, P. (2020). Social protection and jobs responses to COVID-19. *World Bank Publications – Reports 33635*.

Kapur, D. and Nangia, P. (2015). Social Protection in India: A Welfare State Sans Public Goods? *India Review*, 14, 73–90. doi:10.1080/14736489.2015.1001275.

Kishore, A. and Chakrabarti, S. (2015). Is more inclusive more effective? The 'New Style' public distribution system in India. *Food Policy*, 55, 117–130. doi:10.1016/j.foodpol.2015.06.006.

Klenert, D., Mattauch, L., Combet, E., Edenhofer, O., Hepburn, C., Rafaty, R., and Stern, N. (2018). Making carbon pricing work for citizens. *Nature Climate Change*, 8, 669–677. doi:10.1038/s41558-018-0201-2.

Kornek, U., Klenert, D., Edenhofer, O., and Fleurbaey, M. (2021). The social cost of carbon and inequality: When local redistribution shapes global carbon prices. *Journal of Environmental Economics and Management*, 107, 102450. doi:10.1016/j.jeem.2021.102450.

Kozicka, M., Weber, R., and Kalkuhl, M. (2019). Cash vs. in-kind transfers: the role of self-targeting in reforming the Indian food subsidy program. *Food Security*, 11, 915–927. doi:10.1007/s12571-019-00942-x.

Malerba, D., Gaentzsch, A., and Ward, H. (2021). Mitigating poverty: The patterns of multiple carbon tax and recycling regimes for Peru. *Energy Policy*, 149, 111961. doi:10.1016/j.enpol.2020.111961.

Rao, N.D., Sauer, P., Gidden, M., and Riahi, K. (2019). Income inequality projections for the Shared Socioeconomic Pathways (SSPs). *Futures*, 105, 27–39. doi:10.1016/j.futures.2018.07.001.

Ravallion, M. (2019). Guaranteed employment or guaranteed income? *World Development*, 115, 209–221. doi:10.1016/j.worlddev.2018.11.013.

Riahi, K. et al. (2017). The Shared Socioeconomic Pathways and their energy, land use, and greenhouse gas emissions implications: An overview. *Global environmental change*, 42, 153–168.

Samir, K. and Lutz, W. (2017). The human core of the shared socioeconomic pathways: Population scenarios by age, sex and level of education for all countries to 2100. *Global Environmental Change*, 42, 181–192.

Shah, S. (2017). The Conflicted Mahatma: Universal Basic Income in the Government of India's 2016–17 Economic Survey. *Indian Journal of Human Development*, 11, 190–193. doi:10.1177/0973703017734718.

Soergel, B., Kriegler, E., Bodirsky, B.L., Bauer, N., Leimbach, M., and Popp, A. (2021). Combining ambitious climate policies with efforts to eradicate poverty. *Nature Communications*, 12. doi:10.1038/s41467-021-22315-9.

Soergel, B. et al. (2021). A sustainable development pathway for climate action within the UN 2030 Agenda. *Nature Climate Change*, 11, 656–664. doi:10.1038/s41558-021-01098-3.

Steckel, J. C. et al. (2021). Distributional impacts of carbon pricing in developing Asia. *Nature Sustainability*, 4, 1005–1014. doi:10.1038/s41893-021-00758-8.

Taconet, N., Méjean, A., and Guivarch, C. (2020). Influence of climate change impacts and mitigation costs on inequality between countries. *Climatic Change*, 160, 15–34. doi:10.1007/s10584-019-02637-w.

Unnikrishnan, V. and Imai, K.S. (2020). Does the old-age pension scheme improve household welfare? Evidence from India. *World Development*, 134, 105017. doi:10.1016/j.worlddev.2020.105017.

van Soest, H.L. et al. (2019). Analysing interactions among Sustainable Development Goals with Integrated Assessment Models. *Global Transitions*, 1, 210–225. doi:10.1016/j.glt.2019.10.004.

Vogt-Schilb, A.et al. (2019). Cash transfers for pro-poor carbon taxes in Latin America and the Caribbean. *Nature Sustainability*, 2, 941–948. doi:10.1038/s41893-019-0385-0.

Weyant, J. (2020). Some Contributions of Integrated Assessment Models of Global Climate Change. *Review of Environmental Economics and Policy*, 11, 115–137. doi:10.1093/reep/rew018.

World Bank. (2019). *Pathways to Reducing Poverty and Sharing Prosperity in India: Schemes to Systems-Lessons from Social Protection in India.* Retrieved from:

World Bank. (2022). ASPIRE. Retrieved from: www.worldbank.org/en/data/datatopics/aspire.

Yemtsov, R. and Moubarak, A. (2018). *Assessing the readiness of Social Safety Nets to Mitigate the Impact of Reform.*

12

POVERTY, INEQUALITY, AND GROWTH

The East Asian experience

Pooja Balasubramanian

Introduction

Over the past few decades, economic growth has been given substantial importance in lifting individuals out of poverty (Ravallion, 2001; Son and Kakwani, 2008). Some of the prominent studies by Dollar, Kleineberg, and Kraay (2016), show that when average incomes rise, the share of income for the bottom 20/40% also increases, thereby suggesting the pro-poor benefits of economic growth. However, many studies observe that the link between economic growth and poverty is not as straightforward. The poverty growth nexus is even more unclear when we use non-monetary indicators such as the multidimensional index (Bourguignon, 2004) or consider other variables such as inequality in the poverty-growth relationship. In one of his seminal works, Bourguignon (2004) acknowledged that growth alone is not sufficient to reduce poverty, but what is important is how this growth is redistributed. Thus, there is a need to introduce active pro-poor policies in the form of government redistribution that also tackle inequality and move beyond the 'trickle down' narrative that economic growth alone is good to reduce poverty (Seguino, 2019a).

More than a decade after Bourguignon's paper was published; the Agenda 2030 emphasized the need to consider the interrelated nature of Sustainable Development Goals (SDG) and highlighted the synergetic role of reducing inequality and eradicating poverty as an overarching goal, anchored on a strategy of 'Leaving no one behind'. Furthermore, active fiscal policies such as implementing progressive and effective taxation systems and delivering social protection and basic public services to all became crucial in realizing SDG 1 (eradicating poverty) and SDG 10 (reducing inequality).

Based on studies by Braunstein, Seguino, and Altringer (2021), the contribution of this chapter is to identify policy mixes that distinguish two growth strategies: (i) equitable growth that focusses on greater redistribution particularly in the areas of

DOI: 10.4324/9781003254683-12

the care sector i.e health, education and access to social protection; and (ii) inequality enhancing growth that favors lower human capacity generation and focusses on economic growth at the cost of low human development and living standards of people. First, the chapter provides a descriptive overview of trends in policies such as social protection, health care and education (Kwon, Cook, and Kim, 2015; Mosley, Hudson, and Verschoor, 2004), taxation structures (Enami, Lustig, and Aranda, 2018) and global orientation (measured by the level of openness and trade) that are important variables in determining the growth paths followed by different countries.

Second, the chapter considers the poverty and inequality reducing impact of each of the policy instruments, namely social protection programs, investment in health and education, different taxation strategies and finally the level of global orientation. Greater redistribution, with particular focus on investment in public goods such as health care, education and social protection could enable growth to be equitable thereby reducing poverty. However, increasing dependencies on FDI and export-oriented manufacturing along with low investment in building human capabilities (through poor investment in care based social policies) could lead to inequality enhancing growth, with poverty reduction being uncertain. The chapter particularly focusses on the poverty, inequality and growth trends in four low and middle-income countries in East and South East Asia – namely the Philippines, Vietnam, Thailand, and Mongolia, in the decade post-2000. The hypothesis is that countries which implement active fiscal redistributive policies (focussing on the care sector) are more successful in reducing poverty and inequality compared to countries focusing only on economic growth policies driven by trade openness. The chapter finds on the one hand that Mongolia has experienced an increase in economic growth, but this growth is accompanied by increasing inequality. On the other hand, Thailand, although recording moderate growth, experienced both a reduction in poverty and inequality. This also holds for Vietnam. The Philippines experienced slow growth as well as a slow reduction in poverty and inequality. The four countries vary in their approach to social protection systems with Mongolia focusing on one large cash transfer to all their citizens (from the profits earned through the exports of the mining sector) while Vietnam and Thailand have a bundle of different social protection programs targeting different groups. Indirect taxation is either regressive or has no impact on poverty and inequality across the countries. Finally, with respect to global orientation (participation in global trade), Mongolia which is highly dependent on the export of commodities such as mining, redistributed the profits from the export of minerals for establishing a cash transfer program. However, owing to the unstable global demand and fall in prices, Mongolia suffered fiscal instability from 2012 until 2016 and drastically reduced their cash transfer program, restricting it to child benefits alone. Vietnam experienced an increase in both foreign direct investment (FDI) and exports. While global orientation and dependency on international trade exposed the country to greater instability, the wide range of social protection programs in Vietnam could offset possible instabilities.

Trends in Poverty and Inequality in South East Asia

The debt crisis of the 1997s resulted in a contraction of many economies within South East Asia. Despite this, post-2000 the region of East and South East Asia experienced a moderate recovery in economic growth alongside a reduction in poverty. In the period between 2000 and 2016, the Pearson's correlation for the levels in GDP per capita and income poverty in South East Asia is -0.6, higher than countries in Sub-Saharan Africa (SSA) (-0.35) and Eastern Europe and Central Asia (-0.25). On average, the inequality-growth relationship, is positive for South East Asia (0.25) along with SSA (0.51). Higher levels of economic growth is associated with increasing inequality levels. In this chapter we focus on four low and middle-income economies – two of which were directly impacted and 'worst hit' by the Asian financial crisis, namely Thailand and the Philippines and two other countries Mongolia and Vietnam that were indirectly affected. Apart from understanding the policies instruments used by these countries with respect to their social protection systems and taxation structures, the chapter also highlights the heterogeneities across the four countries within the same region in their approach to reduce poverty and inequality.

TABLE 12.1 Trends in long-term changes in poverty, inequality and growth

Country	*Year*	*Absolute Poverty*	*Inequality*	*Growth*
Mongolia	2002–2011	-0.50	0.2	0.07
Vietnam	2002–2014	-0.26	-0.3	0.05
Thailand	2000–2011	-0.38	-0.35	0.04
Philippines	2006–2015	-0.10	-0.31	0.04

Source: The growth and absolute income poverty measures are from the World Development indicators and Povcalnet dataset of the World Bank. Gini (relative income inequality) is taken from UNU-WIDER.

Table 12.1 shows the long-term annualized change in GDP per capita, the squared poverty gap at $1.90, and the GINI (relative inequality). Considering the changes in poverty, inequality, and growth post-2000, the largest reduction in income poverty is for Mongolia and Thailand, followed by Vietnam and the Philippines. Despite experiencing a large increase in GDP per capita and reduction in poverty, Mongolia is the only country among the four nations that experienced a rise in inequality by 0.2. However, although Thailand's change in GDP per capita was not the largest, this country saw a 0.38 reduction in poverty along with a 0.35 reduction in inequality. Vietnam experienced a reduction in poverty and inequality, along with a moderate increase in growth. Finally, the Philippines had not only the lowest GDP per capita, but also the smallest reduction in poverty compared to the other three nations. Thus, we have Mongolia at one end of the spectrum, which experienced an inequality increasing growth while reducing poverty levels. Vietnam also experienced a steady GDP per capita, but unlike Mongolia, not only reduced poverty but also inequality over time. Finally, Thailand and the Philippines had the same change in growth, but Thailand was more successful in reducing poverty and inequality compared to the latter.

What were each of these countries doing differently in terms of their social protection, taxation and growth policies? In the next sections, the chapter observes policies that were implemented in the respective countries leading to varying combinations of poverty, inequality and growth trends over the long run.

Social Protection, Taxation, and Economic Growth

In this chapter, three channels are identified that can either directly or indirectly affect poverty and inequality. The first channel is characterized by the presence of a robust social protection system, encompassing health care and education that can have a direct positive affect on reducing poverty but can also have a long-term impact on growth by investing in human capacities. Complementary to a sound social protection system is the second channel that includes different taxation structures which enable redistribution of resources, not only having a direct positive impact on poverty reduction but also lowering inequalities. A policy mix of the two channels can be defined as growth strategies that are poverty and inequality reducing. The third channel primarily focusses on economic growth (measured as GDP per capita) by increasing the country's dependency on global trade through the promotion of exports and FDI and, investment in capital. Such a profit driven growth strategy is expected to increase the average income of the poorest population groups but the inequality-growth relationship in this case is ambiguous. This third channel is defined as an inequality enhancing growth strategy with an uncertain impact on poverty reduction.

Social protection in South East Asia?

The chapter assesses social protection systems, including access to health and education in East and South East Asia from two key parameters: level of public expenditure on social protection, health and education, and coverage or universalization of different programs (Barrientos, 2010; Barrientos and Hulme, 2008; van de Walle, 1998).

What encompasses social protection systems?

The International Labour Organization conceptualizes social protection as a basic human right, and that comprises of a set of policies and programs designed to reduce and prevent poverty, vulnerability, and social exclusion throughout the life cycle by a mix of contributory schemes (social insurance, social security) and non-contributory tax-financed benefits (including social assistance) (International Labour Organization, 2017). Social assistance includes non-contributory social protection programs in the form of in-kind transfers, cash or conditional cash transfers and subsidies with an aim to reduce chronic and extreme poverty (Barrientos, 2010; Barrientos and Hulme, 2008; Barrientos and Malerba, 2020). Social insurance schemes are based on a principle of solidarity and is funded by contributions from beneficiaries and their employers or subsidized by the government or partly by both the beneficiaries and the government (International Labour Organization, 2017). The objective of social insurance schemes is to protect individuals from risks related to income, health, and climate shocks, particularly those who are poor and vulnerable. More recently, micro insurance has been promoted widely as a complementary social protection tool for people who are excluded from formal social insurance schemes. A third pillar of social protection includes labor market policies through minimum wages, unemployment allowance and pension systems.

The function of social protection policies can further be differentiated between (i) social protection in education and health domains in the form of transfers for childcare and maternity health, access to schooling and public pensions for old age; and (ii) social protection for the working population in the form of unemployment allowance, disability, injury, and public work programs (Niño-Zarazúa, 2019). Social protection programs in education, particularly free or subsidized primary and secondary schooling, school feeding programs, cash transfers conditional on attendance are effective to enhance pro-poor progress in regions where demand is highly income and price elastic (Harttgen, Klasen, and Misselhorn, 2010). Complimentary to public expenditure in social protection programs, increased expenditure in creating universal education systems is essential and can contribute to reducing horizontal inequalities between different social groups from the supply side.

Social protection in health includes non-contributory social health insurance schemes, cash and conditional cash transfer for maternity and childcare. In addition, direct investment in public healthcare can also contribute to the preventative and curative forms of social protection that aim to reduce out of pocket expenditure (OOPE) and user fees for the vulnerable population groups (Mohanan, 2013; Mohanan, Hay, and Mor, 2016; O'Donnell et al., 2008 in Asia; Wagstaff, 2007). However, there is mixed evidence about the role of direct investment in public health. Many studies which observe the impact of public investment in health, find an improvement in child survival rates and life expectancy at birth (You et al., 2015) and increase in utilization of services amongst the poorest quartiles (Kruse, Pradhan, and Sparrow, 2009). The catastrophic effects of HIV have been largely mitigated by the support of state funded social spending (UNAIDS, 2020). However, studies find that an increase in public expenditure on healthcare spending does not matter, as this will not translate to quality services and subsequently demand for public healthcare (Filmer, Hammer, and Pritchett, 2000; McGuire, 2006).

Public expenditure in social protection, health, and education

In this section, I analyze the public expenditure in overall social protection, healthcare and education for each of the four countries. Considering changes in public expenditure for social protection, Thailand ranks the highest, followed by Vietnam and Mongolia, with the Philippines being the lowest. Thailand is one of the few countries in this region that made a switch from contributory insurance schemes to tax funded national systems for social security and health. This helped social protection programs reach the large informal economy (International Labour Organization, 2017). Although direct expenditure on health is low compared to the Philippines or Vietnam, Thailand has been successful in not only ensuring full minimum coverage of health insurance for its population but also focused on the supply side initiatives by expanding healthcare services and improving the service quality (2017). The bar next to public healthcare shows the long-term change in the OOPE that is negative in Thailand compared to the other three nations.

At the other end of the spectrum is the Philippines – where the change in expenditure on social protection programs is the lowest. One of the main reasons is that the Philippines social protection is highly dependent on contributory insurance schemes. The main social protection is the social insurance scheme, which is mandatory for private sector employees, but for workers in the informal sector, it is based on voluntary contributions. The changes in expenditure on the healthcare sector is however, highest in the Philippines along with a marginal increase in OOPE. The increase in healthcare expenditure is primarily driven by harnessing digital technologies to facilitate contribution payments such as smart cards or digital transfers by collaborating with private money transfer agencies.

Mongolia and Vietnam have very similar social protection systems, one that is dependent on contributory insurance schemes and both countries have found it difficult to cover self-employed or hard to reach workers, even though contributions in the insurance programs are subsidized by the governments. Despite these similarities, Mongolia experienced an inequality increasing growth along with poverty reduction, while Vietnam was able to not only reduce poverty but also inequality. Furthermore, Mongolia saw a decline in government health spending (See Figure 12.1) and an

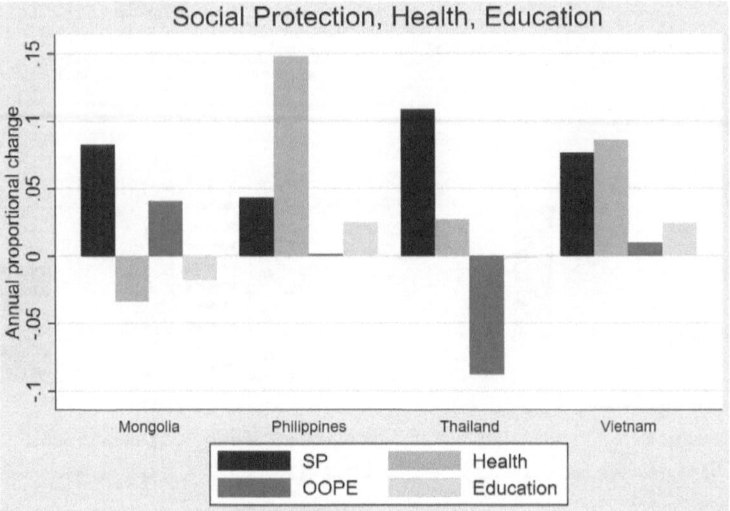

FIGURE 12.1 Annualized long-term changes in public expenditure on social protection, health, education and out of pocket expenditure (OOPE)

Source: Author's own elaboration

increase in OOPE, while in Vietnam there was increased government health spending, reducing OOPE during the same period.

Universal coverage

There has been a longstanding debate on whether social protection programs need to be based on specific targeting or the coverage must be universal. Poverty alleviation programs indeed have some form of targeting across social groups, regions for e.g. reducing the child mortality rate or maternal health programs will focus on children and mothers. Proponents of such broader forms of targeting have pointed out various practical issues like the cost of exclusion (making poverty alleviation programs more regressive), incentive distortion, social stigma and high administration costs that comes with narrow means-based targeting (Brown, Ravallion, and van de Walle, 2016; Klasen and Lange, 2016; Sen, 1995). Finally, many political economy studies find that narrow targeting is not feasible as it might be detrimental

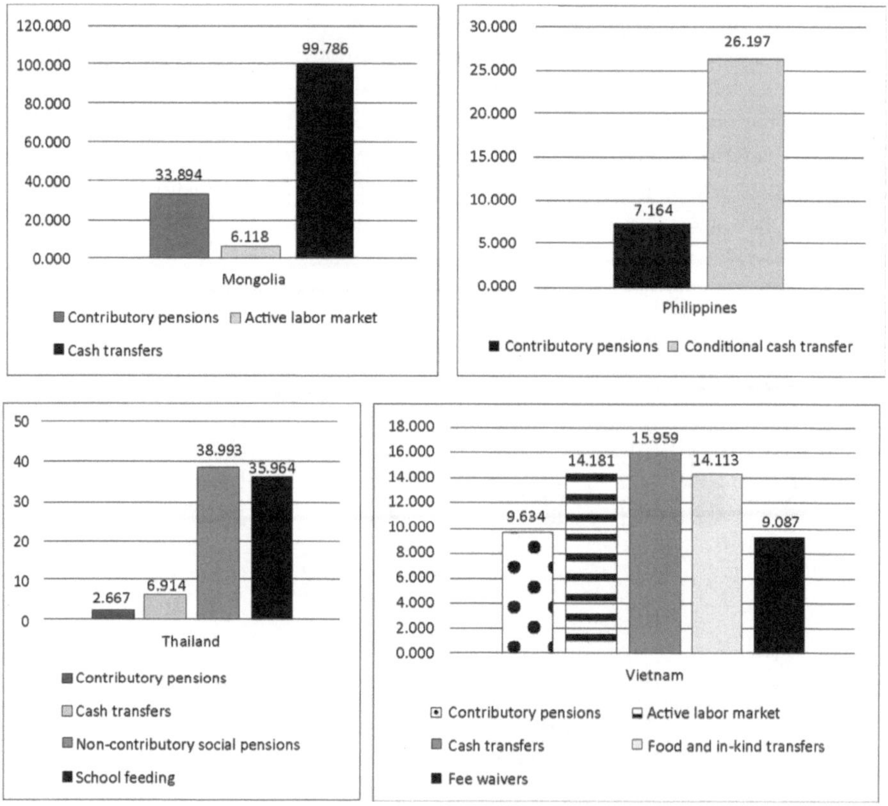

FIGURE 12.2 Percentage of the population covered by different programs providing social assistance

Source: Author's own elaboration

in gaining political support. The main argument against universal coverage is the high public expenditure.

The Figure 12.2 shows the percentage of the population covered by the various social protection programs for the most recent year in each country. The top left panel shows the figures for Mongolia in 2011. Compared to the other three countries the former has been the most successful by reaching out to 99% of the population through their cash transfer program. This is followed by Thailand, the Philippines, and Vietnam. However, in terms of number of programs, Mongolia had only one cash transfer program that was initially funded from the Human Development Fund i.e. profits from the mining industry which was transferred to the entire population. Soon, this transfer could not be fiscally sustained and now it is is only restricted to children (Barrientos, 2019).

The top right panel of Figure 12.2 shows the two social protection programs in the Philippines – the contributory pensions and conditional cash transfer which was targeted at improving the health, nutrition and education of children. In 2012 some 14 million beneficiaries utilized this transfer (Barrientos, 2019). This focus on providing healthcare and education for children can also be seen in Figure 12.1 that shows a high expenditure on healthcare for Philippines (along with negligible OOPE). However, the coverage of the conditional cash transfer in the Philippines is 26% of the eligible population. The bottom right panel of Figure 12.2 shows the coverage of social protection programs for Vietnam. Although the coverage for the eligible recipients is lower than the Philippines, Vietnam has managed to implement a wide range of programs addressing different population groups and is the only country to make discernable efforts to implement active labor market programs. Finally, on the bottom left panel is Thailand which not only has a larger coverage but also includes more than two social protection programs addressing different vulnerable groups.

Impact of social protection on poverty and inequality

Having understood the landscape of social policies, including health and education in the four East Asian countries, I now discuss how successful have the countries been in reducing poverty and inequality. The Asian Development Bank's Social Protection Indicator not only provides the aggregate resources being spent on social assistance and insurance policies but also the poverty and inequality reduction effectiveness of social protection in these countries. Overall in developing countries – less than 20% of the expenditure on social protection is spent on the poor. As at 2015, Vietnam spent the largest amount of their social protection on non-poor beneficiaries i.e. nine times more followed by Mongolia, Thailand, and the Philippines. One of the reasons is that most social protection spending is concentrated on social insurance that has less ability to reach the informal workers and as a result have a weaker distributional impact (according to a study by the Asian Development Bank).

Table 12.2 shows the percentage reduction of inequality and income poverty, owing to change in expenditure on social assistance (excluding social insurance schemes). The change in social assistance spending is a part of the overall expenditure in social protection programs. Overall, I observe the reduction in poverty

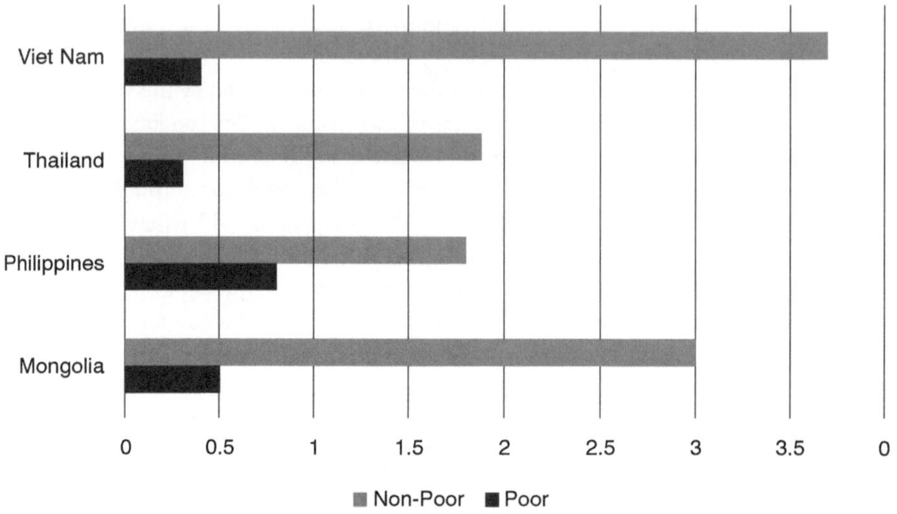

FIGURE 12.3 Percentage of social protection spending on poor and non-poor populations in 2015
Source: Author's own elaboration

TABLE 12.2 Poverty and inequality reducing impact of social assistance

Country	*Year*	*Social Assistance*		
		Gini Inequality Index (% reduction)	*Poverty Headcount (% reduction)*	*Poverty Gap (% reduction)*
Mongolia	2007	6.00	22.35	36.36
Mongolia	2012	10.28	34.75	52.85
Mongolia	2014	4.09	14.83	26.71
Philippines	2013	2.44	12.51	27.33
Philippines	2015	1.51	8.75	15.24
Thailand	2006	0.05	0.14	0.57
Thailand	2009	1.53	7.71	15.34
Thailand	2011	2.79	12.49	23.04
Vietnam	2006	1.45	6.68	13.25
Vietnam	2010	1.35	7.63	13.52
Vietnam	2012	0.38	1.44	3.95

Source: Barrientos, 2018.

(poverty gap and headcount) is larger than reduction in inequality. One of the reasons for this is that although social assistance reaches more beneficiaries, the public expenditure on assistance is less compared to that of social insurance. As a result, absolute poverty might reduce, but this does not affect the distribution extensively. Thailand is the only country that shows a sustained reduction in inequality and poverty. Mongolia, for its part, saw a large reduction in inequality and poverty in 2012, but this reduction fell by about one-half in 2014. One of the main reasons could be that Mongolia stopped their only large-scale cash transfer program using the profits earned from their mining industries after 2012 (Barrientos, 2018).

Taxation systems

While observing the trends in poverty and inequality, the impact of social transfers have to be analyzed alongside various taxation policies (Higgins and Lustig, 2016). Tax constitutes a major source of financing social transfers, but tax revenue is still low across many East and South East Asian countries, with the average tax to GDP ratio being lower than the OECD countries (34%) and even lower than an average of 21 African countries (18.2%)[1]. Furthermore, depending on how fiscal reforms are implemented, the distributional impacts might differ either benefiting the ultra-rich and upper middle class or narrowing inequality (Inchauste and Lustig, 2017; Younger, Myamba, and Mdadila, 2016).

Comparing the four countries, Figure 12.4 shows the annualized long-term change in tax revenues collected by each of the countries. Mongolia (5%) and Thailand (1.9%) show a positive change in domestic resource mobilization (through taxes). Vietnam (-0.03%) and the Philippines (-0.1%) experienced a marginal decrease in the collection of tax as revenue.

What constitutes these tax revenues for each of the countries? On the one hand is the contribution of value added and indirect taxes on consumption. However, revenues

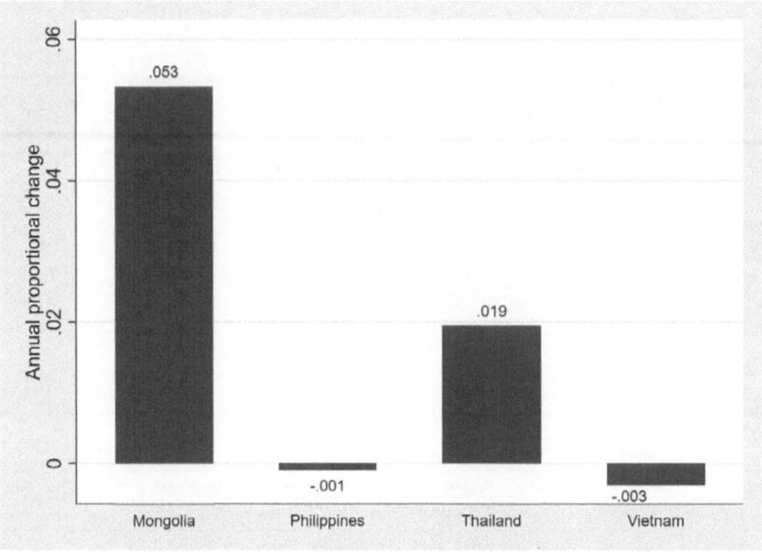

FIGURE 12.4 Annualized long-term changes in tax revenue as a percentage of GDP
Source: Author's own elaboration

from such taxes can increase out of pocket expenditure and thereby have a regressive impact on income poverty, unless governments are able to use the revenue from the indirect taxes for the purposes of public spending on social protection, health and education (Hirvonen, Mascagni, and Roelen, 2018; Muñoz and Cho, 2004). Many countries in East Asia introduced the 'sin taxes' on goods such as alcohol, tobacco, and sugar, which are hazardous for health. It was expected that the revenues from this can be diverted towards healthcare spending. However, this was not sustainable as most of these taxes disproportionately impacted low and middle-income households.

On the other hand, corporate taxes, wealth and inheritance, and property taxes are progressive forms of revenue generation and can be important levers generating revenue for social protection programs (UN Conference on Trade and Development, 2012). There are many reasons for strengthening corporate taxes, which the government can use as revenue for increasing social transfers and lower income poverty. Entrepreneurs do not make investments only thinking about the net profit for one single period, and instead make expectations regarding the future demand of the goods and services they produce. If higher tax revenues from corporate taxes are used as government expenditure in enhancing human capabilities and ensuring the availability of a productive labor supply, this will enable sustained growth in demand. The net profits will rise despite initial tax increase (Braunstein, Bouhia, and Seguino, 2019; Seguino, 2012, 2019a; UN Conference on Trade and

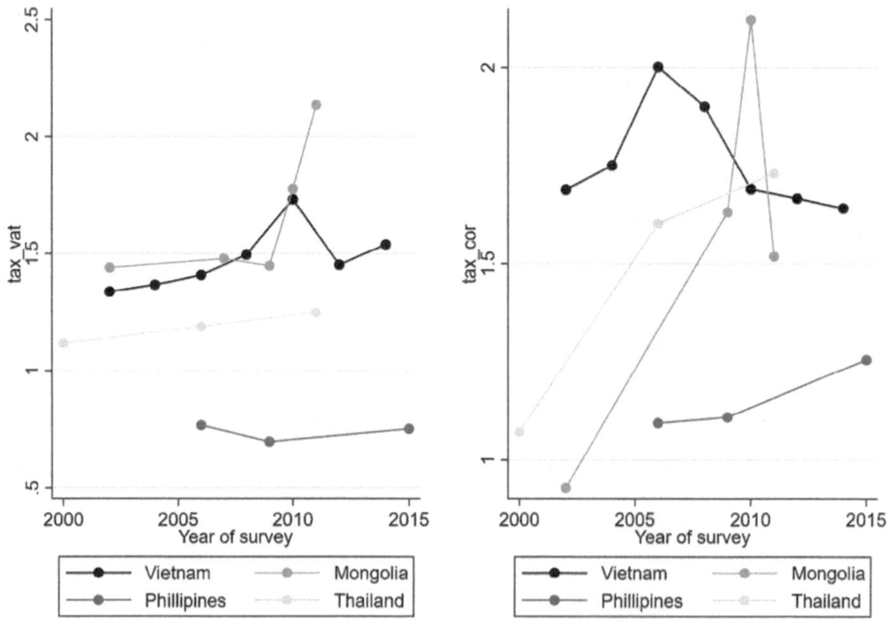

FIGURE 12.5 Absolute values of value added tax (VAT) as a percent of GDP on the left hand side, and corporate tax as a percent of GDP on the right hand side.

Source: Author's own elaboration

Development, 2012). Similarly, inheritance and wealth tax can act as a huge leveler reducing inequality and poverty.

In Figure 12.5, the left panel plots VAT for each country and the right panel denotes revenue from corporate taxes. Mongolia that can be considered as an example of inequality led growth shows one of the highest increases in VAT compared to the other three countries, while corporate taxes increased until 2010 and subsequently had a steep decline. Mongolia is an important example that shows a trend of increasing indirect taxation (that can be fiscally impoverishing) and rise in inequality in the same year span (Corbacho, Cibils, and Lora, 2013; Higgins and Lustig, 2016; Llerena Pinto, Llerena Pinto, and Llerena Pinto, 2015). For its part, Thailand not only reduced income poverty but also experienced a decrease in inequality had a marginal increase in VAT, along with a rise in corporate taxation. Vietnam and the Philippines, which experienced a reduction in inequality, also show either no change or a decrease in indirect taxation respectively.

Impact of tax-transfers systems on poverty and inequality

In the previous section, I described the different modalities of the tax system for the four countries – a transition country like Mongolia experienced an increase in inequality post-2012 and around the same time saw an increase in indirect taxes such as VAT. The revenue from corporate taxes was also unstable, owing to the dependency on taxation from corporates and royalties of one of the main exports in the commodities sector i.e mining. Alternatively, both Thailand and the Philippines observe a flat curve in terms of indirect taxes such as value-added tax (VAT) and higher revenue from corporate taxes over time. In a study by the Centre for Equity Assessment (CEQ), they assess how progressive a fiscal policy is in order to reduce poverty and inequality within the tax-transfer framework (Enami et al., 2018; Inchauste and Lustig, 2017). The tax-transfer system is an important calculation also from the point of view of studying progress in the SDG 10 (Reducing inequality). Among the four countries, this analysis is available only for Mongolia and Vietnam.

In the case of Mongolia, the study by CEQ was conducted on the Household Survey of 2016, which found the tax and transfers system to reduce both poverty and inequality. However, when considering the marginal impact of each of the policies, VAT and excise taxes increase the inequality by one GINI point, and personal taxes and social security contributions reduce inequality between 0.7 and 0.4 GINI points. Despite the redistributive ability of personal income tax, and social security, the public expenditure from these sources in the form of tax-funded contributions were less and the cash transfer in Mongolia was funded using the revenue from taxing corporations and royalties from the export of mineral commodities. This had particularly led to fiscal instability between 2012 and 2016, when the benefits from mining fell (Freije and Yang, 2018).

For Vietnam, the total reduction in GINI from market to final income was 0.35 points, primarily driven by inequality reduction from in-kind and cash transfers in education and health. This is also corroborated in Figures 12.1 and 12.2 which shows Vietnam's increase in expenditure on social protection, health and education and a diverse range of programs covering the beneficiaries respectively. However, tax alone had a very small impact on reducing inequality in Vietnam (World Bank, 2016).

Global orientation

The third channel proposed is the extent of global orientation measured by exports and FDI inflows that are expected to bring about trickle down economic growth, increase average incomes, thereby reducing income poverty and inequality. Studies supporting this argument find that while openness does not significantly impact income poverty in a direct manner, it reduces poverty indirectly through the growth component (Dollar and Kraay, 2002; Foster and Székely, 2008; Kraay, 2006; Ravallion, 2007). This has been echoed in a recent study by Santos, Dabus, and Delbianco (2017) where they find export-driven countries and greater trade openness reduce not only income but also non-income measures of poverty.

There is a direct impact of factors such as FDI inflows and manufacturing exports on enabling an environment of decent employment and domestic wage growth. However, the experience of the 'East Asian Tigers' including countries such as South Korea, Japan and Hong Kong shows that growth rates are not enough. Semi-industrialized economies focusing on large-scale export manufacturing are under extreme pressure to ensure a conducive business environment, and continue to struggle with high poverty levels, poor labor standards and large gender wage gaps (Seguino, 2000). With regard to FDI inflows, the high share of FDI in domestic investment, makes the host country more vulnerable to capital flights. Therefore many semi-industrialized countries, maintain low labor costs; otherwise, the mobility of capital will result in firms to relocate, enabling a race-to-the-bottom competitive environment between different low and middle income countries. Several studies have explored the employment and wage effects of the intense global competition causing firms to adopt flexible and informal work arrangements that are temporary, casual based on unregulated labor contracts. Both these measures of global orientation are likely to have a negative impact on decent employment structures in different countries (Braunstein et al., 2019; Seguino, 2019b; UN Conference on Trade and Development, 2016).

I expect both FDI and increasing dependency on export-oriented products to have positive impact on GDP growth, but the direct impact on income poverty and inequality to be negative or not significant. The increasing dependency of countries on foreign capital flows and unstable global demand is expected to reduce their ability to mobilize revenue for investment in social protection programs and taxation policies. This could hamper expenditure in the provisions of public goods such as universal healthcare and education. Laudage (2020) observes a trade-off between FDI inflows and revenue generated from corporate income taxes in many countries, subsequently having a negative impact on domestic well-being.

Before observing the impact of global orientation on growth, poverty and inequality, it is helpful to discuss the trends of exports and FDI as a percentage of GDP for each of the countries. Mongolia transitioned from a socialist economy and opened up their country to a large increase in FDI (Figure 12.6 A). The only other country that experienced a marginal increase in FDI is Vietnam. Meanwhile, the Philippines and Thailand experienced a decrease in FDI. With regard to global orientation, Thailand shows a long-term decrease in dependency on FDI but a marginal increase in exports. Vietnam experienced a positive increase in both FDI and exports as a percentage of its GDP. Finally, the Philippines shows a decrease in both long-term FDI and exports.

In order to understand the role of FDI and export in reducing poverty and inequality, I undertake a pairwise correlation of the variables respectively. The correlation reveals that on average, both FDI and exports as a percentage of the GDP are not significantly correlated with poverty and inequality. At the country level, in Mongolia, FDI as a percentage of GDP is negative and significantly correlated with reducing income poverty, and in Vietnam, exports as a percentage of GDP are negative and significantly correlated with reducing income poverty. However, these correlations are very weak, as shown in the scatter plot in Figure 12.6 B. For the four countries, there is no association between FDI and exports, as a percentage of the GDP, with inequality.

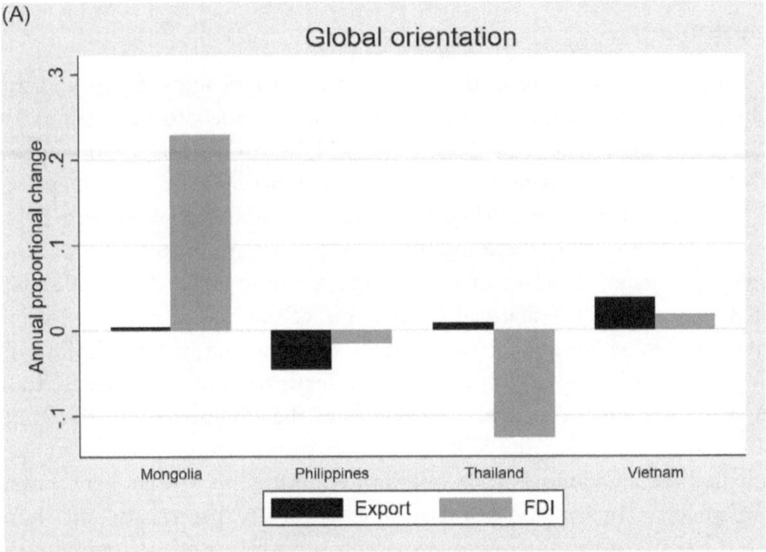

FIGURE 12.6A Annualized long-term changes in FDI and Export as percentage of GDP
Source: Author's own elaboration

(B)

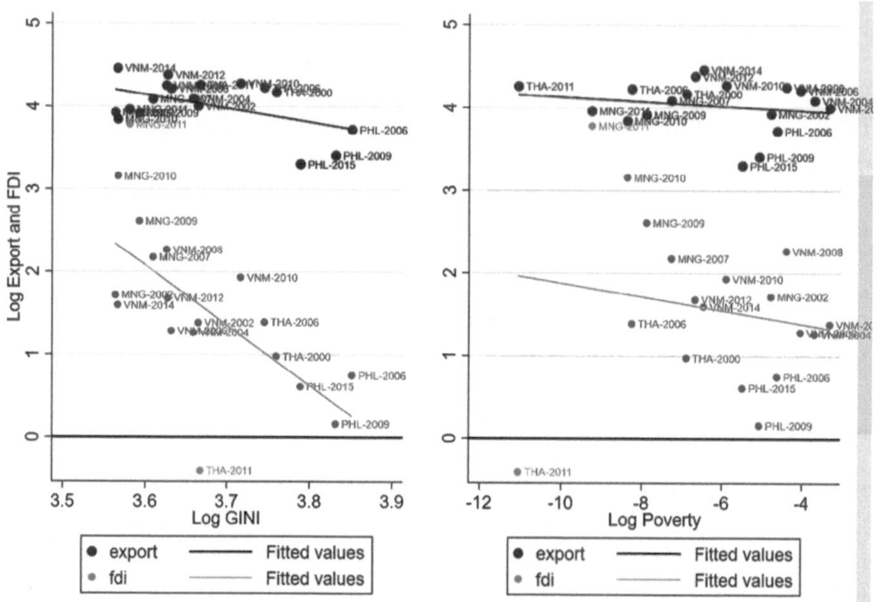

FIGURE 12.6B Scatter plot of the correlation between export and FDI as percentage of GDP with inequality (GINI) and income poverty (squared-poverty gap), respectively
Source: Author's own elaboration

Conclusion

There is a growing literature in the area of interlinkages across SDGs and the need to facilitate the harnessing of synergies and mitigate trade-offs between and within SDGs (Breuer, Leininger, and Tosun, 2019; Niestroy, Hege, Dirth, Zondervan, and Derr, 2019; Organisation for Economic Co-operation and Development, 2019; Tosun and Leininger, 2017). However, the academic work and policy discussions on how to tackle these interlinkages, particularly in terms of policy mixes is limited. The main objective of this chapter is to provide a descriptive picture of the different policy mixes that can address one of the crucial concerns of integrating three SDGs; economic growth (SDG 8), reducing poverty (SDG 1), and reducing inequality (SDG 10) with a specific focus on four countries in South and East Asia – Mongolia, Vietnam, Thailand, and the Philippines. In the context of the COVID-19 pandemic, the SDGs on poverty and inequality received attention, not only because an addition of 88 million to 115 million people were pushed into extreme poverty, but it also revealed the extremely haphazard and short-term response of many governments in providing subsidies and cash transfers through the social protection systems (Target 1.3). It was clear that our existing systems of public services, healthcare, and social protection were not sufficient to provide a safety net to people.

Keeping these aspects in mind, the policies highlighted in the chapter include social protection (also involving health and education expenditure) and progressive taxation policies for the purpose of redistribution on the one hand and exports and FDI as drivers of economic growth on the other, as potential means to reduce poverty and inequality. Based on the conceptual framework of Seguino (2019a), the chapter identifies two growth trajectories, namely (i) inequality enhancing growth; and (ii) inequality reducing growth that can lead to reduction in income poverty and inequality. On the one hand, a transition country such as Mongolia followed an inequality increasing growth path and was largely dependent on unstable gains from the export of minerals within the commodity sector. The large gains from this export were distributed to the entire population in the form of a centralized cash transfer scheme. However, owing to instability in the world market for the minerals, Mongolia experienced fiscal instability and could no longer continue the cash transfer scheme.

Thailand, for its part, can be categorized as an economy that promotes inequality decreasing growth. The universal health insurance scheme complemented by increased expenditure on investment in care related infrastructure has enabled Thailand to reach out to informal sector workers. Thailand, similar to Vietnam has broadened their social protection systems to cover various schemes addressing specific target groups e.g. old age pensions, cash transfers for schooling and school feeding, food and in-kind transfers, to name just a few. As a result, there is no direct impact of taxation, but an indirect effect of tax-funded social protection schemes in reducing both poverty and inequality.

Note

1 Botswana, Burkina Faso, Cameroon, Cabo Verde, Congo, Côte d'Ivoire, the Democratic Republic of the Congo, Egypt, Kingdom of Eswatini, Ghana, Kenya, Mali, Mauritius, Morocco, Niger, Rwanda, Senegal, South Africa, Togo, Tunisia, and Uganda.

References

Barrientos, A. (2010). Social Protection and Poverty. Retrieved from: www.unrisd.org/80256B3C005BCCF9/(httpPublications)/973B0F57CA78D834C12576DB003BE255?OpenDocument&cntxt=1A348&cookielang=es

Barrientos, A. (2018). *Social Assistance in Developing Countries Database.* Retrieved from: https://socialprotection.org/discover/publications/social-assistance-developing-countries-database-0.

Barrientos, A. (2019). The role of Social Assistance in reducing poverty and inequality in Asia and The Pacific. In *ADB Sustainable Development Working Paper Series No* 62.

Barrientos, A. and Hulme, D. (2008). *Social Protection for the Poor and Poorest: An Introduction.* In A. Barrientos and D. Hulme (Eds). Palgrave Macmillan.

Barrientos, A. and Malerba, D. (2020). Social assistance and inclusive growth. *International Social Security Review*, 73, 33–53. doi:10.1111/issr.12244.

Bourguignon, F. (2004). The Poverty-Growth-Inequality Triangle. *Working Paper No.* 125. New Delhi: Indian Council for Research on International Economic Relations.

Braunstein, E., Bouhia, R., and Seguino, S. (2019). Social reproduction, gender equality and economic growth. *Cambridge Journal of Economics*, 44. doi:10.1093/cje/bez032.

Braunstein, E., Seguino, S., and Altringer, L. (2021). Estimating the Role of Social Reproduction in Economic Growth. *International Journal of Political Economy*, 50(2), 143–164. doi:10.1080/08911916.2021.1942963.

Breuer, A., Leininger, J., and Tosun, J. (2019). Integrated Policymaking. In *Choosing an Institutional Design for Implementing the Sustainable Development Goals (SDGs)*. Bonn: Deutsches Institut für Entwicklungspolitik.

Brown, C., Ravallion, M., and van de Walle, D. (2016). A Poor Means Test? Econometric Targeting in Africa. *National Bureau of Economic Research Working Paper Series*, No. 22919. Retrieved from: www.nber.org/papers/w22919.

Corbacho, A., Cibils, V.F., and Lora, E. (2013). *More than Revenue: Taxation as a Development Tool*: Springer.

Dollar, D., Kleineberg, T., and Kraay, A. (2016). Growth still is good for the poor. *European Economic Review*, 81, 68–85. doi:10.1016/j.euroecorev.2015.05.008.

Dollar, D. and Kraay, A. (2002). Growth is Good for the Poor. *Journal of Economic Growth*, 7, 195–225. doi:10.1023/a:1020139631000.

Enami, A., Lustig, N., and Aranda, R. (2018). Analytical Foundations: Measuring the Redistributive Impact of Taxes and Transfers. In N. Lustig (Ed.), *Commitment to Equity Handbook: Estimating the Impact of Fiscal Policy on Inequality and Poverty*: Brookings Institution Press and CEQ Institute, Tulane University.

Filmer, D., Hammer, J.S., and Pritchett, L.H. (2000). Weak Links in the Chain: A Diagnosis of Health Policy in Poor Countries. *The World Bank Research Observer*, 15, 199–224. doi:10.1093/wbro/15.2.199.

Foster, J.E. and Székely, M. (2008). Is economic growth good for the poor? Tracking low incomes using general means. *International Economic Review*, 49, 1143–1172. doi:10.1111/j.1468-2354.2008.00509.x.

Freije, S. and Yang, J. (2018). *Mongolia: Distributional Impact of Taxes and Transfers*. Retrieved from:

Harttgen, K., Klasen, S., and Misselhorn, M. (2010). Pro-Poor Progress in Education in Developing Countries? *Review of Economics and Institutions*. doi:10.5202/rei.v1i1.6.

Higgins, S. and Lustig, N. (2016). Can a poverty-reducing and progressive tax and transfer system hurt the poor? *Journal of Development Economics*, 122, 63–75. doi:10.1016/j.jdeveco.2016.04.001.

Hirvonen, K., Mascagni, G., and Roelen, K. (2018). Linking taxation and social protection: Evidence on redistribution and poverty reduction in Ethiopia. *International Social Security Review*, 71(1), 3–24. doi:10.1111/issr.12159.

International Labour Organization. (2017). *World Social Protection Report 2017–2019: Universal social protection to achieve the Sustainable Development Goals*. Retrieved from: www.ilo.org/wcmsp5/groups/public/—dgreports/—dcomm/—publ/documents/publication/wcms_604882.pdf.

Inchauste, G. and Lustig, N. (2017). *The Distributional Impact of Taxes and Transfers: Evidence from Eight low- and middle-income Countries*. World Bank Group.

Klasen, S. and Lange, S. (2016). How Narrowly Should Anti-poverty Programs Be Targeted? Simulation Evidence from Bolivia and Indonesia. *Courant Research Centre: Poverty, Equity and Growth – Discussion Papers* 213. Retrieved from: https://ideas.repec.org/p/got/gotcrc/213.html.

Kraay, A. (2006). When is growth pro-poor? Evidence from a panel of countries. *Journal of Development Economics*, 80, 198–227. doi:10.1016/j.jdeveco.2005.02.004.

Kruse, I., Pradhan, M., and Sparrow, R.A. (2009). Marginal Benefit Incidence of Public Health Spending: Evidence from Indonesian Sub-National Data. *SSRN Electronic Journal*. doi:10.2139/ssrn.1526565.

Kwon, H.-J., Cook, S., and Kim, Y. (2015). Shaping the national social protection strategy in Cambodia: Global influence and national ownership. *Global Social Policy*, 15, 125–145. doi:10.1177/1468018114543310.

Laudage, S. (2020). *Corporate tax revenue and foreign direct investment: Potential trade-offs and how to address them*. 17, 2020.

Llerena Pinto, F., Llerena Pinto, M.C., and Llerena Pinto, M.A. (2015). Social Spending, Taxes and Income Redistribution in Ecuador. *Commitment to Equity (CEQ) Working Paper Series* 28.

McGuire, J.W. (2006). Basic health care provision and under-5 mortality: A Cross-National study of developing Countries. *World Development*, 34, 405–425. doi:10.1016/j.worlddev.2005.08.004.

Mohanan, M. (2013). Causal Effects of Health Shocks on Consumption and Debt: Quasi-Experimental Evidence from Bus Accident Injuries. *Review of Economics and Statistics*, 95, 673–681. doi:10.1162/rest_a_00262.

Mohanan, M., Hay, K., and Mor, N. (2016). Quality of Health Care in India: Challenges, Priorities, and the Road Ahead. *Health Affairs*, 35, 1753–1758. doi:10.1377/hlthaff.2016.0676.

Mosley, P., Hudson, J., and Verschoor, A. (2004). Aid, Poverty Reduction and the 'New Conditionality'. *The Economic Journal*, 114, F217–F243. doi:10.1111/j.1468-0297.2004.00220.x.

Muñoz, S. and Cho, S.S.-W. (2004). Social Impact of a Tax Reform: The Case of Ethiopia. *IMF Working Papers*, 3, 1. doi:10.5089/9781451875584.001.

Niestroy, I., Hege, E., Dirth, E., Zondervan, R., and Derr, K. (2019). *Europe's approach to implementing the Sustainable Development Goals: Good practices and the way forward*. Brussels: European Union.

Niño-Zarazúa, M. (2019). Welfare and Redistributive Effects of Social Assistance in the Global South. *Population and Development Review*, 45, 3–22. doi:10.1111/padr.12308.

O'Donnell, O. et al. (2008). Who pays for health care in Asia? *Journal of Health Economics*, 27, 460–475. doi:10.1016/j.jhealeco.2007.08.005.

Organisation for Economic Co-operation and Development. (2019). *Policy Coherence for Sustainable Development 2019: Empowering People and Ensuring Inclusiveness and Equality*. Paris: OECD Publishing.

Ravallion, M. (2001). Growth, Inequality and Poverty: Looking Beyond Averages. *World Development*, 29, 1803–1815. doi:10.1016/s0305-750x(01)00072-9.

Ravallion, M. (2007). *Looking beyond averages in the trade and poverty debate*. (pp. 118–144): Palgrave Macmillan.

Santos, M.E., Dabus, C., and Delbianco, F. (2017). Growth and Poverty Revisited from a Multidimensional Perspective. *The Journal of Development Studies*, 55, 260–277. doi:10.1080/00220388.2017.1393520.

Seguino, S. (2000). Accounting for Gender in Asian Economic Growth. *Feminist Economics*, 6, 27–58. doi:10.1080/135457000750020128.

Seguino, S. (2012). Macroeconomics, Human Development, and Distribution. *Journal of Human Development and Capabilities*, 13, 59–81. doi:10.1080/19452829.2011.637376.

Seguino, S. (2019a). Macroeconomic Policy Tools to Finance Gender Equality. *Development Policy Review*, 37, 504–525. doi:10.1111/dpr.12396.

Seguino, S. (2019b). Tools of macroeconomic policy: Fiscal, monetary and macroprudential approaches. *Gender Equality and Inclusive Growth*, 46–75. doi:10.18356/8055dda1-en.

Sen, A. (1995). *Gender Inequality and Theories of Justice*. M.C. Nussbaum and J. Glover (Eds.), (pp. 259–273): Clarendon Press.
Son, H.H. and Kakwani, N. (2008). Global Estimates of Pro-Poor Growth. *World Development*, 36, 1048–1066. doi:10.1016/j.worlddev.2007.10.002.
Tosun, J. and Leininger, J. (2017). Governing the Interlinkages between the Sustainable Development Goals: Approaches to Attain Policy Integration. *Global Challenges*, 1(9), 1700036. doi:10.1002/gch2.201700036.
UNAIDS. (2020). 2020Global AIDS Update – Seizing the moment – Tackling entrenched inequalities to end epidemics. United Nations.
UN Conference on Trade and Development. (2012). *World investment report: Towards a new generation of investment policies*: United Nations.
UN Conference on Trade and Development. (2016). *World investment report: Investor nationality, policy challenges*. United Nations.
van de Walle, D. (1998). Assessing the welfare impacts of public spending. *World Development*, 26, 365–379. doi:10.1016/s0305-750x(97)10064-x.
Wagstaff, A. (2007). Health systems in East Asia: what can developing countries learn from Japan and the Asian Tigers? *Health Economics*, 16, 441–456. doi:10.1002/hec.1180.
World Bank. (2016). Sustaining Success: Priorities for inclusive and sustainable growth. Retrieved from: https://documents1.worldbank.org/curated/en/334491474293198764/pdf/108348-REVISED-PUBLIC-ACS.pdf.
You, D. et al. (2015). Global, regional, and national levels and trends in under-5 mortality between 1990 and 2015, with scenario-based projections to 2030: a systematic analysis by the UN Inter-agency Group for Child Mortality Estimation. *The Lancet*, 386, 2275–2286. doi:10.1016/s0140-6736(15)00120-8.
Younger, S.D., Myamba, F., and Mdadila, K. (2016). Fiscal Incidence in Tanzania. *African Development Review*, 28, 264–276. doi:10.1111/1467-8268.12204.

13

A DECISION-MAKING TOOL FOR SYSTEMS THINKING IN SDG IMPLEMENTATION

Experiences from Sweden's Voluntary National Review 2021

Henrik Carlsen, Nina Weitz and Therese Bennich

Introduction

The establishment of the United Nations High-Level Political Forum on Sustainable Development (HLPF) was mandated in 2012 by the outcome document of the United Nations Conference on Sustainable Development (Rio+20), "The Future We Want". HLPF replaced the Commission on Sustainable Development, which had met annually since 1993. The HLPF is the main UN platform on sustainable development and it has a central role in the follow-up on the Sustainable Development Goals (SDG) at the global level.[1]

As part of UN follow-up and review mechanisms, the 2030 Agenda encourages member states to conduct regular and inclusive reviews of progress at the national and sub-national levels. These so-called voluntary national reviews (VNR) serve as a basis for the regular reviews by the annual HLPF. Those regular reviews are voluntary, state-led, undertaken by both developed and developing countries, and in most cases they involve stakeholder engagement processes. The process facilitates sharing of experiences and mutual lessons learned, with the aim of accelerating the implementation of the 2030 Agenda.

In the 2021 VNR process, Sweden, as one of 42 countries, decided to submit a report. While Sweden is often seen as a forerunner on sustainable development and makes good progress on implementing the SDGs, the 2021 VNR report shows that more actions are needed in order to fully implement the 2030 Agenda (Government Offices of Sweden, 2021). The report especially highlights that economic and social inequalities are growing, that several of the national environmental objectives are still lagging behind and that mental ill-health, especially among younger women, presents a challenge.

In 2016 the Government appointed a Delegation for the 2030 Agenda, tasked with producing a report to serve as a basis for supporting and stimulating Sweden's

DOI: 10.4324/9781003254683-13

efforts to implement the 2030 Agenda. Of specific interest for this chapter is that the Delegation, in its final report (Official Reports of Sweden, 2019), stressed the importance of policy coherence for sustainable development. This was taken forward by the Government in its bill to the Parliament on the 2030 Agenda in 2020 and reflects the importance of coherence, which now constitutes a core objective in Sweden's implementation of the 2030 Agenda.

The Swedish approach to policy coherence is in line with the OECD Recommendation on Policy Coherence for Sustainable Development (Organisation for Economic Co-operation and Development, 2019). The OECD model builds on eight principles for promoting policy coherence, including political commitment and leadership, policy integration, whole-of-government coordination, stakeholder engagement as well as monitoring, reporting and evaluation.

Operationalizing policy coherence means that it is important to identify interlinkages between different SDGs. Sweden's VNR states: "Identifying potential synergies and trade-offs between the goals in a structured way provides a better basis for making decisions that lead to sustainable development" (Government Offices of Sweden, 2021, p. 70). In the course of developing the Swedish VNR, the Government commissioned the Stockholm Environment Institute (SEI) to launch a pilot project utilizing *SDG Synergies* process and tool[2] to survey, assess and analyse synergies and trade-offs between the SDGs at the Swedish national level. This chapter describes and critically assess this application of *SDG Synergies*.

This chapter primarily addresses research question 4 of this volume: What institutional processes and policy strategies can strengthen the capacities of different stakeholders in order to achieve inclusive and participatory sustainability governance? In doing this we attune the generic SDG Synergies process to the context of national Swedish policy-making with regards to sustainable development, and design a participatory process including civil servants from most of the Government's ministries. The primary aim of this study is to investigate whether this participatory and tailor-made process could help the Government in operationalizing policy coherence for the achievement of the 2030 Agenda.

By investigating whether SDG Synergies could provide building blocks of effective governance mechanisms, this chapter furthermore touches upon the first part of research question 2: What governance mechanisms are needed to manage SDG interlinkages and address power asymmetries between different stakeholders and sectors?

Lastly, this chapter also addresses research question 1: What do we know about the most important interlinkages between the SDGs? Although being a pilot study, this chapter reveals what SDG interlinkages are potentially considered most important in the national implementation of the 2030 Agenda in Sweden.

SDG Synergies: A Decision-Making Tool for Synergetic Implementation of the 2030 Agenda

SDG Synergies is a method designed to help decision-makers to understand how SDGs or associated targets (or, for that matter, any set of targets) interact in a given context. It provides a basis for priority-setting by ranking goals as wells as cross-sectoral collaboration by identifying sub-groups of goals. The method looks beyond simple target to target interactions in order to include also indirect, secondary, effects across the whole set of selected goals. The method relies heavily on the legacy of (International Council for Science, 2016) and (Nilsson, Griggs, and Visbeck, 2016). The scientific basis of the method is described in (Weitz, Carlsen, Nilsson, and Skånberg, 2018) and a short description for users is provided by (Weitz, Carlsen, and Trimmer, 2019).

SDG Synergies has been applied in a wide range of different cases from the regional sub-national level to the level of EU, see Table 13.1. The breadth of the contexts in which SDG Synergies has been applied indicates the flexibility of the approach and therefore its utility in many different settings. The acknowledgment of the fact that each analysis of SDG interaction is unique has been an important guiding principle of the development of SDG Synergies. The tool and a description of the process is available for free at www.sdgsynergies.org.

The approach is built around a three-step process of collaborative analysis:

Step 1. Customization: The 2030 Agenda is 'universal', which means it applies to all countries, too all sectors of society and to all levels of governance. As a consequence, every application of SDG Synergies is unique, depending on the coalition of stakeholders, and the wider context in terms of institutional arrangements, natural resources, economic conditions, governance set-ups, technological options available, current policies and practices, and prevailing ideologies. These factors, in turn, shape which targets are perceived as most relevant and important by decision-makers. Given this – and the fact that there are almost 30,000 possible direct interactions between the 169 targets – the first crucial step in applying the SDG Synergies approach is to narrow the scope of the analysis and choose a subset of SDGs or targets to focus on. It is also possible to work with goals that are derived from the global SDGs or the associated targets. In many applications, this turns out to be the most common setting.

This step also includes the important aspect of contextualizing the selected goals. It is obvious that the goals are interpreted differently in different contexts; this is yet another consequence of the Agenda being universal: It applies to all countries at all scales, but in different ways. What main challenges do each goal pose? Do goals address issues that are central to enhancing sustainability in the given context? Are they relevant for a specific country or organization? Which goals to include can then be selected based on the discussions and some pre-established criteria, such as choosing at least one target from each of the 17 SDGs and setting a maximum number of targets.

Step 2. Scoring interactions: The selected targets build a cross-impact matrix (see further Figures 13.1 below), in which each matrix element except the diagonal is used to score the interactions. The guiding question for the scoring is:

If progress is made towards Target A, how does this influence progress towards Target B?

Consistent scoring is facilitated by the use of a scale of different types or strengths of interaction. For this purpose, SDG Synergies relies on the scale proposed by

TABLE 13.1 Examples of studies using SDG Synergies

Case	*Region*	*Purpose*	*Goals/targets*	*Reference*
Towards systemic and contextual priority setting for implementing the 2030 Agenda	Sweden	Using Sweden as a case to develop the methodology	Two targets per SDGs, hence 34 targets	(Weitz et al., 2018)
Piloting the SDG Synergies Approach in Mongolia	Mongolia	To build capacity among government ministries on systems thinking and informing Mongolia's national water management plan	17 targets	(Barquet, Järnberg, Alva, and Weitz, 2021)
SDGs and the environment in the EU: A systems view to improve coherence	EU	A proof of concept of applying the SDG Synergies approach at a regional level, and of its potential usefulness for environmental policy-making in the EU	SDGs 1–11, 16 and 17 on goal level and two targets each for SDGs 12–15	(Weitz, Carlsen, Skånberg, Dzebo, and Viaud, 2019)
The application of soft systems thinking in SDG interaction studies: a comparison between SDG interactions at national and subnational levels in Colombia	Colombia, national and subnational level	To support the work of the national commission on SDGs, related to inter-institutional coordination for the national implementation of the 2030 Agenda and the National Development Plan	National level: 20 targets; subnational level: 15 targets	(Barquet et al., 2021; Hernández-Orozco et al., 2021)
Interactions among the SDGs in Sri Lanka: A systemic assessment	Sri Lanka	To provide input to the national Council for Sustainable Development, and to build capacity of decision-makers to apply SDG Synergies	36 targets	(Barquet et al., 2021; Järnberg, Weitz, Maltais, and Carlsen, 2021)
Sweden's VNR	Sweden	Analysing SDG interactions in Sweden's national implementation of the 2030 Agenda	Swedish national-level interpretations of the 17 SDGs	(Carlsen and Weitz, 2021)
Six Swedish municipalities	Sweden, subnational	Pilot study aiming at laying out the ground for using SDG Synergies in all Sweden's municipalities	Between 8 and 2.	(Carlsen and Weitz, 2021)

Source: Author's own elaboration

Weimer-Jehle (2006), which reads: +3: strongly promoting, +2: moderately promoting, +1: weakly promoting, 0: no influence, -1: weakly restricting, -2: moderately restricting, -3: strongly restricting. In addition to providing quantitative score according to this scale, it is possible to add textual inputs in order to justify or further explain the selected score.

This step in the process is usually performed in a workshop setting. A participatory co-developing approach is usually highly appreciated among stakeholders because it facilitates joint assessments of interaction between goals often situated in different parts of an organization.

An important feature of this systems analysis approach is that the stakeholders should only focus on the direct influence of progress on one target, contingent of progress of another target. This makes the scoring manageable for participants and enables analysis that accounts for indirect impacts in the next step. Also worth noting is that interactions between two targets can be scored differently depending on the direction (e.g. progress on Target *A* could promote progress towards Target *B*, whereas progress to reach Target *B* could restrict progress towards Target *A*), hence the cross-impacts matrix is in general asymmetric.

Step 3. Analysis: The completed cross-impact matrix only shows direct interactions between the selected goals. In this last step of the process, network analysis methods are used to gain a better understanding of how progress towards different targets could affect the whole system. This step goes beyond the direct interactions and identifies patterns, clusters of closely interacting targets, and other network effects. These effects are more difficult to detect because they emerge through several stages of interaction and involve a large number of targets. Identifying these 'emerging effects' is important to get a more realistic view of the real-life behaviour of the connected SDGs and to avoid surprises in implementation further down the line. Another useful type of analysis in this step is to identify clusters of positively interacting targets and how they might interact with other such clusters. Findings from such analyses could serve as a basis for creating cross-sectoral working groups and partnerships.

The process can involve scientific experts, representatives of different sectors of society (business, civil society, government), and a range of other stakeholders. This inclusiveness can help to build bridges and partnerships between actors and sectors, generating shared understanding of the challenges and opportunities, highlighting common interests, and building ownership among stakeholders. Our experiences with SDG Synergies indicate that these outcomes can be just as valuable as the analytical outputs.

Operationalizing Policy Coherence: Applying SDG Synergies for Sweden's VNR 2021

The Government of Sweden has worked on policy coherence for almost two decades. In its Policy for Global Development (PGD) from 2003, the government adopted an overarching, farsighted goal of policy coherence. PGD was set out to work towards the UN Millennium Development Goals, the eight goals preceding the SDGs. The objective of PGD is to ensure coherence between national policy decisions to promote the goals of development policy. PGD furthermore stresses the importance of enhanced synergies. The Government decided to relaunch PGD in the light of the 2030 Agenda and the effort on implementing the SDGs. This involved all ministries being tasked with producing an action plan for their respective work on policy coherence.

In an internal evaluation of its work on PGD, one of the Government's conclusions was the importance of guidance and methodological support in efforts to increase policy coherence. The majority of ministries considered that they did not receive sufficient support and guidance when producing and drafting their action plans.

In an effort to respond to this challenge and operationalize policy coherence in line with the Government's strategy, SEI was commissioned to undertake a pilot project to identify and assess synergies and trade-offs between the SDGs for the Swedish 2021 VNR. The analysis presented in this chapter provides an overview of systemic effects resulting from SDG implementation in Sweden.

Identifying synergies and trade-offs

For the Swedish VNR, it was first decided that the analysis should remain on the SDG level, not involving any of the associated targets. As a first step, the 17 global SDGs were 'contextualized', i.e. interpreted so as to describe what they mean for the Swedish national level. For example, concerning SDG 2 (Zero hunger) hunger is not a major problem. Instead, issues like, overweight and unhealthy food habits are in focus. For another example with regard to SDG 12, the focus in Sweden is on reducing waste and sustainable consumer habits.

Together with the Swedish Ministry of Foreign Affairs the research team then identified civil servants in all Government Ministries responsible for at least one SDG. The scoring exercise was divided into different parts, where each ministry was put in charge for the SDGs under its sectoral responsibility, i.e. a ministry responsible for SDG x was tasked to answer the question: "If progress is made on SDG 1, how does this influence progress on SDG x?", "If progress is made on SDG 2, how does this influence progress on SDG x?" etc.

Focusing only on the direct influence between goals – also referred to as "first order effects"- was key at this stage in order to make the scoring manageable for the participants. In addition to provide quantitative scores (according to the scale ranging from -3, to +3), motivating texts were provided for all 272 (=17*17–17) interactions. Figure 13.1 below show the cross-impact matrix, which visualizes the output from this process step.

Analysis

The analysis focused on gaining a better understanding of how progress towards each of the SDGs could affect the whole system. These effects are generally more difficult to detect because they emerge through several stages of interaction and involve a large number of SDGs. Different techniques from network analysis were used in this step, for example in order to identify 'catalytic' goals that have the potential to accelerate progress across multiple SDGs.

A less complex approach is to simply calculate each SDGs' effect as the row sum (also called out-degree) in the matrix in Figures 13.1. Based on this we see that SDG 17 (Partnership for the goals) is the most influential goal in Sweden, followed by SDG 4 (Quality education), SDG 11 (Sustainable cities and communities), SDG 13 (Climate action) and SDG 16 (Peace and justice).

It has to be noted that the same direct influence can be the result of different underlying dynamics: SDG 8 and SDG 12 have the same total influence on all other goals, but while SDG 12 only has positive influence on other goals, progress on SDG 8 has a weakly restricting influence on three other goals (SDGs 12 to 14).

While the above analysis only includes first-order, direct effects it is also important to understand how influence ripples through the network in order to generate

(A)

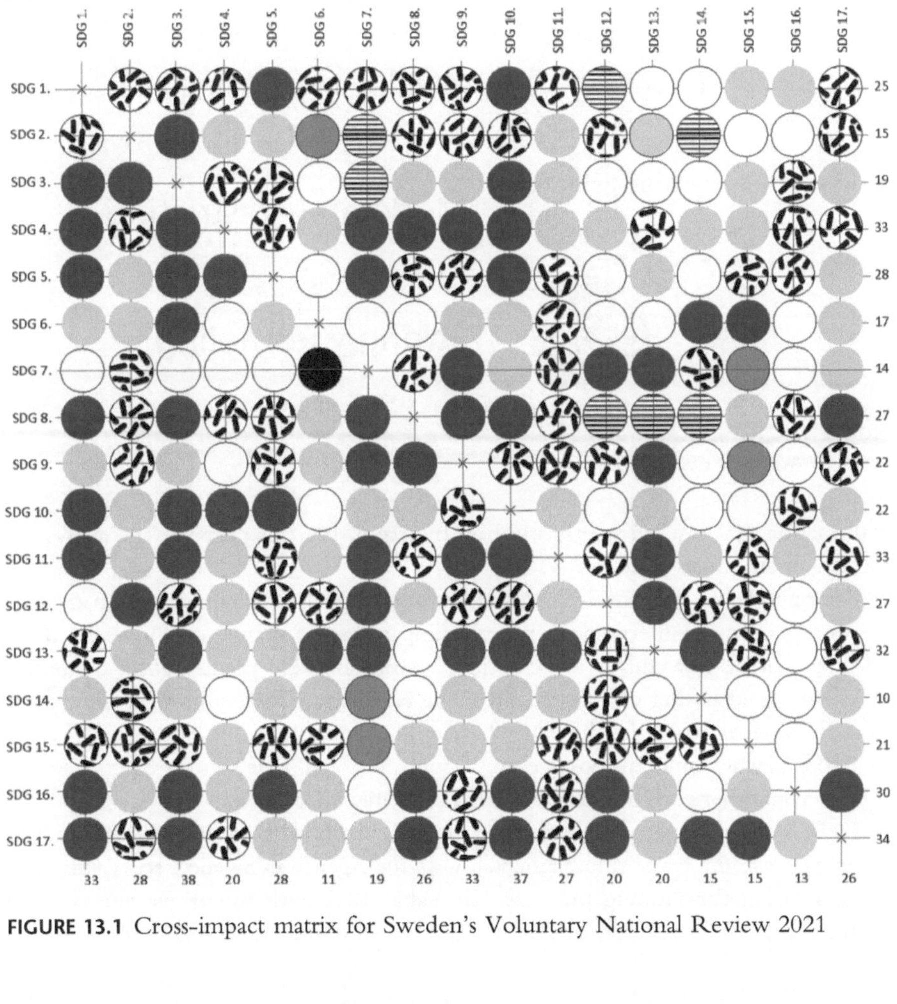

FIGURE 13.1 Cross-impact matrix for Sweden's Voluntary National Review 2021

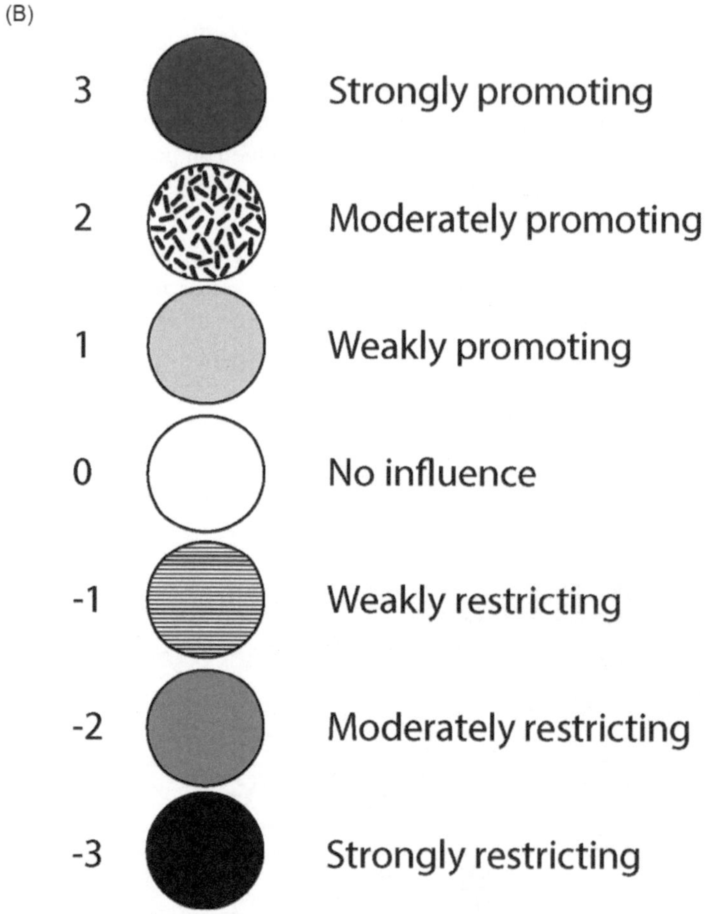

FIGURE 13.1 (Cont.)
Source: Author's own elaboration

information that can guide prioritization of action. For this purpose, the analysis of second-order effects is necessary.

For example, if a goal positively influences another goal, which in turn has many and/or strong positive connections, its systemic impact can be significant. However, if the affected goal has few and/or weak positive connections, the positive effect diminishes quickly without having much systemic effect. Furthermore, many strong positive connections to other goals with the same characteristics generates a high and positive multiplier effect, while a strong positive connection to a goal that in turn exerts negative influence on other goals generates a negative systemic impact.

By including such secondary effects the analysis identified a ranking of the most 'catalytic' SDGs in the national implementation of the Agenda in Sweden, see Table 13.2.

Comparing the ranking presented in Table 13.2 with the above analysis that includes only direct influences we see that SDGs 11 and 17, as well as SDGs 16 and

TABLE 13.2 The most influential SDGs on the system as a whole.[3]

Ranking	*Sustainable Development Goal*
1	*SDG 11: Sustainable cities and communities*
2	*SDG 17: Partnerships for the goals*
3	*SDG 4: Quality education*
4	*SDG 16: Peace and justice*
5	*SDG 13: Climate action*

Source: Author's own elaboration

13 change places. This illustrates how indirect effects influences the overall results of the analysis.

In follow-up meetings with ministries and stakeholder dialogues these results were discussed. The top position for SDG 11 on sustainable cities and communities was attributed to the development over recent decades in Sweden which has seen a widening gap between rural areas and cities, as well as increased differences with regards to living conditions in different areas within cities. The fact that SDG 17 was high ranked did not surprise participants as this goal is chiefly interpreted as a means for having progress on the Agenda as a whole. The education system has been one of the most debated sectors in Sweden over the last decades. The assessment of Sweden's performance in the education sector in international studies such as the OECD's Programme for International Student Assessment (PISA) has decreased and the correlation between school results and educational levels of parents has increased. Against this background, the result that the systemic effect of 'fixing' the educational system is substantial came as no surprise.

The analysis furthermore showed which SDGs are most positively affected by progress overall and hence might need less of explicit support and which SDGs are least positively affected by progress overall and hence might need extra support, see Table 13.3.

For Sweden SDG 1 is predominantly about social security and income gaps, where especially the latter is seen as a challenge with rather rapidly increased gaps. The cross-matrix presented in Figures 13.1 give a hint on why this goal turned out as most positively affected by progress overall since SDG 1 is strongly promoted by eight other SDGs. This is, however, only with regards to first degree influence; the full analysis (as shown in Table 13.3) also takes into account second order influences.

Regarding goals that are 'left behind' it was noted that clean water (SDG 6) indeed has become a relatively new concern in Sweden. It also did not come as a

TABLE 13.3 Most and least positively affected goals by progress overall

Goals most positively affected by progress overall	*Goals least positively affected by progress overall*
SDG 1: No poverty	SDG 6: Clean water and sanitation
SDG 3: Good health	SDG 12: Responsible consumption and production
SDG 5: Gender equality	SDG 14: Life below water

Source: Author's own elaboration

surprise that SDG 12 needs special attention since responsible consumption – like in most high-income countries – will not evolve by itself.

Since it is beneficial for progress overall to have progress on the top-ranked SDGs (cf. Table 13.2), it is important to manage trade-offs associated with those goals. Table 13.4 shows how the five top-ranked goals (SDG 11, SDG 17, SDG 4, SDG 16, and SDG 13) influence the eight goals which have direct negative influence on any other goal (SDGs 1–3, SDGs 7–9, SDG 14, SDG 15; cf. Figures 13.1). The rightmost column shows the total influence from the five top-ranked SDGs. We note that this effect is not, in general, lower for the most high-ranked goals. SDG 16 has a larger systemic effect on the whole set of goals, while the total negative effect is only 25, which is a comparably low number. The exact numbers are not, however, of great importance. It is the general overarching picture which is important.

Implementing the 2030 Agenda might require new collaborations that move beyond the siloed or sector-based approaches that typically dominate governments. As mentioned in section 2 of this chapter, network analysis methods applied to the cross-impact matrix can be used to identify cross-sectoral collaborations based on SDGs that are strongly interconnected. We carried out several such network-based analyses to identify "clusters" of SDGs that could inform the creation of cross-sectoral collaboration. This is rather difficult for a network like the one in Figures 13.1 since the network is very dense, meaning that out of all possible links very few are non-existing, i.e. grey dots. However, the SDG Synergies tool allows for relaxing the density of the network by considering different sub-sets of networks where only links of certain pre-defined strengths are included. Figure 13.2 shows emerging clusters of SDGs (marked by different colors) when only links of strengths +3 (= strongly promoting) are included.

As can be seen, the pink cluster is clearly 'environmentally oriented', with SDG 6 (Clean water), SDG 13 (Climate action), SDG 14 (Life below water), and SDG 15 (Life on land). The green cluster consisting of SDG 7 (Clean energy), SDG 8 (Economic growth), and SDG 9 (Infrastructure) could be interpreted as an

TABLE 13.4 Top-ranked SDGs (11, 17, 4, 16, and 13) influence on goals with negative influence (1–3, 7–9, 14, 15)

Goals with negative impacts on other goals	*SDG1 (-1)*	*SDG2 (-4)*	*SDG3 (-1)*	*SDG7 (-5)*	*SDG8 (-3)*	*SDG9 (-2)*	*SDG14 (-2)*	*SDG15 (-2)*	
Support from SDG 11	+3	+1	+3	+3	+2	+3	+1	+2	-43
Support from SDG 17	+3	+2	+3	+1	+3	+2	+3	+3	-46
Support from SDG 4	+3	+2	+3	+3	+3	+3	+2	+2	-52
Support from SDG 16	+3	+1	+3	0	+3	+2	0	+1	-25
Support from SDG 13	+2	+1	+3	+3	0	+3	+3	+2	-40

Note: The sum of negative interaction on all other goals is shown below each SDG. For example, SDG 2 influences SDG 6 with a -2, and SDG 7 and SDG 14 with a -1. The right most column shows the total influence from the five top ranked SDGs for example in the case of SDG 11 we have; 3*(-1)+ (-4) + 3*(-1) + 3*(-5) + 2*(-3) + 3*(-2) + (-2) + 2*(-2) = -43.

Source: Author's own elaboration

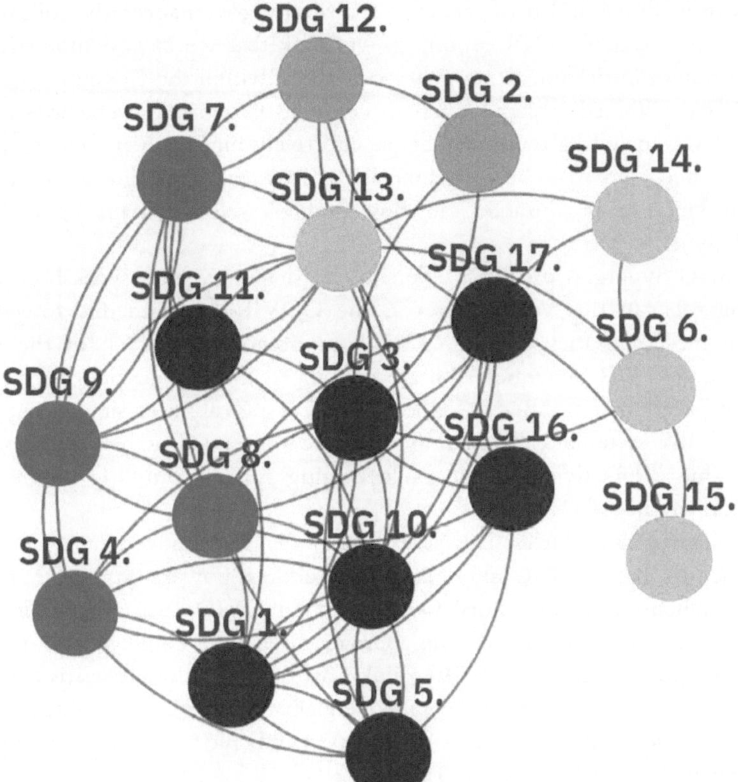

FIGURE 13.2 Four clusters identified with network analysis. Note that only interactions of strengths +3 is included in this analysis. The identification of clusters in SDG Synergies is based on the Louvain method, see (Blonde, Guillaume, Lambiotte & Lefebvre, 2008)
Source: Author's own elaboration

'economic and industry cluster'. Given the importance of the high-tech sector in Sweden, which requires a highly skilled workforce, it is perhaps not surprising to also find SDG 4 (Education) in this cluster. The blue cluster is the biggest and includes three of the top-ranked goals (see Table 13.2): SDG 11 (Sustainable cities), SDG 16 (Peace and justice), and SDG 17 (Partnership). Since these three highly ranked goals are positively tightly linked (via +3 interactions), joint actions on those goals will have a large impact on progress overall.

Conclusion

This chapter primarily addresses research question 4: What institutional processes and policy strategies can strengthen the capacities of different stakeholders in order to achieve inclusive and participatory sustainability governance? By utilizing SDG Synergies in Sweden's VNR reporting, we think that we have demonstrated one efficient way of institutionalizing policy coherence within the Government Offices. The approach 'forced' policy-makers to consider *all* possible interactions between the 17 SDGs, including trade-offs, hence effectively hindering any cherry-picking, which has been shown to be commonplace as a way of focusing only on subsets of goals on which an organization can show progress (see Siegel and Bastos Lima in this volume).

The SDG Synergies process is explicitly participatory with workshops playing a central role. In this work however, the COVID-19 pandemic forced us to organize most activities on-line. This is an obvious drawback of the current application of SDG Synergies. We are convinced that the participatory element whereby people with different expertise and responsibilities sit down together physically and systematically identify and assess interactions between goals will be even more effective in terms of leveraging systems thinking in the implementation of the SDGs.

This chapter also addresses research question 2: What governance mechanisms are needed to manage SDG interlinkages and address power asymmetries between different stakeholders and sectors? Owing to the COVID-19 pandemic, it showed to be difficult to assess whether the approach used here could help in managing asymmetric power relationships. We think however, that the process is democratic in the sense that all scores are counted equally and, owing to systemic effects beyond first order interaction, it is in retrospect difficult to derive for example the list of top-ranked goals shown in Table 13.2.

Lastly, this chapter also set-out to address research question 1: What do we know about the most important interlinkages between the SDGs? This question is extremely hard, if not impossible, to answer in a general, universally applicable manner. While there have been several attempts to do so (e.g. Pham-Truffert, Metz, Fischer, Rueff, and Messerli, 2020) these attempts have, by necessity, remained rather coarse-grained. Owing precisely to the 'universality principle' of the 2030 Agenda, goals and associated targets must be contextualized, and as a result interactions will look different dependent on context. The results provided by the application of the SDG Synergies approach in Sweden reflect the fact that inputs to the process were based on the perceptions of participants working in the Swedish context. As a result, they allow for a first assessment on key interactions between the 17 SDGs at the national Swedish level or more precisely, the systemic effects of such interactions for ranking SDG with regards to their systemic effects. However, these results cannot be extrapolated to different country contexts.

Reflecting on avenues for future research it needs to be pointed out that the SDG Synergies approach does not distinguish between SDGs that are more oriented around sectors, like food, education, health etc., and the SDGs that are more cross-cutting and process oriented, such as SDG 16 and SDG 17. This can be seen both as an advantage and a disadvantage for our method. On the one hand, it is useful to assess content (i.e. sectoral goals) versus process because the processes are fundamentally there to support progress across the entire agenda. On the other hand, it might be useful to only focus on interactions between sectoral goals and use the process-oriented goals as a way of qualifying the assessments of these interactions. How the interactions between sectoral and process-oriented goals can be analyzed should be the subject of future research.

Finally, this study shows the utility of a simple and transparent method for identifying, assessing and analysing SDG interactions in a given context. This was also clearly articulated by participants in the stakeholder dialogues. In this sense, it is in line with the findings of Di Lucia, Slade, and Khan (2021) who investigated whether existing scientific methods are fit-for-purpose in supporting decision-making regarding SDG interactions in Sweden. They find that decision-makers appreciate methods that are simple, flexibly applicable and suited to provide directly actionable and understandable results. They also find that decision-makers are less concerned with the accuracy, precision, and quantitative nature of the knowledge. SDG Synergies as a method and tool to support decision-making processes has been shown to be well in line with these findings.

Notes

1 General Assembly resolutions 70/299 and 75/290B provide further guidance on the follow-up and review of the 2030 Agenda and the SDGs.
2 www.sdgsynergies.org.
3 For a mathematical description of how second order effects are included see Nina Weitz et al. (2018).

References

Barquet, K., Järnberg, L., Alva, I.L., and Weitz, N. (2021). Exploring mechanisms for systemic thinking in decision-making through three country applications of SDG Synergies. *Sustainability Science*. doi:10.1007/s11625-021-01045-3.

Blondel, V.D., Guillaume, J.-L., Lambiotte, R., and Lefebvre, E. (2008). Fast unfolding of communities in large networks. *Journal of Statistical Mechanics: Theory and Experiment, 2008*, P10008. doi:10.1088/1742-5468/2008/10/p10008.

Carlsen, H. and Weitz, N. (2021). Analys av synergier och målkonflikter i Sveriges frivilliga granskning av arbetet med Agenda 2030. doi:10.51414/sei2021.015. Retrieved from: www.sei.org/publications/analys-av-synergier-och-malkonflikter-i-sveriges.

Di Lucia, L., Slade, R., and Khan, J. (2021). Decision-making fitness of methods to understand Sustainable Development Goal interactions. *Nature Sustainability*, 5, 131–138. doi:10.1038/s41893-021-00819-y.

Government Offices of Sweden. (2021). *Report on the Implementation of the 2030 Agenda for Sustainable Development.* Retrieved from: www.government.se/4a5bef/globalassets/government/dokument/regeringskansliet/agenda-2030-och-de-globala-malen-for-hallbar-utveckling/voluntary-national-review-vnr/voluntary_national_review_2021_sweden_report_on_the_implementation_of_the_2030_agenda_web.pdf.

Hernández-Orozco, E., Lobos-Alva, I., Cardenas-Vélez, M., Purkey, D., Nilsson, M., and Martin, P. (2021). The application of soft systems thinking in SDG interaction studies: a comparison between SDG interactions at national and subnational levels in Colombia. *Environment, Development and Sustainability*. doi:10.1007/s10668-021-01808-z.

International Council for Science. (2016). A Draft Framework for Understanding SDG Interactions. Retrieved from: https://council.science/wp-content/uploads/2017/05/SDG-interactions-working-paper.pdf.

Järnberg, L., Weitz, N., Maltais, A., and Carlsen, H. (2021). *Interactions among the Sustainable Development Goals in Sri Lanka: a systemic assessment*. doi:10.51414/sei2021.004. Retrieved from: www.sei.org/publications/interactions-among-the-sustainable-development-goals-in-sri-lanka-a-systemic-assessment.

Nilsson, M., Griggs, D., and Visbeck, M. (2016). Policy: Map the interactions between Sustainable Development Goals. *Nature*, 534, 320–322. doi:10.1038/534320a.

Organisation for Economic Co-operation and Development. (2019). *Recommendation on Policy Coherence for Sustainable Development*. Retrieved from: www.oecd.org/gov/pcsd/oecd-recommendation-on-policy-coherence-for-sustainable-development.htm.

Official Reports of Sweden. (2019). Agenda 2030 Och Sverige: Världens Utmaning - Världens Möjlighet. Retrieved from: www.regeringen.se/rattsliga-dokument/statens-offentliga-utredningar/2019/03/sou-201913.

Pham-Truffert, M., Metz, F., Fischer, M., Rueff, H., and Messerli, P. (2020). Interactions among Sustainable Development Goals: Knowledge for identifying multipliers and virtuous cycles. *Sustainable Development*, 28, 1236–1250. doi:10.1002/sd.2073.

Weimer-Jehle, W. (2006). Cross-impact balances: A system-theoretical approach to cross-impact analysis. *Technological Forecasting and Social Change*, 73, 334–361. doi:10.1016/j.techfore.2005.06.005.

Weitz, N., Carlsen, H., Nilsson, M., and Skånberg, K. (2018). Towards systemic and contextual priority setting for implementing the 2030 Agenda. *Sustainability Science*, 13(2), 531–548. doi:10.1007/s11625-017-0470-0.

Weitz, N., Carlsen, H., Skånberg, K., Dzebo, A., and Viaud, V. (2019). *SDGs and the Environment in the EU: A Systems View to Improve Coherence*. Retrieved from: www.sei.org/publications/sdg-synergies-environment-eu.

Weitz, N., Carlsen, H., and Trimmer, C. (2019). SDG Synergies: An approach for coherent 2030 Agenda implementation. *Stockholm Environment Institute*. Retrieved from: www.sei.org/publications/sdg-synergies-factsheet.

CONCLUSIONS

Anita Breuer, Daniele Malerba, Pooja Balasubramanian and Srinivasa Srigiri

This book set out to explore challenges in implementing the UN 2030 Agenda for Sustainable Development that result from the interdependent and interlinked nature of its 17 Sustainable Development Goals (SDG). Specifically, our aim was to deepen the understanding about governance and policy innovations that are needed to support policy coherence across different policy sectors, government levels, and societal actors in varying country contexts in order to harness synergies and mitigate trade-offs in the process of implementing the SDGs.

To this purpose, this book formulated the following four overarching research questions:

1. What do we know about the most important interlinkages between the SDGs?
2. What governance mechanisms and policy processes are needed to address power and capacity asymmetries between stakeholders form different sectors and levels in order to promote an inclusive, coherent and participatory governance of SDG interlinkages?
3. Which policy mixes to increase policy coherence in the implementation of different interlinked SDGs have proven to be effective, socially just, and acceptable, and leaving no one behind?
4. What political-institutional preconditions are conducive to the establishment of effective governance mechanisms to manage SDG interactions?

In responding to the above questions, the 12 studies collected in this volume adopted different methodological approaches ranging from single and comparative case studies, quantitative national and cross-national analyses, to meta-reviews of literature on SDG interlinkages. Notwithstanding their rich methodological diversity, contributions to this volume were guided by a joint research heuristic that paid particular attention to three central dimensions of SDG implementation namely (i) the role of

DOI: 10.4324/9781003254683-14

political-institutional preconditions; (ii) governance mechanisms as processes of political decision-making; and (iii) policy instruments and policy mixes to steer the implementation of politically set goals.

The picture that emerges from these studies is mixed. Their findings show that the achievement of the 2030 Agenda requires careful consideration of governance mechanisms and policy mixes. On the one hand, it is difficult to fully consider the interlinkages between all SDGs, as action on one goal has implications for many others. On the other hand, studies in this book show that progress on specific SDGs or targets as well as specific policies, governance mechanisms and institutional preconditions are particularly suited to spur positive effects across the 2030 Agenda. Although our conclusions can only be tentative the analyses presented in this book allow for cautious optimism, as they show that a wide spectrum of scientifically based political and methodological approaches is potentially available to accelerate the integrated implementation of the 2030 Agenda. Yet seeing that the time to achieve the SDGs is running short and the speed at which positive changes are being made is still insufficient, a faster and broader uptake and actual application of these approaches is urgently needed.

Based on the findings of the individual studies and using the overarching research questions as structuring elements, the following sections discuss this argument in more detail and highlight related policy implications.

Key Findings and Policy Implications

Regarding the first of our research questions about major SDG interlinkages, based on the evidence and analyses presented in this book, we cannot but confirm some of the main messages coming from the existing literature. It is in fact clear that it is nearly impossible to clearly and systematically understand and present the main interlinkages between the SDGs in a generalized fashion. One reason is that "one size does NOT fit all" and the 2030 Agenda's principle of "universal applicability" of the SDGs (§55) necessitates their contextualization to different country, local and organizational contexts. The way in which the SDGs and their targets interact can vary significantly between these contexts. The challenge thus resides in striking a balance between respecting local context, working at the international level to reform institutions and carefully consider the scale at which actionable policy advice can be directed. Another – but related – reason is that the links between two or more goals are usually mediated by a multitude of variables and factors, which represent the context. This complicates the possibility of actually measuring the interlinkages. An example for such mediating effects, is provided in Chapter 10 by Khan, who uses cases studies and examples across the globe to discuss the interlinkages between education, gender equality, and economic growth and female employment. She shows that increasing education among girls does not automatically translate into closing the gender gap in labour force participation. Instead, the relationship between female education and employment varies between different socio-cultural, economic, and institutional settings. Another problem in relation with interlinkages is that the 2030 Agenda itself is fraught with several conceptual flaws that virtually provoke some negative interactions, most notably between the Agenda's proclaimed economic and environmental goals. In view of this, what is needed is not simply accumulating more knowledge about interlinkages but at the same time a) develop methods that help to translate this knowledge into policy action and b) gain an understanding of the Agenda's inherent conceptual flaws and develop proposals to overcome them. Several chapters in this book make valuable contributions towards these ends.

As the brief overview of the academic "state-of-the-art" on SDG interlinkages research presented by Bennich, Weitz, and Carlsen state in Chapter 2 shows, the body of literature on this topic has grown almost exponentially over recent years and, as a quasi-inevitable consequence, increasingly difficult to navigate. As the authors state, this might ultimately hamper the uptake of academic knowledge about SDG interlinkages by decision makers and practitioners in charge of implementing the SDGs. In order to facilitate the translation of knowledge into practice, they propose a "reading guide", consisting of a set of five questions, to help readers determine the usefulness and transferability of a given study to their specific decision-making context. One of the questions to be addressed is to which specific policy challenge a study responds, distinguishing, for example, between strengthening of policy coherence, policy prioritization, or monitoring and evaluation. Another question to be addressed is which methods and tools are best suited to address specific policy challenges. The authors posit, for example, that studies applying quantitative modelling methods and rely on data sourced from official databases might be particularly useful for the purpose of policy prioritization. In turn, network analysis, cross-impact analysis, and participatory methods that rely on data obtained through stakeholder and expert consultations might be best suited for the purpose of informing the design of governance mechanisms for SDG implementation. A practical application of the latter approach to the case of Sweden is presented by Carlsen et al. in Chapter 13. The methodological steps include contextualization of SDGs and targets and scoring of interactions through a participatory consultation of experts and stakeholders and then analysing the underlying reasons for these interactions. The approach described in this chapter can be a good starting point for integrated implementation of SDGs at the national level and has the potential for wider application in many national contexts. As a result, it provides a replicable and commendable example of how science can support (political) decision makers in addressing the complexity of the 2030 Agenda and facilitate its integrated implementation.

A method to extract and harness existing knowledge about interlinkages between different aspects of a selected subset of SDGs is presented by Allen et al. in Chapter 3. The reduction of poverty (SDG 1) and inequalities (SDG 10) suffered severe setbacks from the COVID-19 pandemic. To learn about how principles of good governance mandated by SDG 16 relate to selected targets of SDGs 1 and 10, the authors conduct a scoping review of scholarly literature. The review's findings reveal empirical evidence of the beneficial impacts of transparency, accountability, and inclusive and participatory institutions on targets of SDGs 1 and 10 from a wide number of quantitative and qualitative evaluations spanning most nations from all parts of the world. Having identified these positive interactions, the authors use systems mapping in order to obtain an overview of explanations provided by individual studies regarding the underlying pathways to impact and causal relationships. Through this approach, they identify several positive feedback loops that represent reinforcing dynamics in complex systems and, as such, are important for identifying key entry points for interventions and accelerators, which can deliver desirable outcomes. The essential contribution of this chapter towards knowledge about SDG interlinkages is twofold. First, it allows policy makers to draw on empirical evidence to argue that investing in accountability, participation and transparency makes interventions for poverty reduction and reducing inequalities more effective and can thus make a significant contribution towards post-pandemic recovery and resilience building. Second, it offers a methodology how to identify enabling SDG interlinkages in specific contexts and transform that knowledge into action: Scoping reviews combined with systems mapping can be used to extract and harness information about interactions between SDG targets that are considered particularly relevant in a given context. When conducted during the early stages of policy or programme development, the results of such analysis can help policymakers prioritize or refocus budgets and mobilize the funding needed to implement related policies and programmes.

Climate modelling is another important methodological approach for analysing SDG interlinkages. It consists in integrating socio-economic and environmental data, and making projections about how these might change in the future. In Chapter 11, Malerba and Emmerling provide an overview of current evidence from Integrated Assessment Modelling (IAM), which are the state-of-the-art modelling tools in the climate community. They conclude that both climate mitigation policies (SDG 13) and climate impacts are increasingly being found to affect poorer households disproportionally, thus potentially hampering progress on SDG 10 (reduced inequalities). However, the authors also point to important gaps that remain in IAM, such as refining models with regard to within-country income distribution and inequality and integrating mechanisms to recycle revenues from CO_2 taxation for social protection measures. Closing these gaps will be paramount in providing decision makers with policy options that are suitable for increasing the social acceptance and thus the political feasibility of climate mitigation policies. The authors' application of such a refined model to the case of India demonstrates that the redistribution of tax revenues can effectively lower the negative inequality effects of climate mitigation policies.

While the above chapters offer methodological solutions to identify relevant SDG interlinkages in varying context settings Chapter 9 by Malerba and Oswald discusses a perceived conceptual flaw in SDG interlinkages thinking. The authors criticize that the 2030 Agenda is still largely based on the paradigm of economic growth, which is explicitly formulated in SDG 8 (Decent Work and Economic Growth) but also implicitly contained in the targets of many other SDGs. More specifically, many of the environmental SDGs formulate relative targets that do not necessarily imply a reduction of material and carbon footprints. Furthermore, the income and income growth levels envisioned by the 2030 Agenda work to the detriment of environmental goals and targets, such as SDG 12 and 13. These flaws entail the risk of effectively locking global development around a failing economic model. Against this backdrop, the authors advocate to give greater consideration to alternative development paradigms, such as degrowth.

In summary, responding to our first research question, chapters in this book confirm that many of the interlinkages between goals and targets, significantly depend on the context and their appropriate management requires context sensitive approaches. Furthermore, it is necessary to gain a deeper understanding of the Agenda's inherent conceptual flaws and develop proposals to overcome them. Responding to this challenge will require a strengthening of the science-policy interface. As Donoghue and Khan (2019) put it: "*policy-makers' understanding of trade-offs and synergies should be grounded in a science-based analysis of interactions across SDG domains.*" (ibid. p. 10). Consequently, if the generation of academic knowledge about SDG interlinkages is to have practical relevance, ways must be found to transform this knowledge into actionable advice for decision makers. This is particularly relevant as the complexity that results from the interlinked nature of the SDGs contrasts starkly with limited implementation capacity - particularly in many developing nations - which will be discussed further below in this concluding section.

Turning to our second research question, chapters in this book provide an array of insights regarding how the potential of governance mechanisms and policy processes can help strengthen the capabilities of actors and stakeholders from different levels and sectors to jointly tackle the task of addressing SDG interlinkages. They do so by both pointing to existing deficits as well as potential ways to overcome these deficits.

Drawing on political science literature and sustainability research Chapter 4 by Breuer, Leininger, and Malerba proposes five criteria for assessing national SDG governance mechanisms regarding their potential to promote coherent and effective implementation of the SDGs. Specifically, these criteria are: political leadership, horizontal coordination across policy sectors, vertical coordination between levels of government, horizontal accountability and control between state actors, and societal accountability and participation. Applying these criteria in a cross-national comparative analysis of 137 countries they find that the majority of countries professes commitment to the 2030 Agenda by the highest level of government in their Voluntary National Reports (VNR). Furthermore, in many countries minimum institutional conditions have been created to allow for some degree of horizontal cross-sectoral coordination in SDG implementation. However, institutional conditions to achieve vertical coordination across levels of government, horizontal control between state actors and social accountability and participation are suboptimal in most countries. The comparative analysis also reveals a dominant role of foreign ministries in national SDG governance mechanisms worldwide, which begs the questions whether the true objective of these mechanisms is to showcase commitment to the SDGs internationally rather than to actually improve policy coordination and accountability between sectors and levels domestically. To achieve the latter, the institutional design of national SDG governance mechanisms would have to be reformed to include ministerial representatives of all policy sectors and to strengthen the formal engagement of the sub-national governments and civil society. Furthermore, parliaments, audit offices and national human rights institutions should be more strongly engaged in order to strengthen horizontal accountability in the process of SDG implementation.

The regional comparative study on national SDG politics and agrifood governance in Latin America's Southern Cone presented by Siegel and Bastos in Chapter 6 reflects some of the above findings. The expansion of large-scale industrial agriculture plays a pivotal role for economic growth in the region. At the same time, it is responsible for a number of severe negative social and environmental impacts, including biodiversity loss, infringements on the human rights of small-scale farmers or indigenous communities, and severe health impacts. The comparison of the cases of Brazil, Uruguay, and Paraguay reveals that in such a setting VNR reporting – as one central element of the national SDG accountability cycle – can be crucial in addressing SDG interlinkages and trade-offs. The central question here is whether the mechanisms and processes for VNR reporting are set-up in a way that contributes to trade-offs between SDGs being recognized and eventually addressed or rather to trade-offs being concealed and neglected. The comparative study shows that in some cases powerful actors in agribusiness have resorted to a strategy of SDG "cherry picking", whereby they claim to contribute towards the entire 2030 Agenda by conveniently selecting an isolated SDG or target in line with their "business as usual". Through this practice, the negative impacts of their activities and on the achievement of other SDGs are concealed, which has demonstrably increased and entrenched marginalization in the study region. Encouragingly, the study also shows that an inclusive, participatory and transparent VNR process can play an important role in increasing accountability and lowering the risks of cherry-picking and selective reporting. However, the study also indicates that pre-existing institutions and power relations, which in turn critically depend on political regime type, might influence the design and effectiveness of VNR processes. This issue will be discussed further down in this concluding section.

Drawing on in-depth country studies the chapters by Srinivasa and Scheumann as well as Dombrowsky in this book further explore the governance challenges related to the coherent and integrated implementation of the 2030 Agenda using the example of natural resource governance in the water–energy–food (WEF) nexus in Ethiopia (Chapter 7) and Jordan (Chapter 8). Both contributions find that the dominant governance modes and deficits of existing governance mechanisms in these countries do not only constrain coordinated and coherent natural resource governance but also threaten the achievement of core ambitions of the 2030 Agenda, particularly the principles of inclusiveness and leave no one behind (LNOB). With regards to interlinkages, and in line with the criticism by Malerba and Oswald in Chapter 9, both studies observe a primacy of economic growth in national development thinking that comes to the detriment of social and environmental goals and the principle of LNOB. Dombrowsky, for example, notes that against a background of water scarcity the Jordanian Water Strategy prioritizes groundwater use for agricultural activities with the highest economic returns over activities of smallholder farmers. Similarly, Srigiri, and Scheumann describe how the pursuit of agriculture-led economic growth in Ethiopia encourages the expansion of irrigated areas for large-scale commercial sugar farming at the expense of alienating vulnerable pastoralists from rangelands and waterholes that are crucial to their livelihoods. They also observe that Environmental Impact Assessment procedures in Ethiopia are frequently intervened by actors from the federal government who push for projects that promise great economic benefits. Furthermore, in both country cases, the authors observe natural resource governance to be dominated by a highly centralized and hierarchical governance mode alongside with fragmented sectoral policies and very limited multi-level coordination. As a direct result, the voices of actors from local and community levels are insufficiently heard and considered in national level planning and policy formulation, particularly when they do not belong to sectors that have been identified as key to promoting economic growth. As an indirect result, the ensuing policies and regulations might be met with insufficient ownership or even outright resistance at lower levels of government or by certain communities, which does not only hamper their implementation but ultimately also poses a threat to societal peace and stability. However, these contributions also make clear that there are no easy fixes to overcoming these deficits. For one thing, as Dombrowsky points out for the example of Jordan, a centralized hierarchical governance mode might be unavoidable to control the allocation of scarce natural resources such as groundwater in order to contain their over use and thus prevent a "tragedy of the commons" situation. The particular challenge here will be to find solutions that are ecologically sustainable on the one hand, but on the other hand meet the demand for fair access to resources in the sense of LNOB. For another thing, both the findings by the regional comparative study by Siegel and Bastos as well as the country studies by Dombrowsky and Srigiri and Scheumann suggest that the observed governance deficits can at least partially attributed to the political context of deficient democratic or autocratic rule. This issue will be discussed further down in this concluding section.

Possible ways to overcome horizontal and vertical coordination deficits are presented in Chapter 5 by Meuleman who elaborates on the concept of multilevel metagovernance. As a framework, the concept is concerned with the creation of actionable mechanisms to spur collaboration between actors from different levels of authority. As one of many practical examples, the author refers to the phenomenon of 'real-time collaborative multilevel governance' as an approach that has emerged in some countries to address situations of national crises or emergency. It implies that representatives of different levels of governance get together to discuss and provide rapid responses to issues of national urgency. As the author contends, such real-time multilevel governance mechanisms are an addition that cannot replace the bottom-up subsidiarity style or the classical top-down hierarchical style of governance. However, in some cases, they proved instrumental in ensuring coordination between levels of government and across policy sectors thus contributing to coherent crisis response and mitigating harmful competition for resources. Maintaining such newly created mechanisms for fast and effective collaboration – at least in stand-by mode – beyond times of acute crises could add to the sustainable resilience of nations and societies as a whole.

In Chapter 13, Carsen, Bennich, and Weitz present SDG Synergies, a decision-making tool to support systems thinking in SDG implementation. The authors provide insights from the practical application of the tool for the preparation of Sweden's VNR in 2021. The tool combines participatory and deliberative dialogues between stakeholders from different sectors with scientific methods such network analysis techniques. Essentially, it aims at gaining a better understanding of how progress towards each of the SDGs affect a system as a whole in a specific context. However, it is important that the application of such approaches does not remain limited to institutional bodies or organizations explicitly tasked with SDG politics, such as national SDG coordination bodies or commissions tasked with the elaboration of VNRs. As long as these approaches are not increasingly picked up in national and sectoral politics more generally, overcoming coordination deficits and power and capacity asymmetries between stakeholders from different levels and sectors will remain wishful thinking.

Regarding our third research question, several chapters in this book shed light on the question how "smart" policy mixes can spur synergies and mitigate trade-offs between interlinked SDGs and targets in an inclusive and socially acceptable manner through the implementation of different policies simultaneously or as a package.

In Chapter 11, Malerba and Emmerling investigate the relationship between climate change mitigation and social policies. Using household data from India, they show that in the absence of complementary policies, climate policies such as a carbon tax can increase inequality. However, the redistribution of tax revenues from a carbon tax through cash transfers can effectively counteract the undesired inequality effects of climate mitigation policies. Nevertheless, the authors emphasize that an adequate design of redistribution mechanisms, such as cash transfers, is also critical. Targeted transfers might be most suited to reduce aggregate poverty and inequality, but could prove insufficient to reach the poorest population groups, thus failing to respect the LNOB principle of the 2030 Agenda.

Chapter 9 by Malerba and Oswald discusses the extent to which continued economic growth is instrumental in achieving social development while simultaneously respecting environmental boundaries. The authors criticize that, despite immense environmental challenges, the SDGs still tend to focus on the goal of continuous economic growth. Using quantitative methods, they demonstrate that the global achievement of higher living standards should be pursued by a combination of growth and global redistribution. They show that reliance on economic growth alone would lead to a situation where already achieved improvements in the decarbonization of energy systems would be reversed and emissions would rise. Against this backdrop, they argue that a shift away from a development paradigm that sees economic growth as a desirable goal in itself is needed. Instead, they call for the adoption of policy mixes that exploit the potential of economic redistribution to reduce the need for aggregate growth and to mitigate trade-offs between collective well-being and environmental constraints.

In Chapter 12, Balasubramanian analyses the potential of different combinations of tax and social policy measures and their implications for reducing poverty and inequality. Comparing evidence from four Eastern and South East Asian countries, she finds that the policy mixes through which these countries pursued their growth strategies have led to an increase or decrease in poverty and inequality. In her analysis, Mongolia for example is shown to have adopted a growth strategy that resulted in an increase in both inequality and poverty. The Mongolian growth strategy was mainly fuelled by the large gains from the export of minerals that were distributed to the entire population in the form of a centralized cash transfer scheme. However, owing to the instability of the global mineral market, the country experienced fiscal instability and had to discontinue the cash transfer scheme. Thailand, by contrast, represents a positive example of a growth strategy that was successful in reducing inequality and poverty. The country introduced a tax financed universal health insurance scheme and broadened its social protection system through various schemes addressing specific target groups such as pensions, cash transfers for schooling and school feeding, as well as food and in-kind transfers. This broad and balanced policy mix proved successful in significantly reducing poverty and facilitating more equitable access to health services.

In summary, in responding to our third research question, the above findings show that policy mixes combining tax-financed social security systems with transfers in kind and cash have been shown to be a promising approach to strengthening healthcare and education and enabling a growth strategy that can reduce poverty and inequality. This has been shown to be especially relevant in relation to environmental taxes, which will be an important part of an adequate policy mix to advance socially just transitions toward a climate-neutral global economy. At the same time, the specific design of environmental fiscal reforms (e.g. combining carbon taxes and social protection) is critical and should be continuously adapted to advance climate and social protection concerns together. In particular, as low-income countries grow and households increase their income and consumption, the focus of cash transfers will have to be refocused at compensating poorer households.

However, more generally, and as discussed in the introduction to this book, the success of a policy mix in achieving a given objective will not only depend on the consistency of its elements but also on the comprehensiveness and inclusiveness of the political decision-making processes to negotiate the mix, which in turn will be critically shaped by political-institutional preconditions and the related governance modes in place.

Furthermore, and as discussed in relation with our research question on SDG interlinkages, in most cases it is nearly impossible to accurately assess the effects of an individual policy or policy mix on the entirety of the SDGs *ex ante*. Consequently, in many cases the design of a policy mix will be dictated by the most proximate goals or interests and its positive or negative collateral effects on other goals will only become apparent *ex post*. Getting policy mixes "right" will therefore continue to be a learning process in the foreseeable future. Ideally, this process should be closely accompanied and supported by academic research.

Finally, turning to our fourth question, several chapters in this book provide insight into how political-institutional preconditions - particularly political regime type and state capacity- affect the effectiveness of governance mechanisms in managing SDG interactions.

The scoping literature review presented by Allen et al. in Chapter 3 reveals that several characteristics considered as key principles of good governance – namely transparency, accountability, and inclusive, participatory decision making –contribute towards the reduction of poverty and inequality. For example, evidence was found that increased electoral accountability results in a better targeting of social expenditure and increased access to basic services. For another example, there is strong evidence that increased institutional transparency through corruption monitoring significantly contributes to the poverty reduction potential of economic growth. While the governance principles of transparency, accountability, and inclusive and participatory of decision-making are not exclusive to democratic regimes they are less frequently found in autocratic contexts than in liberal democracies, which in turn suggests that the latter might provide more favourable preconditions for the achievement of the SDGs related to the social dimension of sustainable development.

The comparative study presented by Siegel and Bastos in Chapter 6 suggest that countries with higher levels of democracy might also be better prepared to address SDG trade-offs than those with lower democracy levels. The authors find that in Uruguay, as an established democracy, grievances over negative health impacts linked to the spread of industrial agriculture were taken up to a large extent through institutional channels for reporting problems that have been put in place by the government. By contrast, in Paraguay and Brazil, as defective democracies, such grievances led to protests by excluded actors outside the institutional framework.

The case study of WEF nexus governance in Jordan, presented by Dombrowsky in Chapter 8 finds the country's autocratic regime to have limited capacity to deal with the complexities of the 2030 Agenda. The author finds that formal rules regarding the allocation of scarce water resources tend to be dominated by a hierarchical governance mode. However, the effectiveness of these rules is partly undermined by social norms of nepotism and favouritism (*wasta*), which are deeply embedded in the Jordanian society and monarchic ruling system. Consequently, the country's autocratic regime does not seem to be particularly conducive towards ecological sustainability. The study also shows that the Jordanian government's confidence in the effectiveness of deliberative decision-making processes remains limited which, in turn, indicates that significant progress towards inclusive and participatory decision-making at all levels is less likely in the context of autocratic rule.

The case study of water, food, and land governance in Ethiopia, presented by Srigiri and Scheumann in Chapter 7 comes to similar conclusions. Within the country's authoritarian regime, economic and social development policies and strategies are planned and executed by means of a central hierarchical governance mode, which does not allow for effective participation of sub-national and non-governmental actors in decision-making. Further aggravating the situation, the ability of existing governance mechanisms for natural resource management in minimizing trade-offs between social, economic and environmental goals is severely constrained by low financial, human, and technical capacities at different levels of the state.

In summary, responding to our fourth research question, contributions in this book provide convincing evidence of democracy as providing enabling conditions not only for the achievement of several individual SDGs but also for addressing and mitigating trade-offs between them. Against the backdrop of an ongoing global trend of democratic regression (BTI, 2022) and a global debate on democracy that is increasingly marked by distrust of democratic institutions, processes and elected representatives this finding has important policy implications. Efforts to support sustainability transformation – be it at national levels or in the context of international development cooperation – will in all likelihood be less successful if they are not accompanied by bold and proactive measures to protect and promote democracy. Furthermore, to some extent, findings in this book reflect the often voiced criticism that the complexity that results from the interlinked nature of the SDGs contrasts starkly with implementation capacity in many developing nations where capacities are often particularly low at sub-national levels. Yet, regional and local governments are at the frontline of delivering progress on the SDGs (e.g., Kindornay and Gendron, 2020).

Working towards integrated implementation of the SDGs will therefore not only require improved coordination between different levels of government and incorporation of the 2030 Agenda into locals plans and policies but also continued financial and capacity support to local governments.

Avenues for Future Research

At the time of writing this book, the deadline to achieve the SDGs by 2030 is fast approaching. Yet evidence has shown that we are a long way from achieving many of the SDGs and their targets. While some targets are on track to be achieved at the global level, this is not the case for the majority of goals. For some of them, such as achieving decent employment (SDG 8) or reducing emissions (SDG 13), there are large gaps that remain and the current pace of progress is highly insufficient. A recent report (Organisation for Economic Co-operation and Development, 2022) also shows that OECD countries on aggregate have met less than 10% of the targets, with scope to strengthen countries' efforts in several key areas such as leaving no one behind, institutions and environmental pressures. In short, given the current status quo and rate of progress, achievement of the SDGs by 2030 is in jeopardy.

Against this backdrop, this book has sought to advance our knowledge on how to close the aforementioned gaps and accelerate progress towards integrated achievement of the 2030 Agenda. In doing so, special focus was put on the different governance dimensions of SDG implementation. Based on key findings, the previous section has pointed to policy implications to accelerate SDG progress. However, it has also brought to the fore several other questions that could not be addressed within the scope of this book. In the following, we will outline major policy and research gaps that should inform future research on the SDGs.

When it comes to SDG interlinkages, the need to improve the policy-science interface is well recognized. As Allen, Metternicht, and Wiedmann (2021) for example point out, scientific methods for evaluating interlinkages between the SDGs have advanced considerably over recent years. Nevertheless, their practical application, for instance in Voluntary National Reviews, is still relatively scarce (Allen et al., 2021). One step towards strengthening the policy-science interface would be the creation of a global online SDG research data repository, in which evidence from all published SDG research would be collected. Studies could then be coded in a way that facilitates the location of evidence related to a specific SDG or country of interest. There is evidence that the demand for such digital resources is increasing. It will be critical that in designing them, researchers understand the needs of policy makers and other stakeholders in order to enable them to retrieve actionable evidence that assists them to understand and address interlinkages in their given context. Another important condition for the creation of such resources is to improve the data situation, particularly in lower income countries. The need to do so is already articulated in SDG target 17.18, which calls for the enhancement of capacity-building support to increase the availability of high-quality, reliable, and disaggregated data. Therefore, target 17.18 should be emphasized more strongly as an enabling target for the entire 2030 Agenda.

Future research on SDG interlinkages should also pay heightened attention to distinguish more clearly between *correlations* and *cause-and-effect relationships* between SDGs and targets. Doubtlessly, correlations are easier to estimate, owing to data and estimation simplicity. However, the information that can be derived from them is insufficient from a policy perspective, which needs to consider causation, i.e. the potential effects of a policy or change in outcome on other goals. In addition, and as has been shown by several contributions to this book, correlations might be affected by mediating factors that are hard to detect through quantitative research alone. Nevertheless, they can provide important initial indications of synergies or trade-offs that exist between SDGs, the more precise nature of which needs to be explored in more depth using complementary methods. To facilitate the translation of SDG research into actionable policy advice, carefully designed *multi-* and *mixed-methods* will need to be increasingly applied.

Furthermore, debates and literature on SDG governance should be more closely linked to other global policy debates. A case in point, for example, is the debate on *just transition* which gained much traction following the 2021 United Nations Climate Change Conference (COP 26). Essentially, the concept of *just transition* is concerned with how to achieve a socially just and acceptable transition to a climate-neutral and climate-resilient global economy. Quite obviously, core issues of just transition are inherent in the principles and goals of the 2030 Agenda, in particular through the principle of leaving no one behind, the principle of indivisibility of the social, economic and environmental dimensions of sustainability, and the strong emphasis on inclusiveness and participation as critical enablers of sustainable development (SDG 16). In actuality, however, these debates often still run in parallel without cross-fertilizing each other. Clearly, future global policy debates concerned with sustainability transformation should make sure to capitalize on the rich evidence and findings generated by the ongoing research on the governance of SDG interlinkages.

Finally, the 2030 Agenda in its current form is far from flawless. For one thing, and as several contributions in this book have made clear (see e.g. Chapters 6, 7, 8, and 9), the incompatibility between its proclaimed economic and environmental aims virtually provokes negative interactions between the related SDGs. This entails the risk that governments and policy makers will opt for the defensive approach to continue business as usual, hoping that some of their actions will, eventually, match one of the 169 targets (Breuer, Janetschek, and Malerba, 2019). For another thing, the weak accountability vision offered by the 2030 Agenda is reflected in weak accountability mechanisms in national institutional SDG architectures, which poses a threat to the effective implementation of the SDGs (see e.g. Chapter 4 in this book by Breuer, Leininger and Malerba). A post-2030 agenda will have to revisit and address these problems. The end of the SDG period is already in sight and negotiations to reach international agreement on a follow-up agenda of such scope will necessarily be complex and protracted. In view of this, the debate on these issues will have to start soon. It is our hope that with this book we have provided food for thought and some inspirations for those who will be engaged in these debates.

References

Allen, C., Metternicht, G., and Wiedmann, T. (2021). Priorities for science to support national implementation of the sustainable development goals: A review of progress and gaps. *Sustainable Development*, 1–18. doi:10.1002/sd.2164.

Breuer, A., Janetschek, H., and Malerba, D. (2019). Translating Sustainable Development Goal (SDG) interdependencies into policy advice. *Sustainability*, 11(7). doi:10.3390/su11072092.

Breuer, A. and Leininger, J. (2021). Horizontal Accountability for SDG Implementation: A Comparative Cross-National Analysis of Emerging National Accountability Regimes. *Sustainability*, 13(13), 7002. Retrieved from: www.mdpi.com/2071-1050/13/13/7002.

BTI. (2022). Global Findings: Resilience wearing thin. Retrieved from: https://bti-project.org/fileadmin/api/content/en/downloads/reports/global/BTI_2022_Global_Findings_EN.pdf.

Donoghue, D. and Khan, A. (2019). Achieving the SDGs and 'leaving no one behind': Maximising synergies and mitigating trade-offs. Retrieved from: https://cdn.odi.org/media/documents/Achieving_the_SDGs_and_leaving_no_one_behind_maximising_synergies_and_mitigating.pdf.

Kindornay, S. and Gendron, R. (2020). *Progressing National SDG Implementation: An independent assessment of the voluntary national review reports submitted to the United Nations High-level Political Forum in 2019.* Ottawa, ON: Canadian Council for International Cooperation. Retrieved from: https://resourcecentre.savethechildren.net/pdf/full-report-w-highlights-and-annex-optimized.pdf.

Organisation for Economic Co-operation and Development. (2022). The Short and Winding Road to 2030: Measuring Distance to the SDG Targets. Retrieved from: www.oecd-ilibrary.org/docserver/af4b630d-en.pdf?expires=1652701717&id=id&accname=ocid77015268&checksum=AF38EB6F28B1F9444B940753AE22FEE1.

INDEX